The Intext Series in BASIC MATHEMATICS

Under the Consulting Editorship of

Richard D. Anderson

Louisiana State University

Alex Rosenberg

Cornell University

Fundamentals of Finite Mathematics

Fundamentals of Finite Mathematics

by Benjamin W. Volker / Andrew S. Wargo

Department of Mathematics, Bucks County Community College

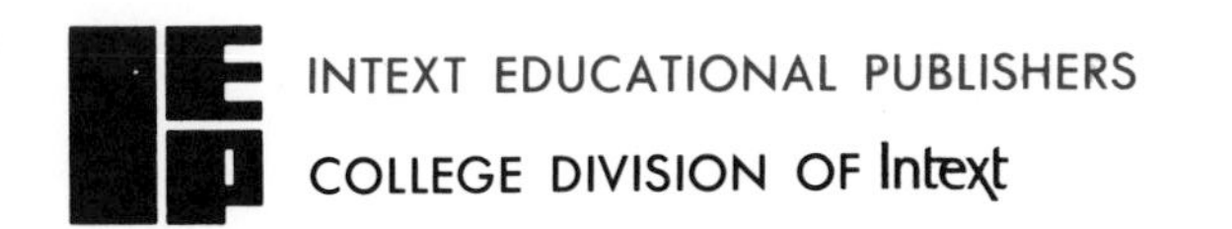

Scranton San Francisco Toronto London

Library of Congress catalog card number 79-177306
ISBN 0-7002-2355-X

Preface

In the past decade the academic community has finally recognized that mathematical course work for the nonscience major need not consist of a typical algebra-trigonometry-calculus sequence. Numerous mathematics courses have come into existence particularly designed for the diverse needs of the nonscience major undergraduate. As a consequence a revitalization has taken place in mathematics course content and instruction. The subject matter for these new courses is generally selected on the basis of its fundamental importance to all branches of mathematics, its broad applicability to many other academic areas, and the interest level such topics may have for the student.

The writing of this text was prompted by the fact that the available texts for this type of course were instructor- rather than student-oriented. We found that students encountered difficulty in reading the text books, that explanations were sketchy, and were overwhelmed by the rapid escalation into extremely complex and abstract problems.

Our goal has been to create a textbook that is written primarily for the student and to demonstrate that mathematics is a lively, interesting, and dynamic field of study. To implement the basic objectives, the authors have been guided by the following philosophy.

1. The text should play an important part in the learning process. The text must be readable, i.e., intelligible to the layman, serve as a study guide, include many explanatory examples and incorporate detailed solutions to many problems so that the student has a working model on which to base his efforts. Also, the presentation must begin on an intuitive level, then gradually build toward a more formal treatment of the various topics.
2. Little or no prior mathematics training should be assumed. The text is essentially self-contained. Some background in high school algebra would be helpful particularly for the study of Systems of Linear Equations, Vectors, and Matrices (Chapter 6) and Linear Programming (Chapter 7), but it is not absolutely necessary.
3. Most of the students who take this type of course have unpleasant recollections of their high school mathematics courses, if not mathematics in general. Therefore, the selection of topics at least initially should be new to the student so as to maximize the possibility of a reawakened interest.

4. Whenever possible, the topics should be capable of a unified treatment stressing the concept of a mathematical system and mathematical reasoning methods. However, flexibility must be incorporated to permit the deletion of certain items without affecting the basic continuity.

With exception of the sections denoted with an asterisk in the Table of Contents, which may be omitted, the material in the first four chapters; Symbolic Logic, Set Theory, Counting Theory, and Probability, is necessary for a study of Statistics (Chapter 5). Chapters 6 and 7 covering Systems of Linear Equations, Vectors, and Matrices and Linear Programming are essentially independent of prior material.

Instructors should find that when the text is used for a one semester course, a considerable number of options are available in selecting topics which fit the background and interest of his particular class. The authors would suggest that based on their experience, three alternate course options may be considered using the chapters indicated as a nucleus with selected topics from other chapters as additive options.

Course A	Chapters 1, 2, 3, 4	Logic, Sets, Counting, and Probability
Course B	Chapters 2, 3, 4, 5	Sets, Counting, Probability, and Statistics
Course C	Chapters 1 (part), 2, 6, 7	Logic, Sets, Systems of Linear Equations, Vectors, Matrices, and Linear Programming

A more in-depth coverage of all topics would permit the use of the text for a two quarter or two semester course.

The text material has been class-tested and revised over a period of two years prior to publication and has proven successful with both large groups and small recitation sections.

We are grateful to our many colleagues who gave us valuable suggestions and constructive criticism. We also express our appreciation to the administration of Bucks County Community College and Gwynedd Mercy College who afforded us the opportunity to class test the manuscript.

Particular gratitude is extended to our hard-working typist Margaret Wilkin and to our colleague Brendan Mulvey who provided us with helpful editing assistance. We also express our appreciation to the publisher and especially to the Mathematics Editor, Charles J. Updegraph, for his sustained interest in the project.

Finally, we wish to thank our wives for their understanding and continuing encouragement.

Benjamin W. Volker
Andrew S. Wargo

Newtown, Pennsylvania
December, 1971

Contents

CHAPTER 3

CHAPTER 4

CHAPTER 5

CHAPTER 6

Fundamentals of Finite Mathematics

The only safeguard against reasoning ill is the habit of reasoning well; familiarity with the principles of correct reasoning and practice in applying these principles.

–JOHN STUART MILL

CHAPTER 1

Symbolic Logic

1.1 Mathematical Systems

Throughout this text one important objective is to make explicit both the structure of a mathematical system and mathematical reasoning methods. Although many of the modern mathematics programs now being used in our elementary and secondary schools place considerable emphasis on mathematical structure; students enrolling in their first college level mathematics course frequently appear not to have a firm grasp of certain fundamental mathematical concepts.

Without attempting to explain why students have had difficulty in absorbing basic concepts or why many students express an obvious dislike for mathematics in general, we propose to take a fresh and different approach to the study of mathematical systems. No extensive knowledge on the part of the student is assumed. Hopefully, this new setting will rekindle some of the reader's latent interest and curiosity for mathematics.

Let us begin with a very basic question, "How can we initiate a thought process aimed at organizing knowledge in a rational manner?" Two broad approaches to reasoning methods are possible. First, we can attempt to draw some generalized conclusions by observing the outcomes of a series of events or the results of repeated experiments. This type of reasoning which is characterized by going from particular events to general conclusions is called **INDUCTIVE REASONING**. Although the inductive method is used in developing new theories, it is plagued by one basic flaw. The observation of a single concrete contradictory result will destroy the most elegant theory. Let us consider some examples of the inductive reasoning process.

Example 1: The product of *1* times *3* is *3*, which is an odd counting number.
The product of *3* times *5* is *15*, which is an odd counting number.
The product of *5* times *7* is *35* which is an odd counting number.
Therefore, the product of any two odd counting numbers is another odd counting number.
The generalization made in this example based on the specific instances can be shown to be true, but not by the inductive process.

Example 2: The formula $n^2 + n + 5$, where n is a counting number, $(1,2,3, \ldots)$ is supposed to produce only prime counting numbers. A counting number is prime if and only if it has only two distinct counting numbers as factors; namely, itself and one. For example, *29* is prime since it has only the factors *29* and *1*. To check the formula, let n assume the values *1*, *2*, and *3*.
Let $n = 1$, then

$$n^2 + n + 5 = (1)^2 + (1) + 5 = 1 + 1 + 5 = 7$$

and for $n = 2$, we have

$$n^2 + n + 5 = (2)^2 + (2) + 5 = 4 + 2 + 5 = 11$$

and for $n = 3$, we have

$$n^2 + n + 5 = (3)^2 + (3) + 5 = 9 + 3 + 5 = 17.$$

Certainly, *7*, *11*, and *17* are prime and the formula appears promising. However, let us consider one more example: Let $n = 4$, then

$$n^2 + n + 5 = (4)^2 + (4) + 5 = 16 + 4 + 5 = 25.$$

The counting number *25* is not prime, since it can be written as $5 \cdot 5 \cdot 1$ in factored form. Thus, the formula fails to produce only prime numbers and the inductive reasoning process does not give rise to a true generalization.

Example 3: It is believed that every even number greater than *2* can be expressed as the sum of two prime numbers. Consider the following typical examples:

$$26 = 3 + 23$$
$$48 = 7 + 41$$
$$156 = 53 + 103.$$

The reader may wish to investigate other examples. This seemingly general property is known as *Goldbach's conjecture.* No one has been able to find an exception to his conjecture. However, even though no contradictory example has been found, the numerous supporting instances do not constitute an acceptable proof of the conjecture.

A mathematician or scientist will not only use induction but also intuition or analogy to initiate a thought process. However, he will not be satisfied until his conclusions can be incorporated into a mathematical system which uses **DEDUCTIVE REASONING**. In the deductive reasoning process, he places primary emphasis not on observations or controlled experiments, but rather on a thought process which is totally abstract. This process begins with accepted generalized properties to which he applies rules of logic to draw necessary conclusions. The apparent paradox is that the mathematician has a greater degree of certainty in the abstract world than he does in the physical world of sense data. Albert Einstein described the situation in the following terms: "As far as laws of mathematics refer to reality, they are not certain; and as far as they are certain they do not refer to reality."

In this text we will rely primarily on the deductive reasoning process as our mode of mathematical thought. Although we have not formally discussed logic and valid reasoning, we will discuss one example of deduction at this time so the reader may gain some feel for the spirit of the activity.

Example 4: Prove that the sum of two positive odd counting numbers greater than *1* is an even counting number.

***Solution*:** A few examples might be considered by the reader to lend a sense of reasonableness to the statement, but we must look for a more general approach in order to develop a proof. Can we find a way to symbolize two distinct odd counting numbers? Every even number has *2* as a factor and every odd number differs from an even number by *1*. We therefore represent an arbitrary odd number in the form $2k_1 + 1$, where k_1 represents any counting number $(1,2,3, \ldots)$ and a second odd number in the form $2k_2 + 1$ where again k_2 is a counting number. The subscript for the *k's* is introduced so that we may consider the possibility that the odd numbers are not necessarily the same number. Having developed some general symbolic form, let us perform the operation of addition on the arbitrary odd numbers and reach a conclusion.

$$\begin{aligned}(2k_1 + 1) + (2k_2 + 1) &= 2k_1 + 2k_2 + 2 \\ &= 2(k_1 + k_2 + 1).\end{aligned}$$

Each line above is justified by a property of our number system. $k_1 + k_2 + 1$ represents a sum of counting numbers which is also a counting number. The term $2(k_1 + k_2 + 1)$ is thus of the form *2* times a counting number which, since it has *2* as a factor, is an even number. Thus, we have proven by deductive methods that the sum of two odd counting numbers greater than *1* is an even counting number.

The preceding example of a deductive process referred to properties of a number system and an operation–addition. These are important characteristics of our conventional arithmetic; however, in a more general sense they also correspond to some features of a mathematical system. Every mathematical system has four essential characteristics: undefined or primitive terms, definitions, axioms (assumptions, postulates), and theorems.

First, we begin with **undefined** or **primitive** terms. We define some object of thought by words which in turn are defined by other words in a seemingly endless process. Somewhere we must agree to accept certain words or terms as essentially undefined. In geometry such terms as "point," "line," and "plane"; in arithmetic, "number"; in algebra, "variable"; in physics, "time"; in chemistry, "charge"; are all examples of these primitive terms. In other words, before we can communicate in any field with each other, we must accept in common certain basic concepts which are understood but not defined.

Once we understand primitive or undefined terms, we form **definitions** which contain undefined terms or previously accepted definitions.

In the further constructing of a mathematical system, we specify **postulates** or

axioms which will become a part of the structure. **Postulates** are **basic assumptions** which are accepted as permissible statements or fundamental rules. For example, in basic arithmetic we assume that:

$$3 + 2 = 2 + 3 \quad \textbf{(Commutative property of addition).}$$

However, the selection of the postulates is a somewhat arbitrary choice of the mathematician. It is this freedom which has removed mathematics from sole confinement to the physical world. For centuries Euclidean geometry was believed to be a "true" model of the physical universe. Yet when one of Euclid's postulates–namely, the famous parallel postulate, which in effect states there is only one line parallel to a given line through an external point–was replaced by either a postulate that said there were no parallel lines or a postulate that said there were an infinite number of parallels to a given line, quite startling but useful non-Euclidean geometries were developed by Riemann (1826–1866), and Lobachevsky (1793–1856). In fact, Einstein's theory of relativity and the physicists' concept of space are explained in terms of the mathematical models provided by non-Euclidean geometries.

The last step in constructing a mathematical system is the development of **theorems.** Theorems are the necessary conclusions obtained from the undefined terms, definitions, and assumptions by applying the rules of logic.

As an illustration of a familiar mathematical system, let us consider the counting numbers $\{1,2,3, \ldots\}$ as a set of elements where we accept "number" as an undefined term. We then define two operations on these elements, viz., "addition" and "multiplication." At this step we introduce axioms. The following three axioms are typical of the type of assumptions which form a part of our number system.

If a, b, and c are counting numbers, then:

(i) $a + b = b + a$ **Commutative property of addition**

(ii) $(a + b) + c = a + (b + c)$ **Associative property of addition**

(iii) $a \cdot (b + c) = a \cdot b + a \cdot c$ **Distributive property of multiplication over addition.**

For example, the Associative Property states that $(2 + 3) + 4 = 2 + (3 + 4)$, which may be illustrated as follows:

$$(2 + 3) + 4 = 5 + 4 = 9$$

$$2 + (3 + 4) = 2 + 7 = 9.$$

With this small portion of the structure made explicit, we can still go on to develop theorems within the system.

Example 5:

THEOREM: If a, b, and c are counting numbers, then

$$a \cdot (b + c) + b = (b + a \cdot b) + a \cdot c.$$

PROOF:

$$a \cdot (b + c) + b = (a \cdot b + a \cdot c) + b \quad \text{Distributive property}$$
$$= b + (a \cdot b + a \cdot c) \quad \text{Commutative property}$$
$$= (b + a \cdot b) + a \cdot c \quad \text{Associative property}$$

At this point it is important that we make a clear distinction between an abstract mathematical system and a model which we may regard as an applied interpretation of the system. In an abstract system, since we begin with undefined terms, the definitions, symbols, assumptions, and theorems are essentially devoid of any concrete meaning. For example, our number system as discussed previously is an abstract mathematical system. We may give an implied interpretation to the system by giving a meaning to the symbols. These symbols may represent "dollars," "apples," or other objects of the same kind. The mathematician operates in a realm of ideas and is not restricted by the physical world. This characteristic of pure mathematics prompted Bertrand Russell (1872–1970), facetiously to observe, "Mathematics is that study in which we never know what we are talking about or what we say is true." It is precisely this freedom of thought from which we derive benefits. In the realm of ideas and symbols we are free of the ambiguities of language and are permitted the advantage of easy symbolic manipulation. Fundamental relationships are exposed and new lines of thought not suggested by any physical interpretation are obtained. Definitions can be altered, assumptions changed, and as a consequence, entirely new lines of inquiry can be developed. Frequently, the motivation for changes within the mathematical system are prompted by an attempt to give an applied interpretation to the elements of the system.

This text will involve the use of induction and intuition to motivate the development of our abstract mathematical system by deductive methods. We will, however, provide the reader with concrete model interpretations so that the applied nature of mathematics will be made explicit.

Problem Set 1.1

A. Read the following statements and be prepared to discuss their apparent truth or falsity.

1. The conclusions made in the inductive reasoning process are never accepted as true.
2. Intuition has no place in the deductive reasoning process.
3. In a mathematical system, postulates are those statements in the system which require proof before they are accepted as part of the system.
4. In any reasoning process, there are statements, implicit or explicit, which we accept with no proof.
5. Although not explicitly stated, the rules of logic are an assumed part of any mathematical system.
6. Deductive reasoning is the process of going from particular events to a general conclusion which in some way identifies a common property of these events.

7. Any mathematical system must have some practical, useful interpretation.

B. 1. Determine whether the following reasoning process is inductive or deductive. Is the conclusion a necessary conclusion? If not, show why the conclusion need not follow from the given statements. Three is a prime number; five is a prime number; seven is a prime number. Therefore, all odd numbers are prime.

2. Show that the sum of a positive even counting number and an odd counting number greater than 1 is an odd counting number. (*Hint:* see Example 4 in this section.)

3. Consider the word "true." In this text "true" is considered an undefined term. Use a dictionary and/or other reference sources to attempt to define the word "true." Do not presuppose any synonyms. Should the word "true" be an undefined term?

4. In the following statements, what are some of the implicit assumptions?
   (a) Man is a rational being.
   (b) All humans have self-respect.

1.2 Validity of Arguments by Euler Circles

Logic is concerned with the principles of valid inference. Historically, Greek philosophers, particularly Aristotle (384–322 B.C.), were the first to make a systematic study of logical reasoning methods. Their analysis frequently took the form of examining syllogisms or arguments which consist of a series of statements or premises and a conclusion which may or may not "follow" from the given premises. Consider the following examples:

Example 1:

All students are diligent.
Bill is a student.
―――――――――――
∴ Bill is diligent.

Example 2:

All exons are bitles.
Zap is an exon.
―――――――――――
∴ Zap is a bitle.

(The straight line is used to separate the premises from the conclusion and the symbol "∴" may be read "therefore.")

In each of the above arguments, the conclusion does seem to follow from the premises even though the second example doesn't make any sense. The Greek logicians were quick to realize that in analyzing arguments it is not the content of the premises and the conclusion, nor their apparent truth or falsity, but the abstract form of the argument that is important which is called an **argument form**. The argument form representing both of the preceding examples is:

Example 3:

All A's are B's.
C is an A.
∴ C is a B.

A visual aid that helps us to illustrate why this structural form is regarded as a valid argument form; i.e., one in which the conclusion must necessarily follow from the premises, is a set of Euler circles, named after the mathematician Leonard Euler (1707–1783).

The first premise is represented by drawing a circle A inside a circle B. This illustrates that any object that is an A is also a B. For the second premise, we have a specific object; namely C, which we designate as a dot inside circle A, showing that C is an A. The conclusion "C is a B" does follow from the premises, since C is also contained inside circle B.

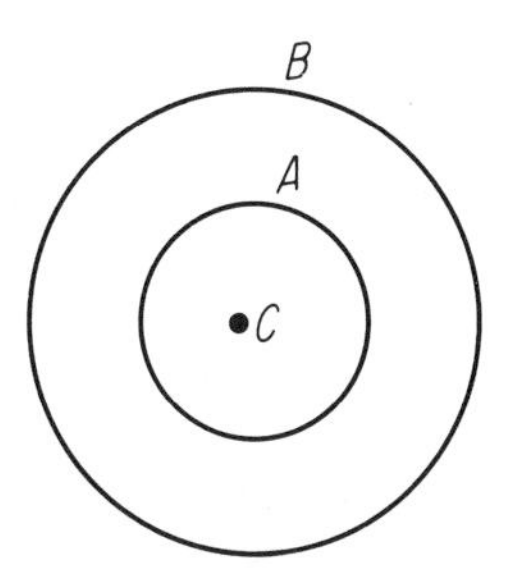

The Euler circle method is useful not only in analyzing valid argument forms, but in showing that an argument is invalid; i.e., the conclusion does not necessarily follow from the premises.

Example 4: Use the Euler circle method to determine whether or not the conclusion in the following argument necessarily follows from the premises.

All women are emotional.
Every woman is fickle.
∴ All fickle persons are emotional.

Solution: The first premise can be represented by drawing a circle W inside a circle E, indicating all women are emotional (Figure 1.1).

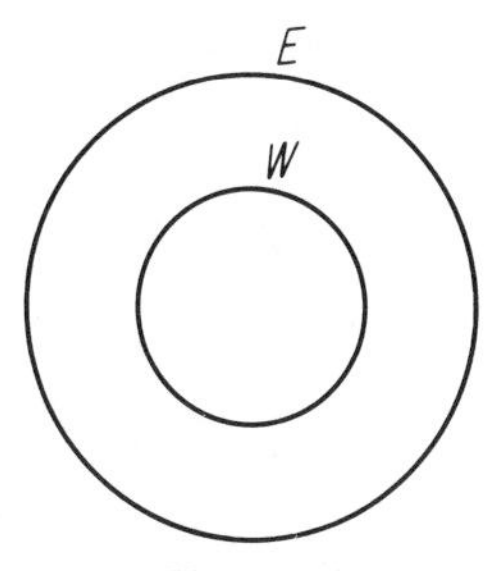

Figure 1.1

The second premise requires that we draw a circle F in such a way that the circle W is inside the circle F, showing that every woman is fickle. However, this drawing may be made in several ways as shown in Figures 1.2, 1.3, and 1.4.

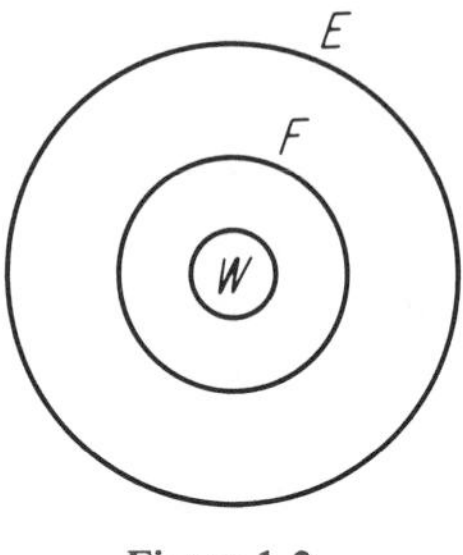

Figure 1.2

Figure 1.2 would appear to show that the conclusion, "All fickle persons are emotional," does follow from the premises, since the circle F is inside circle E. However, Figure 1.3 and Figure 1.4 illustrate that the premises can be satisfied when there is an "f"

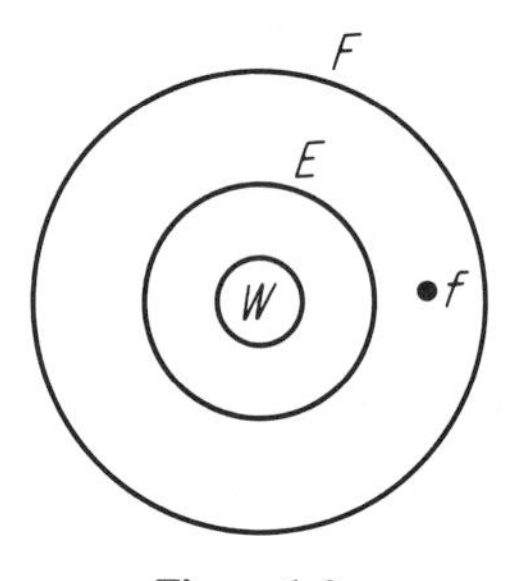

Figure 1.3

who is fickle and not emotional. Thus not all fickle persons are emotional and the conclusion does not necessarily follow from the premises. Our ability to make a drawing which accurately represents the premises and yet indicates that the conclusion does not necessarily follow, demonstrates that the argument is invalid.

Not all argument forms involve affirmative statements about every member of a

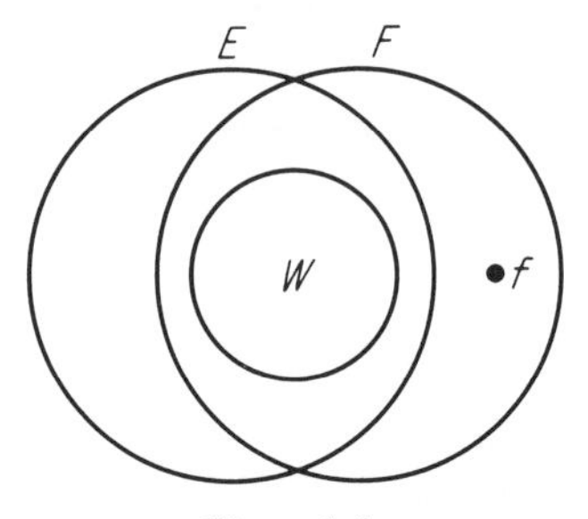

Figure 1.4

class of objects. A premise may inform us of what a class of objects is not, by the use of a negation, or the premise may make a statement referring to some but not all objects of a given class. The following example will illustrate some of the possible statement variations.

Example 5: Use the Euler circle method to analyze the following argument

No students are discourteous.
Some men are students.
∴ Some men are not discourteous.

The first premise may be drawn as two circles which do not overlap, because there are no students (S) who are discourteous (D). The second premise may be drawn by placing a small "m" inside circle S illustrating that there is at least one male student. The conclusion does necessarily follow from the premises, since there is at least one male; namely "m," who is not inside the circle D.

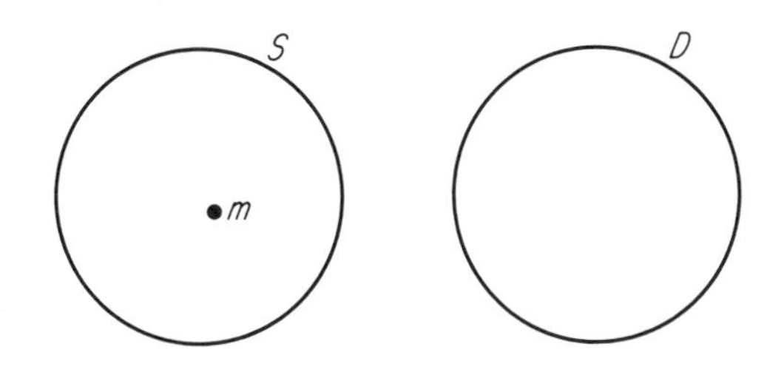

Although the Euler circle method provided students of logic with a useful visual technique for displaying arguments, our knowledge of logic passed on to us by the Greeks remained relatively static until the nineteenth century. Early in the nineteenth century a renewed interest in mathematics and in logic took place. William Hamilton (1805–1865) and Augustus De Morgan (1806–1871) were successful in developing a new symbolic notation for "quantified" statements of the form, "All A's are B's," "Some A's are B's" and their respective denials. However, the most significant contribution to our knowledge of logic was made by George Boole (1815–1864). Boole recognized that in order to develop a complete symbolic representation, it was necessary to symbolize not only statements, but the relations between statements and the logical connectives which join statements together. Boole was successful in creating a system of abstract symbolic logic on which the comprehensive development of modern logic has since been based.

In the next section we will turn our attention to a systematic development of some of the fundamental concepts of symbolic logic.

Problem Set 1.2

A. Determine whether the conclusion in each of the following arguments must necessarily follow from the given premises. (Use Euler circles.)

1. Any statement of the form "All A's are B's," means "If anything is an A then it is a B." For example, "All fish can swim" means "If anything is a fish then it can swim."
   (a) All fish can swim.
   (b) Bill Small can swim.
   ∴Bill Small is a fish.
2. (a) All fish can swim.
   (b) Perch are fish.
   ∴ Perch can swim.
3. A statement of the form "No A's are B's" means "If anything is an A then it is not a B." For example, "No wild birds are domestic." means "If anything is a wild bird, then it is not domestic."
   (a) No wild birds can swim.
   (b) All wild ducks are wild birds.
   ∴ All wild ducks can swim.
4. The statement form "Some A's are B's," means "There *exists* at least one thing that is an A and is also a B."
   (a) Some birds can swim.
   (b) All people can swim.
   ∴ Some people are birds.

B. From the given pair of premises, determine which conclusions must "follow" from these premises. Use Euler circles to aid in determining your results.
(a) All students are learned people.
(b) Some learned people are self-taught.
∴ 1. All students are self-taught.
2. Some students are learned people.
3. All learned people are self-taught.
4. No self-taught person is a student.
5. Some self-taught people are learned.

1.3 Symbolizing of Statements and Connectives

The study of symbolic logic is an analysis of reasoning methods. However, symbolic logic is a mathematical system with its own structure. This structure is by no means unique, and variations in notation, definitions, and assumptions could of course be made. The symbolic logic which we will study, based on the work of George Boole, will preserve the type of logical deductions made in mathematics and will conform to what we might regard as intuitive translations of our ordinary language.

To begin the exposition, we need to identify what will be our primitive terms, such as "number" is a primitive term for arithmetic. Our initial primitive term is "declarative sentence." A statement may be characterized by the following definition:

DEFINITION: A **statement** is any declarative sentence that may be classified as either true or false but not both.

Some declarative sentences that are statements are given in the following example:

Example 1:

1. This book is red.
2. Lyndon Johnson was a president of the United States.
3. In the next decade taxes will go up and purchasing power will decline.
4. If medical research continues at its present rate then the human life span is bound to increase.

The first two statements in Example 1 represent a single idea and can be immediately classified as true or false. However, the last two statements involve two ideas joined together and yet we will assume that the entire sentence can be classified as true or false.

Not all sentences can be assigned a truth value. Some examples of such sentences are:

Example 2:

1. Woodman spare that tree. (Not a statement since it is an imperative sentence rather than a declarative sentence.)
2. Stop, police! (Exclamatory sentence.)
3. Why did you do it? (Interrogative sentence.)
4. This sentence is false. (This is a declarative sentence, but we cannot assign the sentence a unique truth value; since, if it is "true" then it is "false" and if it is "false" then it is "true.")

Since we will be concerned with the truth or falsity of rather complex statements, it is important to make a distinction between the components of the sentence and the overall statement. We will use five basic connectives: "and," "or," "if-then," "if and only if," and "negation" to join the components to form these complex statements.

DEFINITION: A **compound statement** is a statement in which one or more connectives appear.

DEFINITION: A **simple statement** is a statement in which no connectives appear.

We will agree to adopt a notation in which lower-case letters p, q, r, s, t, etc., will represent simple statements and capital letters **P, Q, R, S, T**, etc., will be used to represent compound statements. In the definitions for simple and compound statements we have used the word "connectives" to which we must now give a more precise meaning. The following table lists the basic connectives together with typical English phrasing examples and the symbol for each connective.

Logical Connective	Typical Wording for the Connective Form	Connective Form Symbolized
Negation	not p it is not the case that it is p	$\sim p$
Conjunction	p and q, both p and q p but q	$p \wedge q$
Disjunction	p or q	$p \vee q$
Conditional	if p then q p is a sufficient condition for q q whenever p q if p p only if q q is a necessary condition for p	$p \longrightarrow q$
Biconditional	p if and only if q if p then q and if q then p p is a necessary and sufficient condition for q	$p \longleftrightarrow q$

As the above listing indicates, there are many alternate ways, which we have by no means exhausted, of expressing verbally the same basic connective. In addition, as we translate from words to symbols, we may be required to insert other words and rephrase the sentence until we are satisfied that we have obtained an equivalent logical form. The following examples should help make this point clear.

Example 3:

	Statement	Logical Form
1.	We won't go if it rains.	If it rains then we will not go.

Let "p" represent "it rains" and "q" represent "we will go." The statement symbolized is: $p \longrightarrow \sim q$.

2.	Should the sun shine, we will take the trip.	If the sun should shine then we will take the trip.

Let "p" represent "the sun should shine" and "q" represent "we will take the trip." The statement symbolized is: $p \longrightarrow q$.

3.	It is not the case that Bill or Harry will go with us.	Bill will not go with us and Harry will not go with us.

Let "p" represent "Bill will go" and "q" represent "Harry will go." The statement symbolized is: $\sim p \wedge \sim q$ *or alternately:* $\sim(p \vee q)$.

4.	All cows eat grass.	If it is a cow then it eats grass.

Let "p" represent "it is a cow" and "q" represent "it eats grass." The statement symbolized is: $p \longrightarrow q$.

The biconditional expression, "p if and only if q" represents the conjunction of the two conditionals "if p then q" and "if q then p." In symbolizing statements it may be helpful to observe that "p if q" means "if q then p," symbolized $q \longrightarrow p$ and "p only if q" means "if p then q," symbolized, $p \longrightarrow q$.

Example 4: Let "p" represent "the crops grow" and "q" represent "we fertilize." Symbolize the following statements.
1. The crops grow only if we fertilize.

Solution: $p \longrightarrow q$.

2. The crops grow if we fertilize.

Solution: $q \longrightarrow p$.

3. Fertilizing is a sufficient condition for the crops to grow.

Solution: $q \longrightarrow p$.

4. Fertilizing is a necessary condition for the crops to grow.

Solution: $p \longrightarrow q$.

5. Fertilizing is a necessary and sufficient condition for the crops to grow.

Solution: $p \longleftrightarrow q$.

Example 5: If "p" represents "it is hot," "q" represents "it is humid," and "r" represents "I will go to the city," write out an English sentence translation of the following:
1. $(p \wedge q) \longrightarrow \sim r$.

Solution: If it is hot and humid then I will not go to the city.

2. $p \wedge (q \longrightarrow \sim r)$.

Solution: It is hot, and if it is humid then I will not go to the city.

Example 5, above, shows that the form, and consequently the meaning, of a statement is altered by a change in the way the statement is punctuated; that is, by the way symbols are grouped in parentheses. Now, suppose we consider this same "statement" without any punctuation, i.e., without parentheses. What is its meaning under these conditions: In other words, which of our five logical forms is $p \wedge q \longrightarrow r$? It appears to be either a conjunction or a conditional; but which? Without some prescribed means of making this decision we have to agree that the statement is, at best, ambiguous.

The situation is somewhat analagous to that presented by numerical expressions involving the arithmetic operations of addition and multiplication.

Consider the following example:

$$(2 \times 3) + 4 \neq 2 \times (3 + 4)$$

since

$$(2 \times 3) + 4 = 6 + 4 = 10$$

and

$$2 \times (3 + 4) = 2 \times 7 = 14.$$

Thus changes in the grouping of the numbers may lead to different arithmetic results. However, it is not always necessary to employ grouping symbols since certain basic assumptions are made; namely, multiplication and division, carry precedence over addition and subtraction and we agree to perform the computations from left to right; for example, the expression

$$48 \div 4 + 6 \times 3$$

is equal to 30.

We also recognize that certain combinations of arithmetic symbols are essentially meaningless. For example, the expressions

$$\div - 2$$
$$7 \times 3 + \div$$
$$- \times 2 - 4$$

are meaningless.

For the purpose of clarity and convenience we will make similar assumptions:

A(1) If **P** and **Q** are compound statements, we will regard as meaningful statements only those of the form $\sim \mathbf{P}$, $\mathbf{P} \vee \mathbf{Q}$, $\mathbf{P} \wedge \mathbf{Q}$, $\mathbf{P} \longrightarrow \mathbf{Q}$, and $\mathbf{P} \longleftrightarrow \mathbf{Q}$.

A(2) In the most generalized form, **P** and **Q** when joined by a basic connective are assumed to be **unrelated**; that is, the truth value of **P** is not related to nor dependent upon the truth value of **Q**.

A(3) Statements may be punctuated by grouping symbols, e.g., parentheses ( ), brackets [ ], braces { }, etc. All statement and connective symbols enclosed in a particular grouping symbol are to be considered as one component of the larger statement in which the grouping appears. Thus, when grouping symbols are employed we will agree to perform the operations from the innermost grouping symbols outward.

A(4) If a statement containing two or more connectives is not explicitly punctuated by grouping symbols, the statement will be regarded as having the form associated with the connective having the highest order of dominance in the following list.

Bi-Conditional ($\longleftrightarrow$) dominates all other connectives.

Conditional ($\longrightarrow$) is dominated only by the biconditional.

Conjunction ($\wedge$) and *Disjunction* ($\vee$) are of equal dominance, but are dominated by the conditional and the bi-conditional. *Note*: As a result of the equal dominance of conjunction and disjunction, any statement containing both these connectives, and none of higher dominance, must be punctuated by grouping symbols. Otherwise, the statement is meaningless. *Negation* ($\sim$) is dominated by all other connectives.

Example 6:

$\vee \mathbf{P} \longrightarrow \mathbf{Q}$ is meaningless by A(1).

$(\mathbf{P} \longrightarrow \mathbf{Q}) \wedge \mathbf{R}$ is a conjunction by A(3).

$\sim \mathbf{P} \wedge \mathbf{Q} \longrightarrow \mathbf{R} \vee \mathbf{S}$ is a conditional statement by A(4).

$\sim [\mathbf{P} \wedge \mathbf{Q} \longrightarrow \mathbf{R}]$ is a negation by A(3).

$\mathbf{P} \wedge \mathbf{Q} \vee \mathbf{R}$ is meaningless by the note in A(4).

Example 7: Insert grouping symbols to make the statement form

$$\sim \mathbf{P} \vee \mathbf{R} \longrightarrow \mathbf{S} \longleftrightarrow \mathbf{T}$$

(a) conditional, (b) negation, (c) disjunction.

Solution:

(a) $\sim \mathbf{P} \vee \mathbf{R} \longrightarrow (\mathbf{S} \longleftrightarrow \mathbf{T})$.

(b) $\sim (\mathbf{P} \vee \mathbf{R} \longrightarrow \mathbf{S} \longleftrightarrow \mathbf{T})$.

(c) $\sim \mathbf{P} \vee (\mathbf{R} \longrightarrow \mathbf{S} \longleftrightarrow \mathbf{T})$.

Problem Set 1.3

A. Determine whether each of the following statements is true or false.

1. Symbolic logic has the distinguishing characteristics of a mathematical system.
2. A compound statement is any combination of two or more logical connectives.
3. Some declarative sentences cannot be assigned a truth value.
4. The conditional and biconditional connectives are on the same level of dominance.
5. The sentence, "Should I pass logic, I will take statistics." is a conjunction.

B. Determine which of the following are statements. If it is a statement, classify it as either simple or compound. If it is compound, determine its dominant connective.

1. All mathematics courses are comprehensible.
2. Pay as you go.
3. Students like sports whenever they do not like music.
4. Think about it a while longer.
5. He enjoys logic.
6. If I knew this stuff then it would be easy, and I could pass mathematics.
7. Some people are short.
8. Yankee, go home!
9. Either we go or we stay.
10. Is this the atomic age?

C. Using as few grouping symbols as possible, insert grouping symbols to make the statement form, $\sim p \vee q \longleftrightarrow p \longrightarrow q$, (a) conditional, (b) disjunction, (c) negation.

D. Let "p" represent "he is persistent" and "q" represent "he shall overcome" and "r" represent "he is antagonistic." Translate each of the following into symbolic form.
   1. He is not antagonistic.
   2. It is not the case that he is persistent or that he shall overcome.
   3. He is antagonistic whenever he shall overcome.
   4. He would overcome if he were persistent, and he is not persistent.
   5. If he is persistent then he shall overcome, if and only if he is not persistent or he shall overcome.

E. Let "r" represent "I am involved."
   Let "s" represent "I actively discuss all issues."
   Let "t" represent "I know the answers to all issues."
   Write an English statement that expresses the logical meaning of the following symbolized statement forms:
   1. $\sim s$.
   2. $r \wedge s \longrightarrow t$.
   3. $s \longrightarrow t$.
   4. $r \longleftrightarrow s$.
   5. $r \longrightarrow s \wedge t$.
   6. $\sim s \vee t$.
   7. $\sim (s \vee t)$.
   8. $\sim s \wedge \sim t$.
   9. $(s \vee r) \wedge t$.
   10. $s \vee (r \wedge t)$.

1.4 Truth Tables

Having developed procedures for symbolizing statements and utilizing logical connectives, we now consider the truth values of simple and compound statements.

The simple statement, "This book is red," could be symbolized by the letter "p." The logical possibilities for the truth value of p are: the statement p is true or it is false. In other words, for a simple statement we need consider only two logical possibilities which can be represented in tabular form as follows:

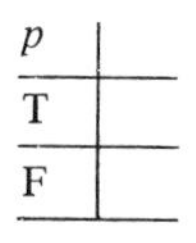

p	
T	
F	

If we were now to consider the negation of p, namely, "$\sim p$," (the book is not red), we can indicate that where previously p was true, $\sim p$ must be false and where p was false $\sim p$ must be true. We will regard the following tabular form as a definition of negation:

DEFINITION: The **negation** of a statement p, denoted $\sim p$, is given by the following truth table:

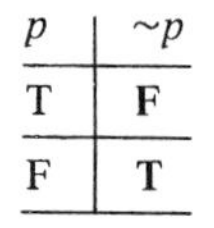

p	$\sim p$
T	F
F	T

To motivate the truth table definitions for a **conjunction** and a **disjunction**, let "This book is red" be represented by p and "This book is black" be represented by q. We must then attempt to assign "true" or "false" to $p \wedge q$ and $p \vee q$ for all logically possible cases. The question which comes to mind is, "How many such logical possibilities are there?" Logically, there are four possibilities; a case where both p and q are true, a case where both p and q are false, a case where p is true and q is false, and finally, a case where p is false and q is true. However, as we continue to consider a compound statement of not only two simple statements, but three or more, we have increasing difficulty to be sure that we have considered all logical possibilities. We need to develop an organized way of presenting the information in a tabular form. We will digress in order to present two concepts which will be useful to us throughout the text, and will give us a systematic way of handling a wide variety of problems.

Fundamental Principle of Counting (F. P. C.)

If a choice for a given action can be made in $\mathbf{n_1}$ ways and this action is followed by a second action that can be performed in $\mathbf{n_2}$ ways, which is then followed by a third action in $\mathbf{n_3}$ ways, and continuing the sequence of actions until the **kth** action can be performed in $\mathbf{n_k}$ ways, then the entire sequence of actions can be performed in:

$$\mathbf{n_1 \times n_2 \times n_3 \times \ldots \times n_k} \quad \textbf{ways.}$$

Example 1: Apply the F.P.C. to our truth table possibilities to determine the number of logical possibilities for a compound statement involving (a) two distinct simple statements, (b) three simple statements, (c) four simple statements, and (d) n simple statements.

Solution:

(a) If our compound statement involves two simple statements, say p and q, then we may first assign true or false to "p" in two ways followed by assigning true or false to "q" in two ways, then there are:

$$2 \times 2 = 2^2 = 4 \text{ logical possibilities.}$$

(b) In a similar manner, if there are three simple statements p, q, r, then we have:

$$2 \times 2 \times 2 = 2^3 = 8 \text{ logical possibilities.}$$

(c) For four simple statements the number of logical possibilities is:

$$2 \times 2 \times 2 \times 2 = 2^4 = 16.$$

(d) For n simple statements, the number of logical possibilities is:

$$\underbrace{2 \times 2 \times 2 \times 2 \times \ldots \times 2}_{n \text{ times}} = 2^n.$$

Although the F.P.C. will inform us of the total number of cases we must consider, we have not yet indicated how we can produce a complete list of the logical possibilities in a systematic manner. A useful device for indicating the logical possibilities is a **tree diagram**. We can best understand the use of a tree diagram by an example.

Example 2: Use a tree diagram to develop the listing of logical possibilities for a compound statement involving (a) two simple statements, (b) three simple statements.

Solution:

(a) Let p and q represent two simple statements. We begin the tree diagram from an initial point where we draw two lines, called branches, representing the two choices for the statement p. We then identify and label each branch with one of the choices. We now consider q and its logical possibilities. We draw the logical possibilities of q as branches from each of the branches of p and label these choices. The set of four logical possibilities may be noted by beginning from the initial point and following the branches to their termination, recording the choices in sequence.

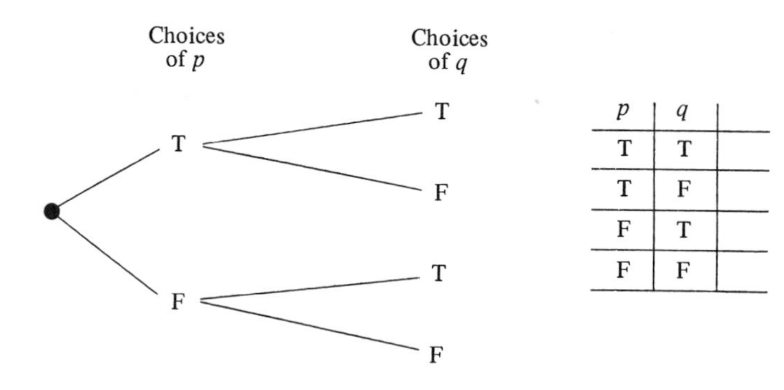

p	q	
T	T	
T	F	
F	T	
F	F	

(b) Extending the tree to include a third simple statement, we obtain the following tree diagram and the corresponding possibilities written in tabular form:

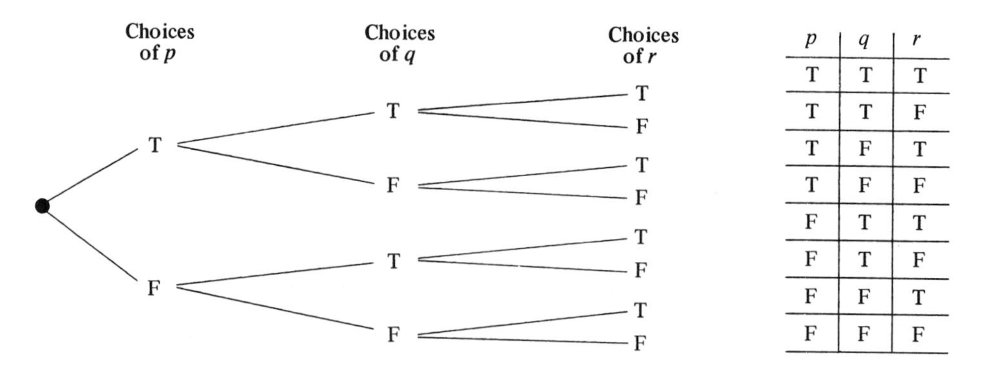

p	q	r
T	T	T
T	T	F
T	F	T
T	F	F
F	T	T
F	T	F
F	F	T
F	F	F

The process described in Example 2 can be extended for any finite number of simple statements. An interesting pattern is evident in the truth table listing which provides us with an automatic method of tabulating the various cases. If two simple statements are involved, we realize by the F.P.C. we must have four logical possibilities; therefore, assign to "p" two "T's" followed by two "F's." Then assign to "q" an alternating sequence of T, F, T, F. If three simple statements are being considered, we use the F.P.C. to mentally compute that eight cases must be considered. Assign four "T's" followed by four "F's" to p, then assign in sequence to q two "T's" followed by two "F's," finally alternate T and F beginning with T for the statement r. In general, we form blocks of "T's" and "F's" by first taking one-half the total number of logical possibilities, one quarter of the total number, etc., until we reach a point where we alternate "T's" and "F's."

We now consider the truth table for a conjunction using as our intuitive example the statement:

"This book is red and this book is black."

Under what conditions does it seem reasonable to consider the entire compound statement true and under what conditions would we regard the compound statement false? If the book is both red and black, we regard the compound statement as true, and if the book is not both red and black we regard the compound statement as false. The example is a typical one for a conjunction and suggests that the following definition will not violate the meaning we attach to "and" in normal discourse.

DEFINITION: The **conjunction** of two simple statements, say p and q, symbolized $p \wedge q$, is given by the truth table:

p	q	$p \wedge q$
T	T	T
T	F	F
F	T	F
F	F	F

The development of a truth table for a disjunction requires a little closer analysis, because in English usage "or" can assume either an exclusive meaning or an inclusive meaning. Consider the following examples:

1. Either the Giants or the Dodgers will win the pennant. (Exclusive–one or the other but not both may win the pennant.)
2. Sue or Mary will be able to help organize a committee. (Inclusive–one or the other or both may help.)

In context there will be a specific meaning attached to "or." However, in an abstract compound statement form, $p \vee q$, p and q represent any simple statements and are assumed to be unrelated (A2). We will therefore interpret "or" in symbolic logic to represent the inclusive sense. We can now discuss a truth value assignment for the four logical possibilities. Consider the statement, "The book is red or the book is black." Our ordinary use of "or" in this sense suggests that the statement may be regarded as true if the book is either red or black and if it is both red and black, and false if the book is neither red nor black. In truth table form, our generalized definition can be expressed in the following manner:

DEFINITION: The **disjunction** of two simple statements, say p, q, symbolized $p \vee q$, is given by the truth table:

p	q	$p \vee q$
T	T	T
T	F	T
F	T	T
F	F	F

We will next turn our attention to the biconditional which as the name suggests involves two conditional statements, "if p then q" and "if q then p."

Example 3: "Democrats will support this piece of legislation if and only if it is to their advantage." This statement contains the two conditionals:

If the Democrats will support this piece of legislation then it is to their advantage.

If it is to their advantage, then the Democrats will support this piece of legislation.

Consideration of the above example suggests that being a Democrat in support of the legislation is tantamount to saying that it is to their advantage and vice versa. The basic ideas are functionally equivalent, suggesting that when the biconditional is true, both simple statements must have the same truth value. The biconditional is false if the simple statements have opposite truth values. This intuitive discussion leads us to define the biconditional as follows:

DEFINITION: The **biconditional** of two simple statements, say p, q, symbolized, $p \longleftrightarrow q$, is given by the truth table:

p	q	$p \longleftrightarrow q$
T	T	T
T	F	F
F	T	F
F	F	T

We have postponed until last consideration of the conditional, since the truth-table definition requires more analysis if we wish an intuitive justification. The conditional consists of a **hypothesis**, also referred to as the **antecedent**, joined in the if-then form with a **conclusion** or what is also called a **consequent**. Consider the following example which we will break up and analyze into the four logical possibilities for the conditional statement:

If birth control information is made readily available, then our social behavior will surely change.

Let us consider in each case when we might have been a truth teller and when we would have been telling a lie.

Suppose both our hypothesis and our conclusion are true, i.e., birth-control information is made readily available and our social behavior does indeed change, then we would identify the entire conditional statement as true.

Now if the hypothesis is true and the conclusion is false, then birth-control information being readily available would not be followed by a change in social behavior. Under these conditions a false truth value would be assigned to the conditional statement.

Consider next the case where both the hypothesis and the conclusion are false. Birth-control information is not made readily available and our social behavior does not change. We have no evidence on which to assign false to the conditional statement and since a statement must be assigned either a true or false, we will agree to assign true to the conditional when both the hypothesis and the conclusion are false. We can rationalize this choice on the basis that we are giving the benefit of the doubt to the if-then statement and will recognize it as true, since we cannot clearly decide it is false. At this point, our discussion has led us to consider three of the four logical possibilities shown in tabular form:

p	q	$p \longrightarrow q$
T	T	T
T	F	F
F	T	
F	F	T

The blank spot where the hypothesis is false and the conclusion is true needs to be assigned either a true or false. Suppose we were to make the assignment to this case "false." Would we not be saying the truth table and the meaning of a conditional is the same as the truth table and meaning of the biconditional? In mathematics and in our ordinary use of the language we never intend that in all cases the hypothesis of any if-then statement has the same logical meaning as the conclusion. As an illustration, the statement

If A and B are right angles, then A and B are equal angles

cannot be distorted to have the same meaning as:

If A and B are equal angles, then A and B are right angles

Otherwise, we would be stating, A and B are right angles if and only if A and B are equal angles. We are left with the only possible choice, namely, assign true to the conditional statement when the hypothesis is false and the conclusion is true. Returning to the example: If we do not make birth-control information available, but social behavior does change, we have no evidence to suggest that the conditional is false. Since we must make a truth value assignment we will regard the conditional as true in this case.

We complete the truth tables for the connectives with the following definition:

DEFINITION: The **conditional** of two simple statements, say p, q, symbolized, $p \longrightarrow q$, is given by the truth table:

p	q	$p \longrightarrow q$
T	T	T
T	F	F
F	T	T
F	F	T

If the reader has absorbed the main points of the discussion so far, we make three remarks which may help to avoid some erroneous conclusions on the reader's part.

Remarks

1. The intuitive discussions which motivated the truth table definitions of each basic connective should not be construed to be a formal justification of these definitions. We have kept one eye on the construction of a mathematical system of symbolic logic and the other eye on possibly significant applications of the model that will preserve the reasoning methods employed in the study of mathematics and will not violate the meaning we ordinarily attach to the connectives as part of our English language. We were, however, under no compulsion to make the truth-value assignments as we did. Other systems in which different truth-value assignments are made provide an interesting field of study though, perhaps of limited utility.
2. The reader is cautioned not to introduce any extramathematical concepts into his interpretation of the connectives, such as the idea of causality. Recall that we have assumed that in an abstract form the statements represented by p and q are unrelated. In no sense have we suggested that a hypothesis causes a certain conclusion. In our system a conditional statement of the form, "If the moon is made of volcanic ash, then there are ten steps in the entrance to the Union Library," is as permissible as any other conditional statement. Mathematicians would be very hard-pressed to develop some rules which would restrict conditional statements to those which we would regard as meaningful. At issue is not the question of whether a specific conditional statement is meaningful or not, but rather does the mathematical model developed lend itself to a significant interpretation?
3. The symbols p, q, r, etc., are variables which we have used to represent simple statements and are not in themselves either true or false. The compound statement form, $p \wedge q$, is again not true or false until we specify the truth value of "p" and the truth value of "q." There is a clear distinction between the form of a statement and its truth value, much in the same sense that the algebraic form $x + 3 = 5$ is neither true nor false until a numerical value is specified for the variable. As we now attempt to determine the truth value of some rather complex statements, it will be convenient for us to have the truth tables for each connective summarized in one table as follows:

		Conjunction	Disjunction	Biconditional	Conditional
p	q	$p \wedge q$	$p \vee q$	$p \longleftrightarrow q$	$p \longrightarrow q$
T	T	T	T	T	T
T	F	F	T	F	F
F	T	F	T	F	T
F	F	F	F	T	T

Although the word has somewhat fallen into disrepute, any student who wishes to master further material must **memorize** the above table. When a person can reproduce a multiplication table or a truth table for a connective in an automatic manner, his mind is free to consider more challenging problems. The following observations may assist the reader in a more rapid assimilation of the basic truth tables.

1. A conjunction $p \wedge q$ is true only when both p and q are true.
2. A disjunction $p \vee q$ is false only when both p and q are false.
3. The biconditional $p \longleftrightarrow q$ is true only when p and q have the same truth value.
4. The conditional $p \longrightarrow q$ is false only when p (hypothesis) is true and q (conclusion) is false.

We will now consider problems where we are interested in determining the truth values of complex statements.

Although there are many different approaches to developing a truth table for complex expressions, we will use a procedure which is both convenient and follows naturally from our rules of dominance.

1. List all of the statement variables and when required, their respective negations in column form alphabetically, starting at the extreme left.
2. Determine the number of rows required and list all of the logical possibilities.
3. If grouping symbols are involved, work within the innermost grouping symbols outward in accordance with the rules of dominance (weakest to strongest). Develop in column form all the required truth values.
4. The final result will be the column of truth values under the dominant connective for the complex statement.

Example 4: Determine under what conditions the following compound statement is true or false.

If I am either happy or miserable, and I am happy then I am not miserable.

Solution: Let "p" represent "I am happy" and "q" represent "I am miserable." The conditional statement may be symbolized as follows:

$$(p \vee q) \wedge p \longrightarrow \sim q.$$

The two statement variables p, q will require a four row truth table where we first list the truth value possibilities for p, q and $\sim q$, and then proceed to form the required truth values under each connective.

The numbers ①, ②, and ③ indicate the sequence of operations. Note that column ② is formed by joining with a conjunction column ① with the truth value column for p row by row, and column ③ is formed by joining with a conditional

p	q	$\sim q$	$(p \vee q) \wedge p \longrightarrow \sim q$		
T	T	F	T	T	F
T	F	T	T	T	T
F	T	F	T	F	T
F	F	T	F	F	T
			①	②	③

column ② with the truth value column for $\sim q$ row by row. The conditional statement $(p \vee q) \wedge p \longrightarrow \sim q$ is true for all cases except when "p" is true and "q" is true.

Although we will not discuss valid argument forms until Sec. 1-7, we point out in passing that if we regard $(p \vee q) \wedge p$ as our hypothesis of the conditional and $\sim q$ as the conclusion, then the conditional is not always true for all possible truth value assignments for p and q and is, therefore, an invalid reasoning process. The following example illustrates the faulty reasoning.

If $2 + 2 = 4$ or $3 \times 2 = 6$, and $2 + 2 = 4$, then $3 \times 2 \neq 6$.

In other words, knowing that one part of a disjunction is true does not permit us to conclude that the other part is false.

Example 5: Develop a truth table for $(p \longrightarrow q) \vee (q \longrightarrow r) \longleftrightarrow (p \longrightarrow r)$.

Solution:

p	q	r	$(p \longrightarrow q)$	$\vee$	$(q \longrightarrow r)$	$\longleftrightarrow$	$(p \longrightarrow r)$
T	T	T	T	T	T	T	T
T	T	F	T	T	F	F	F
T	F	T	F	T	T	T	T
T	F	F	F	T	T	F	F
F	T	T	T	T	T	T	T
F	T	F	T	T	F	T	T
F	F	T	T	T	T	T	T
F	F	F	T	T	T	T	T
			①	④	②	⑤	③

The truth value for the compound statement appears in Column ⑤. Examples 4 and 5 involved statements which were assumed to be unrelated. In the next section we will turn our attention to statements which are related and analyze compound statements where all cases may not be logically possible.

Problem Set 1.4

A. Determine whether each of the following statements is true or false.

1. In logic, a compound statement is an expression which is either true or false, but not both.

2. The negation of a simple statement is always false.
3. The truth-value of a compound statement is uniquely determined if the truth value of each simple statement is known and the definitions of the connectives assumed.
4. The "Fundamental Principle of Counting" is a basic assumption.
5. The conjunction of two simple statements is false when both simple statements are false.
6. The truth-value of a conditional is true when the antecedent is true and the consequent is false.
7. If an expression has three statement variables, say p, q, and r, and if we know the truth value of r is always true, then there are five logical possibilities; since there are two possibilities for p, two for q, and one for r.
8. In everyday usage, the word "or" is used only as a disjunction.
9. The conditional statement is a "causality" relationship between the consequent and the antecedent.
10. A biconditional statement form, $p \longleftrightarrow q$, has the same truth value row by row as does the conjunction of the conditionals $p \longrightarrow q, q \longrightarrow p$.

B. Develop truth tables for each of the following complex expressions:

1. $\sim\sim p$.
2. $\sim(\sim p \vee q)$.
3. $\sim(p \longrightarrow q)$.
4. $p \longrightarrow q \longleftrightarrow \sim q \vee \sim p$.
5. $(p \longrightarrow q) \wedge (q \longrightarrow r) \longrightarrow (p \longrightarrow r)$.
6. $p \longrightarrow q \wedge r$.
7. $p \longrightarrow q \vee r$.
8. $p \wedge q \longrightarrow p$.
9. $(p \vee q) \wedge \sim p \longrightarrow q$.
10. $(p \longrightarrow q) \wedge p \longrightarrow q$.
11. $p \wedge (q \vee r)$.
12. $(p \wedge q) \vee r$.

C. How many rows are there in a truth table of a compound statement involving the statement variables p, q, r, s, and t?

D. If p and q are logically true and r and s are logically false, determine whether the following compound statements are true or false.

1. $\sim q \longrightarrow p \vee q \longleftrightarrow q \vee p$.
2. $[p \wedge (q \vee r)] \vee \sim[(p \wedge q) \vee (p \wedge r)]$.
3. $[(p \wedge q) \longrightarrow r] \longleftrightarrow [r \longrightarrow (p \longleftrightarrow q)]$.
4. $\sim(q \vee p) \wedge r \longrightarrow \sim(q \vee \sim r) \wedge \sim p$.
5. $\sim s \wedge (p \longleftrightarrow r) \longrightarrow (q \longrightarrow r \vee \sim s) \longleftrightarrow r$.

1.5 Related Statements

In previous sections we have considered simple statements joined together by basic connectives to form compound statements. We have assumed the simple statements to be unrelated, leading us to examine all logically possible cases. However, when we substitute specific statements for the variables p, q, r, etc., all of the cases may not be logically possible. Consider as an example the following statements:

Let p represent "Jim is the father of Harry."
Let q represent "Harry is the father of Jim."

If p and q were unrelated, we would list the four possible cases

	p	q
①	T	T
②	T	F
③	F	T
④	F	F

but is it not clear that for this example that case ① is impossible, since it is not possible for each to be the father of the other, i.e., for p and q to both be true? However, the other three cases are logically possible. When a situation like this occurs, we say that p and q are related statements.

DEFINITION: Statements **P** and **Q** are **related** if* one or more of the cases for which truth values are to be assigned to two statements **P** and **Q** are not logically possible.

There are many relations that we might consider, however, we will restrict our attention to three relations of fundamental importance: **implication**, **equivalence**, and **inconsistency**.

Before we define or discuss in detail these three relations we will introduce two terms which will help clarify our discussion.

DEFINITION: A statement which for all possible cases is true is said to be **logically true** or a **tautology**.

Example: The statement, "The book is red or it is not red," is a tautology. In general, the statement form, $p \vee \sim p$, is a tautology.

DEFINITION: A statement which for all possible cases is false is said to be **logically false** or a **self-contradiction.**

Example: The statement, "The book is red and it is not red," is a self-contradiction. In general, the statement form, $p \wedge \sim p$ is a self-contradiction.

We will now define implication, equivalence, and inconsistency.

DEFINITION: **P implies Q**, symbolized **P** $\Longrightarrow$ **Q**, means that whenever **P** is true **Q** must be true.

*Although definitions may be written as an if-then statement, definitions will always be considered as "if and only if" statements.

Note: in accordance with the definition, if **P** is false **P** implies **Q** irrespective of the truth value of **Q**.

> **DEFINITION: P and Q are equivalent**, symbolized **P**$\Longleftrightarrow$**Q**, if and only if whenever **P** is true **Q** must be true and whenever **P** is false **Q** must be false.

> **DEFINITION: P and Q are inconsistent** if and only if there is no case when **P** and **Q** are both true.

For each of the three relations the definition will exclude one or more of the logically possible cases for the two statements **P** and **Q**. When **P** implies **Q**, the definition requires that whenever **P** is true, **Q** must also be true; there is therefore no instance when **P** is true and **Q** is false. If we join **P** and **Q** with the conditional connective, the compound statement **P** $\longrightarrow$ **Q** would be logically true. This perhaps explains why the symbol for implication ($\Longrightarrow$) is similar to the symbol for the conditional connective ($\longrightarrow$). Do not confuse the two, however, implication is a relation between statements, whereas the conditional connective forms new compound statements by joining two statements together. This situation is quite similar to that which we find in typical arithmetic operations. When we symbolize $2 + 3 = 5$ we are indicating how we obtain "5" by combining together 2 and 3 under the operation (connective) of addition. We could also observe that there is a relation between 2 and 3; namely 3 is greater than 2 ($3 > 2$).

The relation of equivalence will occur if whenever **P** is true so is **Q**, and if whenever **P** is false so is **Q**. The statements **P** and **Q** could not have opposite truth values, which means we would be excluding two cases; one where **P** is true and **Q** false, and the other when **P** is false and **Q** is true. Under these conditions, if **P** and **Q** were joined with the biconditional connective, the compound statement, **P** $\longleftrightarrow$ **Q**, would be logically true. Again the choice of symbols for equivalence; namely, $\Longleftrightarrow$, is analogous to, but not the same as the biconditional connective. When statements are equivalent we are sure that they have the same truth table, row by row and may be substituted for each other. This property is of sufficient importance to our future work that we state it as an axiom.

> ***SUBSTITUTION AXIOM:*** When two statements, **P** and **Q**, are equivalent, they have the same logical meaning and may be substituted for each other in any expression.

The definition of inconsistency excludes the case where **P** and **Q** are both true. If there is at least one case where **P** and **Q** are both true, we regard the statements **P** and **Q** as a **consistent** pair of statements.

A considerable amount of mental effort is spent in the search for nontrivial tautologies and in considering the implications of certain thoughts or actions. In the deductive reasoning process, whether it be in mathematics, in the sciences, or in any other activity, we will ultimately produce statements which if not in the *if-then* form originally may be translated in an equivalent manner to that form. When this transition is accomplished, we are faced with the question, does the hypothesis imply the conclusion? If the hypothesis is accepted as false, then the hypothesis implies any conclusion we wish to make, but we obtain little benefit from examining false hypotheses. What is the situation if we are ready to accept the hypothesis as true? We then seek to establish that there is no case in which the conclusion could be false when the hypothesis is assumed true. Volumes could be (indeed, have been) written to discuss schemes of deducing conclusions from hypotheses which are assumed true. We will deal with this problem in terms of symbolic logic in the sections which follow. However, one additional observation should be made at this time. If we know that we have produced a string of theorems $(t_1, t_2, t_3, \ldots, t_n)$ where we have established that

$$t_1 \Rightarrow t_2 \Rightarrow t_3 \Rightarrow \ldots \Rightarrow t_{n-1} \Rightarrow t_n$$

then we must, from a mathematical system point of view, raise a fundamental question. Is the entire string of theorems consistent, i.e., is there at least one case where all the theorems are true? It would indeed be a disastrous situation to discover that we have developed in the same system two theorems which contradict each other and thus could not both be true. A mathematical system is extended by showing that a new theorem is implied by previously developed theorems. Showing that a mathematical system is consistent is a topic of considerable complexity and beyond the scope of this text, however, it is one which continues to be a formidable and lively challenge to contemporary mathematicians.

Example 1: Given the statements $p, q, p \wedge q$, and $p \vee q$, discuss any relations which hold between pairs of these statements.

Solution: Develop a truth table for each statement and determine whether or not the basic definitions are satisfied.

p	q	$p \wedge q$	$p \vee q$
T	T	T	T
T	F	F	T
F	T	F	T
F	F	F	F

(a) $p \wedge q \Rightarrow p$ and $p \wedge q \Rightarrow q$, since whenever $p \wedge q$ is true so are p and q.
(b) $p \wedge q \Rightarrow p \vee q$, since whenever $p \wedge q$ is true so is $p \vee q$.
(c) $p \Rightarrow p \vee q$ and $q \Rightarrow p \vee q$, since whenever p and q are true so is $p \vee q$.
(d) None of the statement pairs are equivalent, since none of the truth tables are identically the same row by row.
(e) All of the statement pairs are consistent, i.e., there is at least one case where they are both true.

Example 1 illustrates that the conjunction of any two simple statements implies either of the simple statements. For example, "The book is both red and black" implies

"The book is red." Example 1 also illustrates that any simple statement implies the given statement in disjunction with any other statement. For example, "The book is red" implies "The book is red or black." The question of whether or not a statement or series of statements imply other statements is what a valid reasoning process is all about. In Sec. 1-7 we will discuss in considerable detail valid argument forms.

Example 2: Show that $p \wedge q \longrightarrow p$ is a tautology and give a sentence example for this tautology.

Solution: Since in all logically possible cases the conditional statement is true, the statement is a tautology. Example: "If a number is even and a multiple of three then it is even."

p	q	$p \wedge q$	$\longrightarrow p$
T	T	T	T
T	F	F	T
F	T	F	T
F	F	F	T
		①	②

Example 3: Show that $\sim [(p \vee q) \wedge \sim p \longrightarrow q]$ is a self-contradiction.

Solution: Since in all cases the compound statement is false, the statement is a self-contradiction.

p	q	$\sim p$	$\sim$	$[(p \vee q)$	$\wedge$	$\sim p \longrightarrow q]$
T	T	F	F	T	F	T
T	F	F	F	T	F	T
F	T	T	F	T	T	T
F	F	T	F	F	F	T
			④	①	②	③

A conditional statement of the form, $p \longrightarrow q$, is very important in mathematics where we attempt to deduce a conclusion q from a given hypothesis p. Situations frequently arise where it is convenient to investigate variations of a given conditional statement. For example, if we interchange the two statements p and q to form the conditional statement, $q \longrightarrow p$, we have a new statement called the **converse** of the given conditional statement, $p \longrightarrow q$. Two other useful variations on the conditional statement, $p \longrightarrow q$, are the **inverse**, $\sim p \longrightarrow \sim q$, and the **contrapositive**, $\sim q \longrightarrow \sim p$.

Example 4: Develop truth tables for the conditional $p \longrightarrow q$, the converse of the conditional, $q \longrightarrow p$, the inverse of the conditional, $\sim p \longrightarrow \sim q$, and the contrapositive of the given conditional, $\sim q \longrightarrow \sim p$. Discuss any implications or equivalences that appear to hold between these compound statements. The reader can verify that the final truth tables are of the form:

Conditional	**Converse**	**Inverse**	**Contrapositive**
$p \longrightarrow q$	$q \longrightarrow p$	$\sim p \longrightarrow \sim q$	$\sim q \longrightarrow \sim p$
T	T	T	T
F	T	T	F
T	F	F	T
T	T	T	T

Observe that: (1) the conditional $p \longrightarrow q$ is equivalent to the contrapositive [i.e., $(p \longrightarrow q) \Longleftrightarrow (\sim q \longrightarrow \sim p)$] and they imply each other, and (2) the converse $q \longrightarrow p$ and the inverse $\sim p \longrightarrow \sim q$ are equivalent [i.e., $(q \longrightarrow p) \Longleftrightarrow (\sim p \longrightarrow \sim q)$] and they also imply each other. However, the conditional and the converse are not equivalent nor are the conditional and the inverse equivalent.

Example 5: Write the converse, inverse, and contrapositive of:
(a) If he is over 21 years of age then I'll eat my hat.

Solution:

Converse: If I'll eat my hat then he is over 21 years of age.
Inverse: If he is not over 21 years of age then I will not eat my hat.
Contrapositive: If I will not eat my hat then he is not over 21 years of age.

(b) I will attend every mathematics class of this course whenever the topics in the course are fully explained.

Solution: First, let "p" represent "I will attend every mathematics class of this course," and let "r" represent "The topics in the course are fully explained." Then the given conditional can be symbolically represented as: $r \longrightarrow p$. In symbols, we can represent the converse, inverse, and contrapositive of the given conditional respectively as: $p \longrightarrow r$, $\sim r \longrightarrow \sim p$, and $\sim p \longrightarrow \sim r$. It is now relatively easy (we leave it to the reader) to write an English statement for the converse, inverse, and contrapositive of the given conditional.

Problem Set 1.5

A. Determine whether each of the following statements is true or false.

1. The negation of a tautology is a self-contradiction.
2. A conditional statement form is always an implication.
3. A biconditional statement form is equivalent to a double conditional statement form, i.e., $(\mathbf{P} \longleftrightarrow \mathbf{Q}) \Longleftrightarrow (\mathbf{P} \longrightarrow \mathbf{Q}) \wedge (\mathbf{Q} \longrightarrow \mathbf{P})$.
4. If **P** and **Q** are related and **P** and **Q** never have the same truth value, then **P** and **Q** are consistent.
5. The converse and contrapositive of the conditional, $p \longrightarrow q$, are equivalent.

B. In each of the following, a pair of related statements is given. Determine which are equivalent and/or which one implies the other.

1. $\mathbf{P}, \mathbf{P} \vee \sim\mathbf{Q}$.
2. $\sim\mathbf{P} \vee \mathbf{Q}, \mathbf{P} \longrightarrow \mathbf{Q}$.
3. $\sim(\mathbf{P} \vee \mathbf{Q}), \sim\mathbf{P} \wedge \sim\mathbf{Q}$.
4. $(\mathbf{P} \longrightarrow \mathbf{Q}) \wedge (\mathbf{Q} \longrightarrow \mathbf{R}), \mathbf{P} \longrightarrow \mathbf{R}$.
5. $\mathbf{P}, \mathbf{P} \wedge \sim\mathbf{Q}$.
6. $\mathbf{P}, \mathbf{P} \longrightarrow \mathbf{Q}$.

C. Write the converse, inverse, and contrapositive of the conditional, "I pass this course whenever I study."

D. Write the following sentences in *if-then* form.

1. A person does not study whenever the person is a student.
2. All women are fickle.
3. No mathematics books can be read easily.
4. He will not go if you go.
5. Physical fitness is a necessary condition for exercising regularly.

E. Find the truth value of the following expressions given that p is logically false, q is logically false, r is logically true, and u is logically true.

1. $(p \longleftrightarrow r) \longrightarrow [(u \wedge r) \longrightarrow (p \vee q)]$.
2. $\sim p \wedge [(q \vee u) \wedge \sim (p \longleftrightarrow q)]$.
3. $(q \longrightarrow u) \longleftrightarrow (\sim q \vee r) \wedge r$.
4. $(\sim p \vee r) \longrightarrow (u \longrightarrow p) \longleftrightarrow \sim (p \vee r)$.
5. $p \wedge (q \vee r) \longrightarrow u \longleftrightarrow \sim (p \vee q) \longrightarrow \sim r$.

F. Change the following expressions to equivalent expressions to contain only the two connectives, disjunction and negation. [*Hint*: Use the equivalences: $(p \longrightarrow q) \Longleftrightarrow \sim p \vee q, \sim(p \wedge q) \Longleftrightarrow \sim p \vee \sim q, (p \longleftrightarrow q) \Longleftrightarrow (p \longrightarrow q) \wedge (q \longrightarrow p)$ and $p \Longleftrightarrow \sim \sim p$.]

1. $p \wedge q$.
2. $(p \longrightarrow q) \longrightarrow r$.
3. $(p \wedge q) \longrightarrow (p \wedge r)$.
4. $(p \wedge q) \vee (r \longleftrightarrow s)$.
5. $(p \longleftrightarrow q) \longrightarrow (r \longleftrightarrow s)$.

G. The British humorist C. Northcote Parkinson has stated that "Expenses rise to meet income," and "Work expands to fill the available time." These statements may be paraphrased to mean, "If income is available then expenses will rise to meet it," and "If time is available, work will expand to fill it." Show that these statements imply, "If income and time are available, then expenses will rise to meet income and work will expand to fill the available time." (*Hint*: Symbolize each statement and assume the given statements are true.)

1.6 Properties of Connectives

In the previous section we observed that a conditional statement and its converse are not equivalent. In other words, we may not in general interchange the hypothesis and conclusion of an if-then statement and obtain an equivalent expression. Mathematicians would describe this situation by stating that the operation is not commutative. In conventional arithmetic we have operations such as addition and multiplication which do satisfy a commutative property. For example,

$$2 + 3 = 3 + 2 \quad \text{and} \quad 2 \times 3 = 3 \times 2$$

and in general, for any numbers a and b,

$$a + b = b + a \quad \text{and} \quad a \times b = b \times a.$$

There are many other properties of arithmetic operations with which the reader is undoubtedly familiar, although the formal names for such properties may be unknown to him. Considerable insight can often be gained into a mathematical system with which one is quite familiar (such as arithmetic) if he investigates other mathematical systems to see if similar properties are preserved. We shall list some familiar arithmetic properties and then attempt to make some correspondence with our system of symbolic logic.

Example	General Property	Formal Name
1. $2 + 3 = 3 + 2$	$a + b = b + a$	Commutative (+)
2. $2 \times 3 = 3 \times 2$	$a \times b = b \times a$	Commutative (×)
3. $2 + (3 + 4) = (2 + 3) + 4$	$a + (b + c) = (a + b) + c$	Associative (+)
4. $2 \times (3 \times 4) = (2 \times 3) \times 4$	$a \times (b \times c) = (a \times b) \times c$	Associative (×)
5. $2 \times (3 + 4) = (2 \times 3) + (2 \times 4)$	$a \times (b + c) = (a \times b) + (a \times c)$	Distributive (×) over (+)

By induction and analogy we might experiment to see whether such properties "carry over" to the connectives $\sim, \vee, \wedge, \longrightarrow$, and $\longleftrightarrow$. We leave to the reader the verification (by truth tables) that the following properties hold in our mathematical system of symbolic logic.

General Property	Formal Name
1. $p \vee q \Longleftrightarrow q \vee p$	Commutative (disjunction)
2. $p \wedge q \Longleftrightarrow q \wedge p$	Commutative (conjunction)
3. $p \longleftrightarrow q \Longleftrightarrow q \longleftrightarrow p$	Commutative (biconditional)
4. $p \vee (q \vee r) \Longleftrightarrow (p \vee q) \vee r$	Associative (disjunction)
5. $p \wedge (q \wedge r) \Longleftrightarrow (p \wedge q) \wedge r$	Associative (conjunction)
6. $p \longleftrightarrow (q \longleftrightarrow r) \Longleftrightarrow (p \longleftrightarrow q) \longleftrightarrow r$	Associative (biconditional)
7. $p \wedge (q \vee r) \Longleftrightarrow (p \wedge q) \vee (p \wedge r)$	Distributive (conjunction over disjunction)
8. $p \vee (q \wedge r) \Longleftrightarrow (p \vee q) \wedge (p \vee r)$	Distributive (disjunction over conjunction)

The above list by no means exhausts all the properties that could be verified, yet we have enough to observe some rather surprising results. The commutative and associative properties hold for three of the connectives, whereas we have such properties for only two arithmetic operations: multiplication and addition. We have a distributive property of conjunction over disjunction, as well as disjunction over conjunction. Should this not seem rather remarkable, consider the situation in arithmetic if we were to reverse the operations of multiplication and addition in the distributive property as illustrated in the following example:

$$2 + (3 \times 4) \stackrel{?}{=} (2 + 3) \times (2 + 4)$$

$$2 + 12 \stackrel{?}{=} 5 \times 6$$

$$14 \neq 30.$$

The above counterexample demonstrates that in arithmetic addition does not distribute over multiplication.

A study of similar basic properties in different mathematical systems is typical of the spirit that pervades contemporary mathematics.

Problem Set 1.6

A. In Problems 1-5, determine whether the pair of expressions are equivalent:

1. $p \wedge (q \vee r), (p \wedge q) \vee (p \wedge r)$.
2. $p \longrightarrow (q \wedge r), (p \longrightarrow q) \wedge (p \longrightarrow r)$.
3. $(q \wedge r) \longrightarrow p, (q \longrightarrow p) \wedge (r \longrightarrow p)$.
4. $p \longleftrightarrow (q \longrightarrow r), (p \longleftrightarrow q) \longrightarrow (p \longrightarrow r)$.
5. $p \longleftrightarrow (q \wedge r), (p \longleftrightarrow q) \wedge (p \longleftrightarrow r)$.
6. Prove or disprove that the conditional is commutative.
7. Prove or disprove that the conditional is associative.
8. Does the biconditional distribute over disjunction? Also, does disjunction distribute over the biconditional?
9. Does disjunction distribute over the conditional?
10. Does the conjunction distribute over the conditional?

1.7 Arguments

When we began our study of symbolic logic, we stated that logic is the study of valid inference and that the overall form or pattern of an argument is of primary importance rather than the truth or falsity of either the premises or the conclusion. We are now ready to give a precise definition of validity of an argument and to discuss methods which are used to determine whether an argument is valid or invalid. First, let us consider some examples of arguments and attempt to determine intuitively whether in each example the conclusion "follows" from the given premises, i.e., is the argument valid? The reader is advised to consider each example carefully before reading further.

Example 1:

1. If dogs bark then horses meow.
2. Horses do meow.

∴ Dogs bark.

Example 2:

1. Horses fly or cats swim.
2. Cats don't swim.

∴ Horses fly.

Example 3:

1. If the moon is made of cream cheese, then there are mice on the moon.
2. The moon is not made of cream cheese.

$\therefore$ There are no mice on the moon.

The argument presented in Example 1 and Example 3 is invalid and the argument in Example 2 is valid. Observe that Example 1 which we have labeled as invalid has a "true" conclusion and Example 2 which we have labeled as valid has a "false" conclusion. The truth or falsity of these particular premises or the specific conclusion does not determine the validity or invalidity of an argument. We cannot depend on some intuitive feel for the form or structure of an argument as a test of validity. We need to state a definition of what we mean by a valid argument and to indicate how we can apply the definition to a specific argument.

DEFINITION: An **argument is valid** if and only if the conjunction of the premises implies the conclusion. Equivalently, an argument is valid if and only if whenever all the premises are true, the conclusion is also true.

In order to be certain that an argument is valid, we must verify that in all the logically possible cases where the conjunction of the premises is true, the conclusion is also true. Since the conjunction of the premises can be true only when each and every premise is true, our work is somewhat simplified. If in any of the logically possible cases considered any premise is false, the conjunction must be false, insuring in that instance that the implication relation holds.

Now we will consider three methods which we will use to determine if the conjunction of the premises implies the conclusion.

METHOD I TRUTH TABLES: To illustrate the use of truth tables in testing an argument for validity or invalidity, let us reconsider the three examples given at the beginning of this section. Example 1 may be represented in the symbolic form as follows:

Example 1A:

(1) $p \longrightarrow q$.
(2) q.

$\therefore p$.

p	q	$(p \longrightarrow q)$	$\wedge\ q$	p
T	T	T	T	T
T	F	F	F	T
F	T	T	Ⓣ	Ⓕ
F	F	T	F	F
		①	②	

Notice that in the third row the conjunction of the premises is "true" and the conclusion "false." One such case is sufficient to verify that the argument is not valid. The third row also provides us with the information that the implication does not hold under the condition that "p" is false and "q" is true. A particular argument of this type will indicate the invalid form.

(1) If $2 + 2 = 5$, then $2 \times 3 = 6$.
(2) $2 \times 3 = 6$.
$\therefore 2 + 2 = 5$.

Another variation on the truth-table method is to join the conjunction of premises and the conclusion with a conditional connective. If the conditional statement formed is logically true, then there is no instance where the premises are all true and the conclusion false; thus the argument is valid. If the conditional statement is not logically true, there is at least one instance where the premises are all true and the conclusion false; thus the argument is invalid. We will use this alternate method in testing Example 2 for validity or invalidity.

Example 2A:

(1) $p \vee q$.
(2) $\sim q$.
$\therefore p$.

p	q	$\sim q$	$(p \vee q)$	$\wedge \sim q$	$\longrightarrow$	p
T	T	F	T	F	T	T
T	F	T	T	T	T	T
F	T	F	T	F	T	F
F	F	T	F	F	T	F
			①	②	③	

Since the conditional statement is logically true, the argument is valid.

The third example will be tested by illustrating that our task can be expedited by considering only those cases where the conjunction of the premises is true.

The symbolic form of Example 3 is shown below:

Example 3A:

(1) $p \longrightarrow q$.
(2) $\sim p$.
$\therefore \sim q$.

p	q	$\sim p$	$\sim q$	$(p \longrightarrow q)$	$\wedge \sim p$	$\longrightarrow$	$\sim q$
T	T	F	F	T			
T	F	F	T				
F	T	T	F	T	Ⓣ	F	Ⓕ
F	F	T	T	T	T	T	
				①	②		

In the third row we have a false under the conditional connective indicating the argument is invalid.

The third row corresponds to the case where p is false and q true. A substitution instance of this type illustrating the invalid form is as follows:

(1) If $2 + 3 = 7$, then $5 \times 2 = 10$.
(2) $\underline{2 + 3 \neq 7.}$
$\therefore 5 \times 2 \neq 10.$

The truth-table method can, of course, be extended to arguments involving more complex premises but as we get to 32, or 64 rows, etc., in the truth table, our efforts although fruitful are tedious. The next two methods which we will consider are more sophisticated and will simplify our testing procedure.

METHOD 2 PROOF BY REDUCTION: In a proof by reduction, we attempt to show that the given argument is invalid by assigning false to the conclusion and then in a step-by-step sequence, attempt to assign true to all the premises. If we succeed then the argument is invalid. If we are unable to assign a false truth value to the conclusion and true to all the premises, then the argument is not invalid. If the argument is not invalid, then it must be valid. By this method we are simply **reducing** the number of rows in a truth table that we investigate to only those cases where the argument could possibly be invalid. Since this can occur only when the conclusion is false, we arbitrarily make the conclusion false by assigning to the statement variables in the conclusion the necessary truth values which will make the conclusion false. Once an assignment of truth values is given to the variables appearing in the conclusion, then these same truth values are assigned to any of these variables appearing in the premises. Then, beginning with the simplest premise, we assign to the other variables a truth value that will make each premise true, if possible. Whatever truth-value assignment is made to a variable, it must be consistently applied throughout all of the premises. At each step it is important to make certain the truth value assignment is necessarily the one chosen. If an optional truth-value assignment could have been made, then more than one case may have to be analyzed. We then continue the process described until all of the logically possible cases have been examined.

Let us illustrate this method by again considering the three examples that appeared at the beginning of this section.

Example 1B:

(1) $p \longrightarrow q$.
(2) $\underline{q.}$
$\therefore p.$

Step 1: Assign F to the conclusion.

(1) $p \longrightarrow q$.
(2) $\underline{q.}$
$\therefore p$ (F).

Step 2: In premise (1) we must also assign the same truth value to p, namely "F":

$$\begin{array}{ll} & \text{(F)} \\ (1) & p \longrightarrow q. \\ (2) & \underline{q.\qquad} \\ & \therefore p \text{ (F)}. \end{array}$$

Step 3: In order to make premise (2) true, we must assign "T" to q.

$$\begin{array}{ll} & \text{(F)} \\ (1) & p \longrightarrow q. \\ & \text{(T)} \\ (2) & \underline{q.\qquad} \\ & \therefore p \text{ (F)}. \end{array}$$

Step 4: In premise (1) we must also assign "T" to "q" as determined in Step 3:

$$\begin{array}{lll} & \text{(F)} \quad \text{(T)} & \\ (1) & p \longrightarrow q & \text{True.} \\ (2) & \underline{q \text{ (T)}} & \underline{\text{True.}} \\ & \therefore p \text{ (F)} & \text{False.} \end{array}$$

The truth-value assignments have made premise (1) true and premise (2) true, and the conclusion false. Thus, the argument is **invalid**. This corresponds to the third row in the truth table of Example 1A. This process need not, of course, be written out in a detailed step-by-step form as we did for purposes of clarity but may be shortened considerably as illustrated in the next example.

Example 2B:

$$\begin{array}{ll} (1) & p \vee q. \\ (2) & \underline{\sim q.} \\ & \therefore p. \end{array} \qquad \begin{array}{llll} & \text{(F) (F)} & & \\ (1) & p \vee q & 3 & \text{False.} \\ & \text{(T) (F)} & & \\ (2) & \underline{\sim \; q} & 2 & \underline{\text{True.}} \\ & \therefore p \text{ (F)} & 1 & \text{False.} \end{array}$$

It is impossible for us to assign false to the conclusion and true to both premises. Since the argument is not invalid, it is valid. The numbers in the column at the right indicate the order of sequence in assigning T's and F's. Finally, let us consider Example 3.

Example 3B:

$$\begin{array}{ll} (1) & p \longrightarrow q. \\ (2) & \underline{\sim p.} \\ & \therefore \sim q. \end{array} \qquad \begin{array}{lll} \text{(F)} \quad \text{(T)} & & \\ p \longrightarrow q & 3 & \text{True.} \\ \text{(T)} \quad \text{(F)} & & \\ \underline{\sim \quad p} & 2 & \text{True.} \\ \text{(F) (T)} & & \\ \therefore \sim \; q & 1 & \text{False.} \end{array}$$

The truth-value assignment shown will make each premise true and the conclusion false. The argument is invalid.

Consideration of a slightly more complex example will demonstrate the strength and the simplicity of this method.

Example 4: Symbolize and test for validity, using the Proof by Reduction method, the argument:

> If it rains today, we will not go on a picnic. We will go on a picnic or be forced to stay indoors. It rained today. Therefore, we will be forced to stay indoors.

Solution: The argument is symbolized and the truth value assignment shown below:

	Statement	Truth values	Step
(1)	$R \longrightarrow \sim P$	(T) (F) (T)	4 False.
(2)	$P \vee S$	(T) (F)	3 True.
(3)	R	(T)	2 True.
	$\therefore S$	(F)	1 False.

Since we were unable to assign true to all the premises and false to the conclusion, the argument is valid.

Now it might appear that the proof by reduction method is completely adequate for testing all arguments; however, as the number of premises and statements increases, we may not find that a unique assignment of T's and F's can be made. Although we could still work through all the possibilities which occur to us we run the risk of overlooking an instance that might have been quite crucial to our analysis.

METHOD 3 VALID ARGUMENT FORMS (RULES OF INFERENCE): This method will provide us with a technique that permits us to either deduce the given conclusion by valid argument forms or reduce a complex set of premises to a manageable number of premises for application of either Method 1 or Method 2.

The rationale for the method lies in the fact that in testing an argument to determine if a string of premises implies a certain conclusion, we may be able to observe that some of the premises imply another statement which may now be used in place of these premises. In other words, if $p_1 \wedge p_2 \wedge p_3 \wedge \cdots\ p_n \Longrightarrow C$ and $p_1 \wedge p_2 \Longrightarrow D$, then $D \wedge p_3 \wedge \cdots \wedge p_n \Longrightarrow C$. For example, suppose that we have a list of premises in which we have included two premises $p \longrightarrow q$ and p together with other premises and a conclusion:

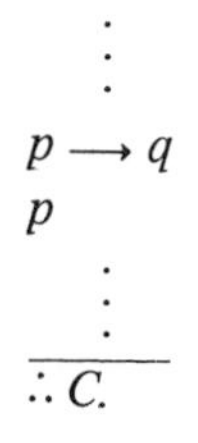

From these two premises, we may conclude the statement q. This is justified, since

$[(p \longrightarrow q) \wedge p] \Longrightarrow q$ is a valid argument form. (The reader should verify that this is indeed a valid argument.) We may now replace $p \longrightarrow q$ and p with the statement q and use q together with any of the other premises to further simplify the overall argument. A listing of the basic valid argument forms which we will employ to simplify our testing for validity is included. The reader should verify that each of these forms is indeed a valid argument. The rather formidable list can be more easily absorbed if the reader will make up a specific example of each of the given valid argument forms. For example, the disjunctive syllogism:

$$\begin{array}{l} p \vee q \\ \underline{\sim p} \\ \therefore q \end{array}$$

can be illustrated by the specific argument,

"The book is red or it is black.
It isn't red.
∴ It is black."

VALID ARGUMENT FORMS
(Rules of Inference)

1. **Disjunctive Syllogism**

$$\begin{array}{l} P \vee Q \\ \underline{\sim P} \\ \therefore Q \end{array}$$

2. **Addition**

$$\begin{array}{l} \underline{P} \\ \therefore P \vee Q \end{array}$$

3. **Modus Ponens**

$$\begin{array}{l} P \longrightarrow Q \\ \underline{P} \\ \therefore Q \end{array}$$

4. **Simplification**

$$\begin{array}{l} \underline{P \wedge Q} \\ \therefore P \end{array}$$

5. **Modus Tollens**

$$\begin{array}{l} P \longrightarrow Q \\ \underline{\sim Q} \\ \therefore \sim P \end{array}$$

6. **Hypothetical Syllogism**

$$\begin{array}{l} P \longrightarrow Q \\ \underline{Q \longrightarrow R} \\ \therefore P \longrightarrow R \end{array}$$

7. **Conjunction**

$$\begin{array}{l} P \\ \underline{Q} \\ \therefore P \wedge Q \end{array}$$

8. $P \Longleftrightarrow \sim\sim P$. Thus we have $\dfrac{P}{\therefore \sim\sim P}$ and $\dfrac{\sim\sim P}{\therefore P}$ **(Double negation)**

9. $(\sim P \vee Q) \Longleftrightarrow (P \longrightarrow Q)$ **(Material implication)**

10. $(P \longrightarrow Q) \Longleftrightarrow (\sim Q \longrightarrow \sim P)$ **(Contraposition)**

11a. $\sim(P \vee Q) \Longleftrightarrow \sim P \wedge \sim Q$ **(De Morgan's laws)**

11b. $\sim(P \wedge Q) \Longleftrightarrow \sim P \vee \sim Q$

12. $(P \longleftrightarrow Q) \Longleftrightarrow (P \longrightarrow Q) \wedge (Q \longrightarrow P)$ **(Material equivalence)**

13a. $(P \vee Q) \Longleftrightarrow (Q \vee P)$ **(Commutation)**

13b. $(P \wedge Q) \Longleftrightarrow (Q \wedge P)$

14a. $[P \vee (Q \vee R)] \Longleftrightarrow [(P \vee Q) \vee R]$ **(Association)**

14b. $[P \wedge (Q \wedge R)] \Longleftrightarrow [(P \wedge Q) \wedge R]$

It is important to point out that each of the given valid argument forms is a completely general form and is not restricted to the given letters of nonnegated statement forms. For example,

$$\begin{array}{l} \sim r \rightarrow \sim t \\ \underline{\sim r \qquad\quad} \\ \therefore \sim t \end{array}$$

is a direct application of Modus Ponens and

$$\begin{array}{l} \sim s \vee \sim t \\ \underline{\quad t \qquad\quad} \\ \therefore \sim s \end{array}$$

is a direct application of Disjunctive Syllogism.

Let us illustrate the application of this method by reconsidering Example 4.

Example 4A:

(1) $R \rightarrow \sim P.$
(2) $P \vee S.$
(3) $\underline{R. \qquad\quad}$
$\therefore S.$

Premise (1) and (3) may be combined to produce a new premise (4) as follows:

(1) $R \rightarrow \sim P.$
(3) $\underline{R. \qquad\quad}$
(4) $\therefore \sim P$ (Modus Ponens).

Now we take premise (2) with premise (4) to give us the desired conclusion; i.e.,

(2) $P \vee S.$
(4) $\underline{\sim P.}$
$\therefore S$ (Disjunctive Syllogism).

In practice, we can condense the format by listing in an adjacent column the new lines which we form and the supporting reason without writing out each subargument. This approach is used in the last two examples given in this section. One must have considerable practice in using this method before he gains confidence.

Example 5: Symbolize and use the valid argument forms to show that the following argument is valid.

(1) If programmed texts are beneficial to students, then we should use them.
(2) <u>It is not the case that we should use them or the students will not study.</u>
∴ Programmed texts are not beneficial.

Solution: The argument may be symbolized as follows:

(1) $P \rightarrow Q.$
(2) $\underline{\sim (Q \vee \sim S).}$
$\therefore \sim P.$

To get started, line (2) may be rewritten as line (3) as follows:

(3) $\sim Q \wedge S$ De Morgan's law.
(4) $\underline{\sim Q \qquad}$ Line 3, Simplification.
$\therefore \sim P$ Lines 1 and 4, Modulus Tollens.

Example 6: Show that the conclusion, $\sim P$, follows from the premises 1 through 6 by the use of valid argument forms.

Solution:

1. $P \longrightarrow Q$.
2. $S \longrightarrow \sim R$.
3. $T \longrightarrow R$.
4. $Q \longrightarrow U$.
5. $\sim U \vee T$.
6. S.

$\therefore \sim P$.

7.	$P \longrightarrow U$	lines 1–4	(Hypothetical Syllogism).
8.	$U \longrightarrow T$	line 5	(Material Implication).
9.	$P \longrightarrow T$	lines 7–8	(Hypothetical Syllogism).
10.	$P \longrightarrow R$	lines 9–3	(Hypothetical Syllogism).
11.	$\sim R$	lines 2–6	(Modus Ponens).
	$\therefore \sim P$	lines 10–11	(Modus Tollens).

Problem Set 1.7

A. Symbolize the following arguments. Also, determine the validity of each of them by use of the definition of validity, i.e., by the truth table method.

1. Mathematics is easy if you should study logic.
   Mathematics is not easy.
   ∴ You should not study logic.
2. I passed all courses this semester.
   ∴ If I do not study, then I will pass all courses this semester.
3. I passed all courses this semester.
   ∴ If I do study then I pass all courses this semester.
4. The sun is shining or it is raining.
   It is raining. Therefore the sun is shining.
5. I will go only if it is fair weather.
   It is fair weather. Therefore I will go.
6. It is cold and I am freezing.
   Hence, it is cold.
7. If the temperature is 80°F or more, I will go swimming.
   The temperature is 80°F or more.
   Hence, I will go swimming.
8. The school team will win this game or I will eat my hat.
   I will not eat my hat.
   Therefore, the school team will win this game.
9. If all the systems in the space ship are reliable, then the moon shot will be a success. If the moon shot will be a success then the astronauts will have a safe journey to and from the moon. Therefore, if all the systems in the space ship are reliable, then the astronauts will have a safe journey to and from the moon.
10. This school cafeteria has good food. Hence this school cafeteria has good food and I will recommend the school cafeteria to everyone.
11. Pepper pot soup is a gourmet dish or you do not know good food. Hence you do not know good food.
12. If he is a student then he is self-taught. He is self-taught. Therefore he is a student.
13. If you read this book, you will enjoy the movie version of this book. You did not read this book. Hence you will not enjoy the movie version of this book.

B. Determine the validity of the following arguments, using the definition of validity:

1. $B \longrightarrow C.$
   $B \longrightarrow D.$
   $\therefore B \longrightarrow (C \wedge D).$
2. $A \longrightarrow B.$
   $\sim (B \vee C).$
   $\therefore \sim A.$
3. $\sim D \vee E.$
   $\sim F \longrightarrow \sim E.$
   $\therefore D \longrightarrow F.$
4. $A \wedge (B \wedge C).$
   $\sim A \vee \sim B.$
   $\therefore A \wedge C.$
5. $P.$
   $\therefore Q \longrightarrow P.$
6. $P \longrightarrow Q.$
   $\sim P.$
   $\therefore \sim Q.$
7. $P \vee Q.$
   $P.$
   $\therefore \sim Q.$
8. $A.$
   $\therefore A \wedge B.$
9. $C.$
   $\therefore C \vee B.$
10. $P \longrightarrow Q.$
    $Q.$
    $\therefore \sim P.$

C. Test the following arguments for validity using the method of Proof by Reduction.

1. The room is cold. Hence the room is cold or tomorrow is Friday.
2. If you enjoy logic then you should enjoy mathematics. You do not enjoy logic or you enjoy music. Therefore, you should not enjoy mathematics.
3. $P.$
   $\therefore P \wedge Q.$
4. $\sim P \vee Q.$
   $Q.$
   $\therefore \sim (P \longleftrightarrow Q).$

   (**Note:** There are two cases to consider.)
5. If the phone rings, then I will answer it. I won't be able to watch the television program whenever I will answer the phone. The phone rings. I will not answer the phone or I will go to a movie. Therefore, I won't be able to watch the television program, but I will go to the movie.
6. $P \longrightarrow \sim Q.$
   $Q \vee R.$
   $P.$
   $\therefore R.$

D. Use the valid argument forms to show validity of the following arguments or argument forms.

1. $P.$
   $\sim P \vee Q.$
   $Q \longrightarrow R.$
   $\therefore R.$
2. $P \longrightarrow (Q \vee R).$
   $\sim R \wedge \sim Q.$
   $R \longrightarrow P.$
   $\therefore \sim R.$
3. $P \longrightarrow Q \vee R.$
   $\sim (S \vee T) \longrightarrow P.$
   $T \longrightarrow \sim U.$
   $U \wedge \sim Q.$
   $U \longrightarrow \sim S.$
   $\therefore R.$

4. My time will be well spent whenever I take physics or calculus. I take physics. Therefore, I will not take psychology or my time will be well spent.
5. If you read the Chronicle you will be bored, and if you read the Courier you will be disgusted. You read the Chronicle and the Courier. Hence, you will be both disgusted and bored.
6. He does not drink or he dances. If he smokes then he does not dance. Therefore, if he drinks then he does not smoke.
7. If x is an element of set A, then x is an element of set B. The element x is in set A. Hence, the element x is in set B.
8. If I work, I am prosperous. If I don't work, I have a good time. Therefore, I am prosperous or I have a good time.
9. $(K \vee P) \vee A.$
   $\sim P \wedge \sim A.$
   $\therefore K.$
   [*Hint*: $(K \vee P) \vee A \Leftrightarrow K \vee (P \vee A)$.]
10. $P \vee (R \longrightarrow Q).$
    $(P \longrightarrow R) \vee \sim S.$
    $S \wedge \sim R.$
    $\therefore R \longrightarrow Q.$
11. A variation of the proof by reduction is to assume the negation of the conclusion as an additional premise. If by such an assumption you can deduce a self-contradiction ($r \wedge \sim r$ for example) then there is no way that this self-contradiction could be consistent with the other premises and the conclusion. If the assumption of the negation of the conclusion has led to this self-contradiction, the assumption must be false, meaning the argument is valid. Use this method on Probs. 1, 3, and 5 in this section.

1.8 Quantification

Up to this point we have discussed in considerable detail simple statements joined by basic connectives to form compound statements. However, we have not discussed in detail the structure of a given statement such as, "All squares are rectangles," in which a property of being a rectangle is conferred on all those items designated as squares. In a similar manner, when we state, "Some triangles are equilateral," we are affirming there must be at least one triangle that is equilateral. The words "all" and "some" as used in these statements indicate the quantity of those items to which a stated property applies. Such words as "all," "some," "none," or "one" (and others that are equivalent in meaning) which describe "how many?" are called **quantifiers**. Consider the following examples of quantified statements.

Example 1:
(a) Some politicians are dishonest.
(b) Every nation is expansionistic.
(c) No individual is perfect.

(d) Nothing is weightless.
(e) Something smells.
(f) Everything is wonderful.
(g) There is a person who does not understand.
(h) At least one student will pass.

Now each of the examples given may be paraphrased by using in some combination only the words "all," "some," and "no" (or their grammical equivalents). For example, (b), (g), and (h) may be rephrased as follows:

(b) All nations are expansionistic.
(g) Some persons are not understanding.
(h) Some student will pass.

We have four basic patterns that we wish to discuss each of which we will designate with a capital letter A, E, I, O, as follows:

A: All P are Q.
E: No P are Q.
I : Some P are Q.
O: Some P are not Q.

Types **A** and **E** are statement forms which refer to all of the objects in a given class. These are called **universal quantification** statement forms.

Types **I** and **O** are statement forms which refer to some but not all of the objects in a given class. These are called **existential quantification** statement forms.

Types **A** and **I** which assert a property about a given class are said to be **affirmative.**

Types **E** and **O** which deny something about a given class are said to be **negative.**

As indicated **A**, **E**, **I**, and **O** are statement forms and may not be designated as either true or false; i.e., P and Q are variables and do not assume a true or false truth value unless a specific class of objects is substituted for the variables.

Although the statement forms are not in themselves true or false, we can consider questions as to the meaning we will give to a specific instance of say "**A**" that we are willing to designate as "true." For example, "All men are primates," means that it is true for all men or that there are no men for whom the statement is false. Thus, statement form **A** may be identified under a substitution instance which is "true" with the general description "all true" or equivalently "none false." In a similar fashion, we can describe other variations in the truth or falsity of the universal quantifier together with an equivalent interpretation as follows:

Table 1.1

Universal Quantifier	Equivalent Interpretation
All true	None false
All false	None true
Not all true	At least one false
Not all false	At least one true

Some very important information regarding the negation of each of these various possibilities may be gained by close examination of each case. The phrase "all true"

and the phrase "not all true" are the negation of each other, but since "not all true" is equivalent to "at least one false," the negation of "all true" is equivalent to "at least one false." A similar analysis can be made of "all false" and its negation "not all false" giving rise to the following comparative listing:

Table 1.2

Universal Quantifier	Negation
All true	At least one false
All false	At least one true
Not all true	None false
Not all false	None true

An examination of the negation column when compared to each listing in the Universal Quantifier column suggests that each universal quantifier is changed to an existential quantifier and the truth value of the statement it quantifies is changed to the opposite truth value. Since each column is the negation of the other, we can also interpret the negation of the existentials in the negation column in terms of universal quantifiers with a corresponding change in the truth value of the statement it quantifies. This relationship is of sufficient importance to be stated as a rule.

RULE OF NEGATION: In order to negate a quantified statement involving a single quantifier, change the quantifier from universal to existential or from existential to universal and negate the statement which it quantifies.

All of the foregoing discussion can be represented in the following tabulation that considerably aids the memorization of the fundamental relationships.

Table 1.3

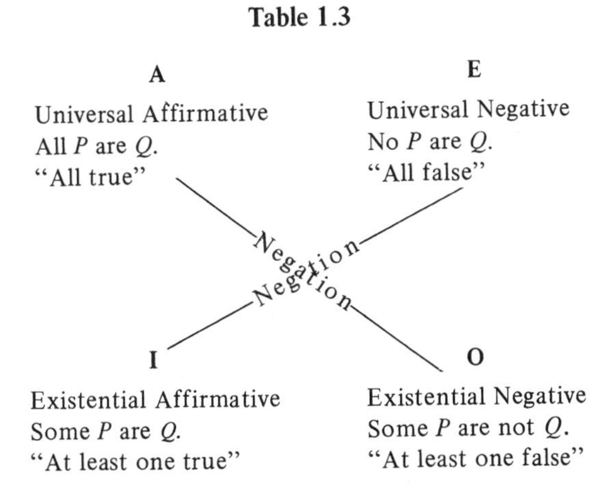

Example 1: Write the negation of the following quantified statements:

(a) All lovers enjoy music. (A)

Negation: Some lovers do not enjoy music. **(O)**

(b) Some men are not hairy. **(O)**
Negation: All men are hairy. **(A)**

(c) No airplanes are flying. **(E)**
Negation: Some airplanes are flying. **(I)**

(d) Some actors are stage-struck. **(I)**
Negation: No actors are stage-struck. **(E)**

Example 2: Write the negation of the following mathematical expressions:

(a) Some rational numbers are repeating decimals. **(I)**
Negation: No rational numbers are repeating decimals. (E)

(b) All the elements of set A are contained in set B. **(A)**
Negation: There is at least one element of set A that is not contained in set B. **(O)**

(c) No scalene triangles are equiangular. **(E)**
Negation: Some scalene triangles are equiangular. **(I)**

With this brief introduction to quantification, we want to determine under what conditions we are able to demonstrate that arguments involving quantifiers are valid. Earlier in this chapter we considered some basic syllogisms of the type which involved two premises and a conclusion. For example:

Example 3:

All mathematicians are absent minded.
All logicians are mathematicians.
∴ All logicians are absent minded.

This argument may be symbolized:

All M's are A's.
All L's are M's.
∴ All L's are A's.

The two premises and the conclusion are all type A statement forms. The validity of this particular form may be demonstrated by the use of Euler circles as follows:

The M circle must necessarily be drawn inside circle A, and the L circle must

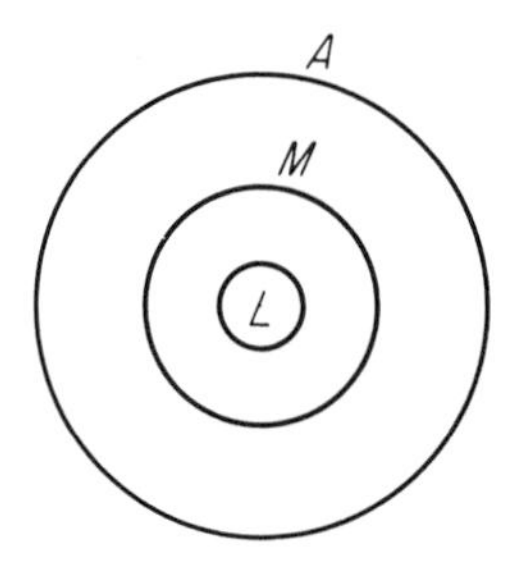

necessarily be drawn inside circle M. The conclusion is supported by observing that all of circle L is inside circle A.

It would be incorrect to conclude, however, that as long as both premises and the conclusion are of the A statement form, we must have a valid argument. A simple interchange of the M's and A's in the first premise will give rise to the following invalid form:

Example 4:

All A's are M's.
All L's are M's.
∴ All L's are A's.

By Euler circles the A circle is drawn inside circle M.

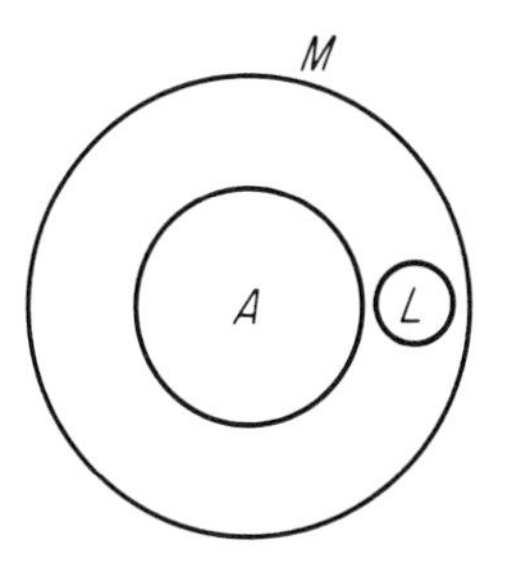

The circle L is drawn inside circle M, but does not necessarily have to be drawn inside circle A. Therefore, the conclusion does not follow as a necessary conclusion. Thus, the argument is invalid.

It is interesting to note how many logically possible cases there are to consider for a syllogistic argument consisting of two quantified premises and a quantified conclusion. We can apply the Fundamental Principle of Counting to find the exact number. The first premise may be one of the four types **A**, **E**, **I**, or **O**, but in each of these forms we can interchange the subject and the predicate in two ways (as we did in our example with M and A). The same choices are available to us for the second premise and the conclusion must be one of the four basic types **A**, **E**, **I**, or **O**. The conclusion is not considered to be reversible for grammatical reasons so as to preserve the syllogistic form. The total number is $\underline{4} \times \underline{2} \times \underline{4} \times \underline{2} \times \underline{4} = 256$.

Fortunately, the 256 distinct possibilities have been thoroughly analyzed so that we do not have to repeat the investigation. Out of the 256 possibilities only 15 are valid argument forms. We could list the 15 valid argument forms and attempt to memorize and apply them in a mechanical fashion; however, we will stress an analysis of each syllogistic argument by an expanded use of Euler circles. To systematize this technique we will now consider some general diagrams and their application to specific syllogisms.

We will use a shading technique to represent the concept of emptiness meaning there are no instances of that type, and the symbol x to represent "some" or "at least one." The four basic types **A**, **E**, **I**, and **O** may be diagrammed as follows:

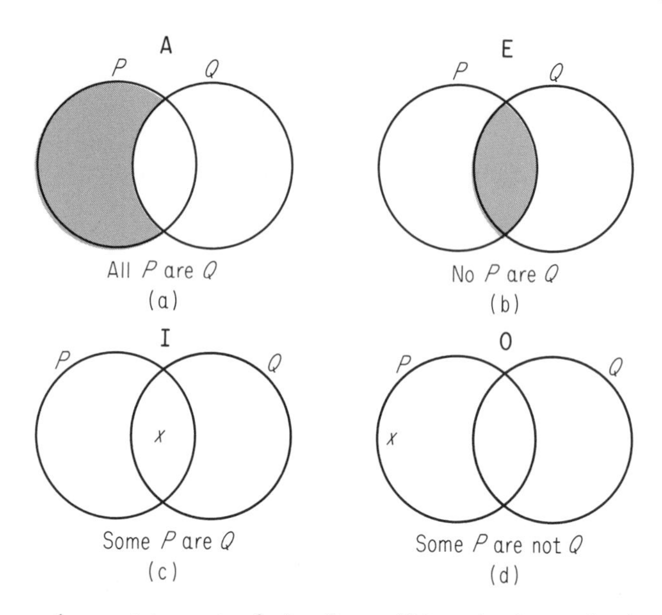

For example, a statement of the form, "No animals are hairless," would be diagrammed

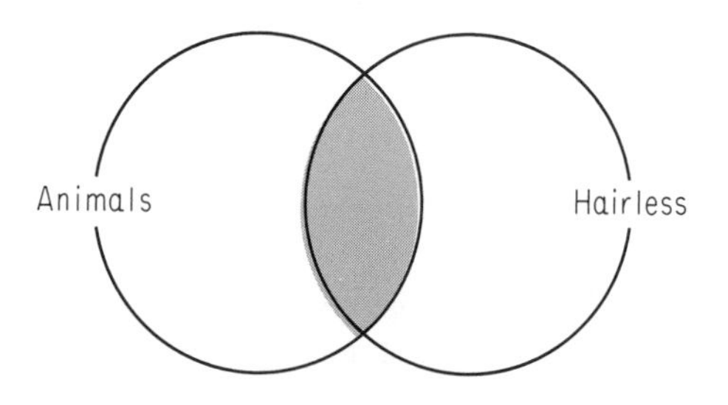

The shaded region tells us that the region which is common to both circles is empty, meaning there are no instances where something is both "animal" and "hairless."

Let us now consider some examples to which we can apply the diagrammatic approach for testing validity. The reader is cautioned once again, however, not to confuse the truth or falsity of the premises and the conclusion with the validity or invalidity of the argument form.

Example 5:

No politicians are naive.
Some women are politicians.
∴ Some women are not naive.

Step 1: Let "*P*" represent politicians, "*N*" represent naive, and "*W*" represent women.

Step 2: Draw three intersecting circles in the most general way possible as illustrated.

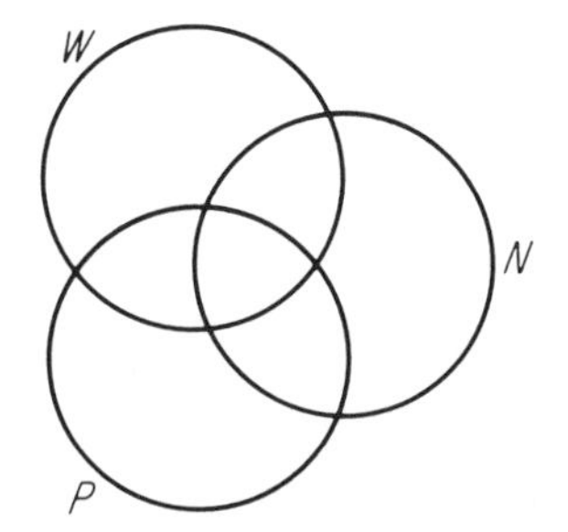

Step 3: Diagram any "universal" premise of the A or E type first. In this example, the first premise is of the E type which we diagram:
Shade the region that circle P and circle N have in common, i.e., there are no politicians who are naive.

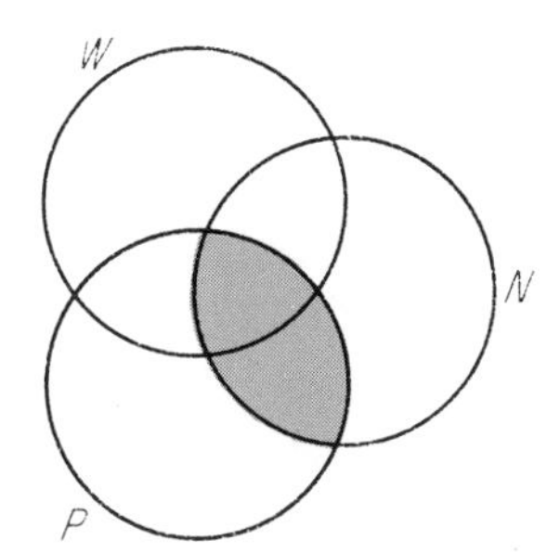

Step 4: Diagram the second premise. Since this is of the I type, we place an "x" in the unshaded region where the circles W and P overlap.

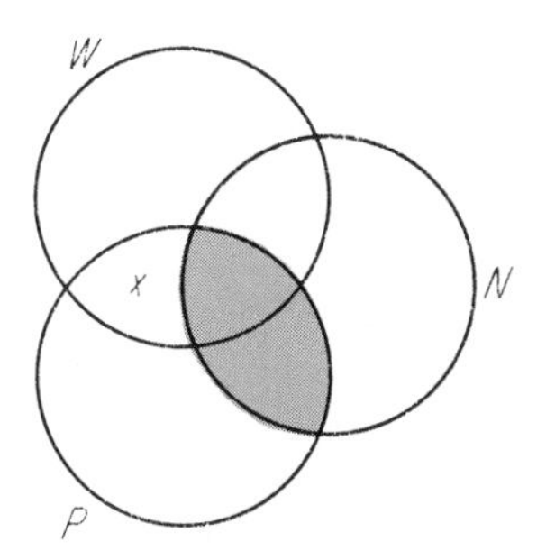

Step 5: The argument conclusion must now be examined in terms of the diagram. The argument will be valid if the diagram correctly represents the conclusion. Since there are some or at least one "x" that lies outside the circle N, the conclusion "Some women are not naive," is accurately diagramed and the argument is valid.

Example 6:

Every athlete is well-conditioned.
No musicians are athletes.
∴ No musicians are well-conditioned.

Step 1: Shade the part of circle A that is not in W, i.e., there are no athletes who are not well-conditioned.

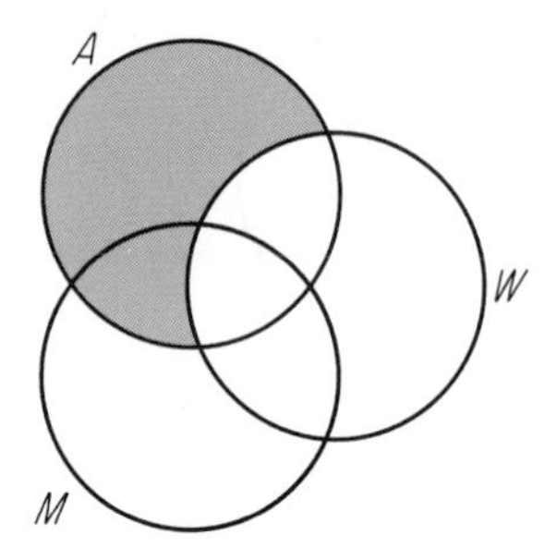

Step 2: Shade what is left of the region common to A and M indicating there are no musicians who are athletes.

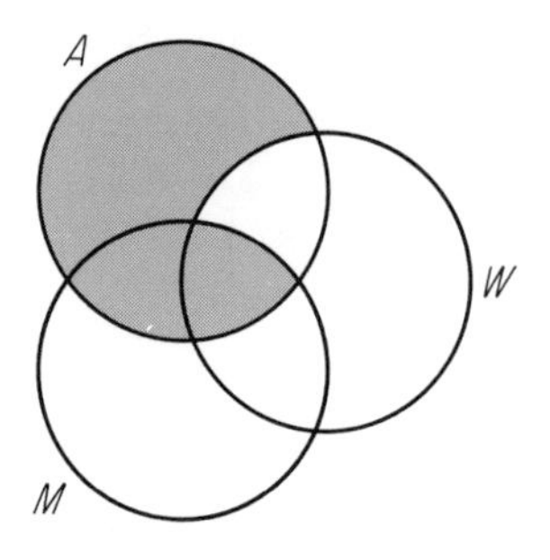

Step 3: Examine the conclusion to determine if it is diagrammed correctly. The conclusion, "No musicians are well conditioned" would require that the entire region common to circles M and W should be shaded. Since this region is not completely shaded the argument is invalid.

We will close our discussion of quantification and valid argument forms involving quantifiers at this point. We have by no means exhausted the topic. Argument forms involving more than two premises might be considered in which case we would break the premises up into pairs to obtain intermediate conclusions which could in turn be combined with other premises, continuing in this manner until we can clearly demonstrate whether the argument is valid or invalid. All of the quantification forms can be completely symbolized and combined with our previous symbols for connectives to afford us a complete set of abstract notation together with rules for combining these symbols. Methods have also been developed for changing the given forms to equivalent forms, and then applying a systematized technique for testing the validity of complex argument forms. Readers who have found this section interesting and wish to extend their knowledge will find books listed in the Suggested Reading List at the end of the chapter that will aid them in further study.

Problem Set 1.8

A. Write the negation of each of the following statements.

1. Some integers are odd numbers.
2. Every set is a subset of itself.
3. No rationals are nonrepeating decimals.
4. Some element of set A is not an element of set B.
5. Some natural numbers have a multiplicative inverse.

B. Prove the validity or the invalidity of each of the following arguments by the use of Euler circles.

1. No student is stupid.
   Jones is not a student.
   ∴ Jones is stupid.
2. Some politicians are liberals. All liberals are shrewd. Therefore some politicians are shrewd.
3. All people who have ambition and intelligence are successful. Some ambitious people are not successful. Some intelligent people are not successful. Therefore, some people have neither ambition nor intelligence.
4. Every human being has his faults. Therefore, nobody is perfect. (*Hint:* Write an equivalent negation of the given premise.)
5. All learning is painful. Anything painful should be avoided. Hence, learning should be avoided.

1.9 Some Applications of Logic in Mathematical Proofs

At the beginning of our study of logic we indicated that not only is logic the cement that binds together the various parts of a mathematical system, but via the methodology of symbolic logic we could have studied logic from a mathematical system point of view. Our treatment of argument forms has exposed the conditions under which we will agree to consider the reasoning methods correct, i.e., an argument to be valid.

The patterns of thought which we have ultimately designated as valid argument forms are employed over and over again in all branches of mathematics, but the logical justifications are not explicit and do not inform a student which particular rules of logic are being used. For example, consider the following theorem:

Prove that for a, b, c, d nonzero positive integers, if $\frac{a}{b} > \frac{c}{d}$ then $ad > bc$.

	Statement	Reason	Logical Form
1.	If $\frac{a}{b} > \frac{c}{d}$ then $bd \cdot \frac{a}{b} > bd \cdot \frac{c}{d}$	Theorem on inequalities	$P \longrightarrow Q$

Statement	Reason	Logical Form
2. If $bd \cdot \frac{a}{b} > bd \cdot \frac{c}{d}$ then $da > bc$	Theorems of algebra	$Q \longrightarrow R$
3. If $da > bc$ then $ad > bc$	Commutative property	$R \longrightarrow T$
4. If $\frac{a}{b} > \frac{c}{d}$ then $ad > bc$	Conclusion	$P \longrightarrow T$

The "statement" and "reason" column would be all that would appear in almost any conventional mathematics text. However, the abstract logical form of the proof is added so that the reader may recognize by two successive applications of the Hypothetical Syllogism valid argument form we can obtain the desired conclusion.

We will now give our attention through the use of examples, to a further consideration of the role of logic in mathematical proofs. The examples have been chosen because they require a minimum amount of prior mathematical knowledge and yet they utilize logical reasoning methods in a nontrivial setting. We will, of course, be permitted to use in our proofs, definitions, assumptions, and previously proven theorems that form a part of a particular branch of mathematics from which the examples are drawn.

Example 1: Prove that the square of an odd number is odd.

Discussion: By the set of odd numbers we mean the set $\{\ldots -5,-3,-1,1,3,5 \ldots\}$ where the three dots indicate the sequence of numbers goes on in the same pattern. But for a proof to be general, we must represent in some symbolic form an arbitrary element in the set of odd numbers. An appropriate choice is to represent an arbitrary element by $2a + 1$ where a is any integer. For example, if $a = 0$ then $2a + 1 = 1$: if $a = 1$ then $2a + 1 = 3$; if $a = -1$ then $2a + 1 = -1$. Conversely, it is also true that if a number can be expressed in the $2a + 1$ form where a is an integer, then the number must be odd. Let us now put together a proof.

PROOF:

Statement	Reason	Logical Form
1. For any arbitrary number, say x, x is odd if and only if x can be expressed in the form $2a + 1$, where a is an integer.	Definition of odd integer	$P \longleftrightarrow Q$
2. If $x = 2a + 1$ then $x^2 = (2a + 1)^2 = 4a^2 + 4a + 1 = 2(2a^2 + 2a) + 1$	Theorems of algebra	$Q \longrightarrow R$
3. If $x^2 = 2(2a^2 + 2a) + 1$, where $2a^2 + 2a$ is an integer, then x^2 is of the form $2a + 1$, hence odd.	Definition of odd integer	$R \longrightarrow S$
4. Therefore, if x is an odd number, then x^2 is an odd number.	Statements 1–3	$P \longrightarrow S$

The logical form of the proof is as follows:

1. $P \longleftrightarrow Q$.
2. $Q \longrightarrow R$.
3. $R \longrightarrow S$.

$\therefore P \longrightarrow S$.

Let us reconstruct the proof using valid argument forms to see the underlying structure.

		Line	
1. $P \longleftrightarrow Q$	4. $(P \longrightarrow Q) \wedge (Q \longrightarrow P)$	1	Material equivalence
2. $Q \longrightarrow R$	5. $(P \longrightarrow Q)$	4	Simplification
3. $R \longrightarrow S$	6. $(P \longrightarrow R)$	2, 5	Hypothetical syllogism
	7. $\therefore P \longrightarrow S$	3, 6	Hypothetical syllogism

Having shown that this is a valid argument demonstrates that the conjunction of the premises implies the conclusion, i.e., if the premises are true the conclusion must also be true. The type of proof represented in this example is called a **direct** proof, since we started with the given hypothesis and in a step-by-step manner proceeded toward the conclusion. Other proofs where we begin with the conclusion or perhaps the negation of the conclusion are called **indirect** proofs.

Example 2: Prove that the additive identity, namely 0 is unique, i.e., there is only one element 0 such that for all a where a is a real number $a + 0 = a$.

PROOF:

Statement	Reason	Logical Form
1. Assume the negation of the conclusion, namely the identity element is not unique. Let 0 and 0′ represent two different identities.	Hypothesis	$\sim P$
2. If 0 and 0′ are identities, then $a + 0 = a$ and $a + 0' = a$.	Property of identity	$\sim P \longrightarrow Q$
3. If $a + 0 = a$ and $a + 0' = a$ then $a + 0 = a + 0'$	Property of equality	$Q \longrightarrow R$
4. If $a + 0 = a + 0'$ then $0 = 0'$	Cancellation property	$R \longrightarrow S$
5. But if $0 = 0'$ they are not different, which contradicts our assumption.		$S \longrightarrow \sim\sim P$ $S \longrightarrow P$
6. Our assumption must be false, which means the identity element is unique.		P

Let us look at the logical form:

		Line	
1. $\sim P$	6. Q	1, 2	Modus Ponens
2. $\sim P \longrightarrow Q$	7. R	6, 3	Modus Ponens
3. $Q \longrightarrow R$	8. S	7, 4	Modus Ponens
4. $R \longrightarrow S$	9. P	8, 5	Modus Ponens
5. $S \longrightarrow P$	10. $\therefore \sim P \wedge P$	1, 9	Conjunction

But $\sim \mathbf{P} \wedge \mathbf{P}$ is a self contradiction.

There is no way in which the conclusion $\sim \mathbf{P} \wedge \mathbf{P}$ could be consistent with the other premises. Our assumption of $\sim \mathbf{P}$ must be false, therefore, $\mathbf{P}$ must be true. This is a typical example of an indirect proof where the desired conclusion is reached by assuming the negation of the conclusion. If we can show that this assumption leads to a self-contradiction, then our assumption must be false, which makes the desired conclusion true.

Another method of proof frequently employed by mathematicians is to prove a statement equivalent to the given statement if it is more convenient to do so. One of the most common examples of this is to prove the contrapositive of a given conditional statement.

Example 3: Prove: If a triangle is scalene (no sides are equal) then none of its angles are equal.
The contrapositive of this conditional may be expressed: If any of the angles of a triangle are equal then the triangle is not scalene (two or three sides are equal).

PROOF:

Statement	Reason	Logical Form
1. If any of the angles of a triangle are equal, then either two or three angles could be equal.	Hypothesis	$P \longrightarrow (Q \vee R)$
2. If two angles of a triangle are equal, then two sides are equal.	Theorem on isosceles triangle	$Q \longrightarrow S$
3. If three angles of a triangle are equal, then three sides are equal.	Theorem on equilateral triangles	$R \longrightarrow T$
4. If any of the angles of a triangle are equal, then two or three sides are equal (i.e., not scalene).	Statements 2 and 3	$P \longrightarrow (S \vee T)$

The logical form may be represented in the following manner:

		Line	
1. $P \longrightarrow (Q \vee R)$	4. $R \vee (Q \longrightarrow S)$	2	Addition
2. $Q \longrightarrow S$	5. $(R \vee Q) \longrightarrow (R \vee S)$	4	Distributive property
3. $R \longrightarrow T$	6. $(Q \vee R) \longrightarrow (S \vee R)$	5	Commutative property
	7. $P \longrightarrow (S \vee R)$	1, 6	Hypothetical syllogism
	8. $S \vee (R \longrightarrow T)$	3	Addition
	9. $(S \vee R) \longrightarrow (S \vee T)$	8	Distributive property
	10. $\therefore P \longrightarrow (S \vee T)$	7, 9	Hypothetical syllogism

Quite often in the study of mathematics when we have failed to prove a given statement or we may suspect the theorem does not hold, we may well consider if we can disprove the statement. The straightforward approach to disproving a statement is to find a counterexample, i.e., to give a specific instance where the statement is false. One counter example is sufficient to demonstrate that the statement is not always true.

Example 4: Prove or disprove that $\sqrt{a^2 + b^2} = a + b$, where a and b are any real numbers.

Counter example:

$$\sqrt{3^2 + 4^2} \stackrel{?}{=} 3 + 4$$
$$\sqrt{9 + 16} \stackrel{?}{=} 3 + 4$$
$$\sqrt{25} \stackrel{?}{=} 3 + 4$$
$$5 \neq 7.$$

We have exposed the reader to some of the techniques of proof and a disproof by counterexample. In our approach to proving a theorem, we have exposed the logical framework which is generally implicit. Considerable practice and study are required to develop a skill in proving theorems. The essence of mathematics is, however, to create new "truths" from old "truths" by whatever methods of proof that may be available. Although a mastery of mathematics must entail acquired proficiency in writing mathematical proofs, an appreciation of the reasoning process employed may be gained by exposing the logical structure.

Problem Set 1.9

In the following proofs develop both a written form of the proof and a symbolic logic form of the proof with logical justifications.

A. 1. Assuming the following definition: If a, b, c are nonzero integers, such that $a \cdot b = c$, then a and b are called *factors* (or *divisors*) of c, and c is called a *multiple* of a or b. Prove for a, b, and c integers, that if a is a factor of b and if b is a factor of c, then a is a factor of c.

2. Let a, b, and c be real numbers. We define: a is less than $b (a < b)$ if and only if there is a positive real number, say c, such that $a + c = b$. Assume the properties of multiplication for real numbers, i.e., the product of two positive real numbers is a positive real number and the product of two negative real numbers is a positive real number, and the product of a negative number and a positive number is a negative number. Prove that, if $a < b$ and if c is negative, then $bc < ac$.

3. The equality relation has the following assumed properties on a set of real numbers:

a_1. For any real number a; $a = a$ (Reflexive property)

a_2. For any real numbers a, b; if $a = b$ then $b = a$. (Symmetric property)

a_3. For any real numbers a, b, c; (Transitive property)
if $a = b$ and $b = c$, then
$a = c$.

a_4. Any quantity may be substituted in a given expression for any equal quantity and an equivalent expression is obtained.

Prove that if $a = b$, then $a + c = b + c$, using any of the equality relations.

4. Prove or disprove the statement, x is an even integer if x^2 is an even integer. Use the definition, x is an even integer if and only if there is an integer k such that $x = 2k$.
5. Prove or disprove the statement, if x is an even integer, then x^2 is an even integer.
6. Prove or disprove that the expression $x^2 + x - 5$ may be used as a formula for reproducing prime numbers where x is any integer.
7. Prove or disprove. If two triangles are similar then the triangles are congruent.

1.10 Concluding Remarks

We have now completed our formal study of logic. The reader will not be surprised when he discovers that the fundamental ideas developed in this chapter will be interwoven with and will support all of the mathematics covered in the rest of this text. Our goal in this first chapter has been to acquaint the reader with the concepts of a mathematical system and mathematical reasoning methods.

Before we direct our attention to a new though yet related subject, we believe a few comments should be made regarding the forward thrust given to mathematics following upon the work of George Boole.

Building on the symbolic notation of Boole, many mathematicians, but notably Charles Peirce (1839-1914) and Gottlob Frege (1848-1925), extended the symbolism to include propositional functions and laid the groundwork for quantification calculus. These efforts transformed logic into a completely abstract system in much the same way that conventional algebra is generalized arithmetic in a completely abstract form.

As we have seen, logic is the means by which we extend all mathematical systems, yet the fact that symbolic logic is in itself a mathematical system, affords it a rather unique but not unquestioned position. Is logic then a part of mathematics or is mathematics a part of logic? These questions are in the final analysis directed at the very foundations of mathematics. What is mathematics and what is its proper role in the history of thought?

At this time we can only inform the reader that no completely acceptable answers have been given to the fundamental questions which have been raised. Most mathematical observers however regard this situation as a further indication of the dynamic vitality of the discipline. In the past, confusion regarding issues of such fundamental importance have often stimulated periods of significant mathematical creativity. In the twentieth century we have witnessed a time of great mathematical achievement. It has been estimated that more mathematics has been invented in the last fifty years than in the previous five thousand years, and the contributions to the field are coming at an ever accelerated pace.

No student, no matter how brilliant, can hope to master all of mathematics. However, any student can develop an understanding of what constitutes a mathematical system and can understand some of the logical reasoning methods employed in mathematics. The reader who has reached this point in the text has achieved a basic understanding and is ready for more advanced work.

Review Exercises

1.1 State definitions for the following:
(a) valid argument.
(b) tautology.
(c) P is equivalent to Q.
(d) self-contradiction.
(e) statement.

1.2 Complete the following table so that it conforms with the truth values in the table and that all logical possibilities are entered in the table.

a	b	$a \longrightarrow b$	$a \wedge b$	$a \vee b$	$a \longleftrightarrow b$
		F			
			T		
					F

1.3 Insert grouping symbols to make the expression, $\sim r \wedge s \longrightarrow t \longleftrightarrow r$,
(a) a negation.
(b) a conditional.
(c) a conjunction.
Use as few grouping symbols as possible.

1.4 Find the truth value of the following expressions if r is true, s is false, and t is false.
(a) $\sim r \wedge s \longrightarrow t \longleftrightarrow r$.
(b) $\sim r \wedge (s \longrightarrow t) \longleftrightarrow r$.
(c) $\sim r \wedge s \longrightarrow (t \longleftrightarrow r)$.
(d) $\sim r \wedge (s \longrightarrow t \longleftrightarrow r)$.

1.5 Construct a truth table for the following expression: $(\sim p \vee q) \wedge \sim q \longrightarrow \sim p$. For the given conditional,
(a) Does the antecedent imply the consequent?
(b) Does the consequent imply the antecedent?
(c) Are the consequent and antecedent equivalent?

1.6 Consider the following conditional statement: "If the product of two real numbers is zero then one or the other of the numbers is zero." Write the (a) inverse, (b) converse, and (c) contrapositive of the given conditional.

1.7 Determine whether the conditional connective distributes over disjunction, i.e., is $p \longrightarrow (q \vee r)$ equivalent to $(p \longrightarrow q) \vee (p \longrightarrow r)$?

1.8 Symbolize the following arguments and then determine their validity by (a) truth table, and (b) by proof by reduction.

(a) If the weather is sunny and warm then I will play tennis. The weather is not sunny and it is not warm. Therefore, I will not play tennis.

(b) If I go skiing then I will break a leg. I will not break a leg. Therefore, I will not go skiing and I will enjoy myself.

1.9 Use the valid argument forms given in the text to show validity of the following arguments.

1. $A \longrightarrow B \vee C.$
$D \longrightarrow \sim C.$
$\underline{D \wedge \sim B.}$
$\therefore \sim A \vee E.$

2. $\sim (D \wedge E) \vee (F \wedge G).$
$D \vee \sim K.$
$K \wedge \sim P.$
$\underline{\sim E \longrightarrow P.}$
$\therefore F \wedge G.$

3. $t \longrightarrow (r \vee s).$
$p \vee t.$
$r \longrightarrow m.$
$\sim s \vee \sim n.$
$\underline{\sim (n \longrightarrow m).}$
$\therefore p.$

4. $\sim (m \vee n).$
$p \longrightarrow m.$
$\underline{\sim s \vee n.}$
$\therefore \sim (\sim p \longrightarrow s).$

5. $\sim (p \wedge q).$
$r \longrightarrow q.$
$\underline{r \vee s.}$
$\therefore p \longrightarrow s.$

6. $\sim p \vee t.$
$\sim r \longrightarrow \sim t.$
$r \longrightarrow (s \vee q).$
$\underline{\sim (p \longrightarrow q).}$
$\therefore s.$

7. $\sim (p \longrightarrow q).$
$s \vee q.$
$\sim t \longrightarrow \sim s.$
$t \wedge p \longrightarrow r.$
$\underline{\sim r \vee m.}$
$\therefore m.$

8. $p \vee q \longrightarrow t.$
$\sim (s \wedge t).$
$m \longrightarrow s.$
$\sim r \vee m.$
$\underline{\sim n \longrightarrow p.}$
$\therefore r \longrightarrow n.$

9. $\sim p \vee q.$
$q \longrightarrow t.$
$\sim s \longrightarrow \sim t.$
$s \longrightarrow m.$
$\underline{\sim (m \wedge \sim p).}$
$\therefore p \longleftrightarrow t.$

10. $(p \vee r) \longrightarrow (s \longrightarrow t).$
$p \vee m.$
$t \longrightarrow n.$
$\sim n \longrightarrow s.$
$\underline{\sim n.}$
$\therefore m.$

Suggested Readings

Christian, Robert R., *Introduction to Logic and Sets*. Blaisdell, New York, 1965.

Copi, Irving M., *Symbolic Logic*. Macmillan, New York, 1954.

Culbertson, James T., *Mathematics and Logic for Digital Devices*. Van Nostrand, Princeton, N. J., 1958.

Eves, Howard, and Carrol V. Newsom, *The Foundations and Fundamental Concepts of Mathematics*. Holt, Rinehart and Winston, New York, 1965.

Dinkines, Flora, *Mathematical Logic*. Appleton-Century-Crofts, New York, 1964.

Lipschutz, Seymour, *Theory and Problems of Finite Mathematics*. Schaum-McGraw-Hill, New York, 1966.

Michalos, Alex C., *Principles of Logic*. Prentice-Hall, Englewood Cliffs, N. J., 1969.

Newman, James R., *The World of Mathematics*. Simon & Schuster, New York, 1956.

Stoll, Robert R., *Sets, Logic and Axiomatic Methods*. W. H. Freeman, San Francisco, Calif., 1961.

Today it is possible to derive almost all of contemporary mathematics from a single source, the theory of sets.

–N. BOURBAKI (1954)

CHAPTER 2

Set Theory

2.1 Introduction

In the previous chapter we briefly described the historical development of symbolic logic after the pioneering work of George Boole. In broad terms, the modern direction of symbolic logic with its abstract form and specialized notation has been an effort to mathematize logic. In a parallel development the mathematical community has increasingly chosen to express their written contributions in a structured deductive form in which the principles of logic become the explicit vehicle for proofs. A most significant mathematical effort in this regard was made by Giuseppe Peano (1858-1932) in 1895. With three undefined terms and five axioms together with the rules of logic, Peano was able to develop arithmetic in a completely deductive form.

In order to make some of the logical relations in his system precise, Peano was required to make a distinction between a member of a class and the class itself, and to make explicit the relation of a subclass to the class which contains the subclass. He was aided in this respect by the prior work of Georg Cantor and Gottlob Frege on the "theory of aggregates," a branch of mathematics which we now designate as set theory.

The same basic concepts of class membership (set theory) were used by Bertrand Russell and Alfred N. Whitehead* to show that all of Peano's axioms are direct consequences of logical relations. This gave rise to their thesis that all of the mathematics is a part of logic.

Today set theory permeates all of the branches of mathematics and is taught at varying levels from kindergarten through graduate school. The concepts of set theory are used to unify subject matter, to clarify basic concepts and to provide a common foundation on which to base more advanced study.

In this chapter many of the basic concepts of set theory will be introduced, the close association between sets and logic will be demonstrated, set theory will be discussed as an abstract mathematical system, and finally, a contemporary application of set theory will be given.

**Principia Mathematica* (1910–1913).

2.2 Basic Concepts of Sets

Our basic concept of a set refers in an intuitive way to a collection of objects. However, we regard **set, element,** (or **member** of) to be our basic undefined (primitive) terms. An important characteristic of all sets is that they must satisfy two important criteria.

1. One criterion is that a set must be **well defined**, by which we mean that for any given arbitrary element we can determine whether or not the element is a member of the set.
2. A second criterion is a **uniqueness** property, by which we mean that we are able to distinguish one element from another in the set. (In other words, the elements are uniquely represented without repetition.)

Example 1: The following collections are some typical examples of sets.
The set of books in a library.
The set of Presidents of the United States.
The New York Mets baseball team.
The counting numbers between 1 and 10.

Example 2: The following collections would not satisfy the well-defined property of a set. (Why?)
All tall men.
The underprivileged Americans.

Example 3: The following collections would not satisfy the uniqueness property. (Why?)
A collection of *all* the letters in the word Mississippi.
A collection of *all* the digits in the number 101, 122.

A number of statement forms regarding sets occur so frequently that a symbolic notation has been developed. The statement, "a is an element (or member) of set S," is symbolized $\boldsymbol{a \in S}$. The statement, "c is not an element of A," is denoted $\boldsymbol{c \notin A}$. Capital letters such as A, B, and C are frequently used to denote sets and lower case letters such as a, b, and c as well as numerals and other symbols are used to represent elements of a set.

Basically, there are two ways of specifying the elements in a set. One way is to list the elements within braces, separating elements by commas, where we regard the order in which we list the elements as irrelevant. For example, the set A consisting of the elements 2, -7, π, 3/4, and e can be written:

$$\{2,-7,\pi,3/4,e\}$$

as well as

$$\{2,\pi,e,3/4,-7\}.$$

The process of listing all of the elements is known as the **roster method**.

The roster method is particularly useful when the number of elements in a set is small; otherwise this method can be quite cumbersome. For example, the task of listing by the roster method the elements in the set of all the books in the world would be an extremely unpleasant chore. In such situations where the number of elements in a set is a large finite number, we devise some rule which enables us to completely describe the objects in the set. Furthermore, we can, if it is convenient, use a rule to describe a set which has any number of elements or which is infinite.* This method which uses a rule or statement about the elements in a given set is known as the **descriptive method**.

We will assume throughout that unless specifically denoted, the sets which we will consider will be finite. Intuitively, a set is finite if we can count the number of elements in the set.

The descriptive method is of fundamental importance and a specialized "set-builder notation" is used to specify sets by this method. The symbol $\{\mathbf{x} | \mathbf{P}(\mathbf{x})\}$ will represent the set of all elements (where x denotes any arbitrary element) such that the statement **P** is true for each element in the set.

Example 4: Specify by both the roster method and the descriptive set builder notation, the set of vowels in the English language.

Roster: $\{a,e,i,o,u\}$.

Set Builder: $\{x | x \text{ is a vowel}\}$.

Example 5: N represents the set of counting numbers or positive integers, $\{1,2,3, \ldots\}$, where the "three dots" represent the continuing unending sequence of counting numbers after 3. List in an equivalent roster form the set given by $\{x | x \in N \text{ and } x \text{ is less than } 8\}$.

Roster: $\{1,2,3,4,5,6,7\}$.

We frequently express a mathematical statement in a set builder form and designate those elements which make the statement true as solutions and the set of all such solutions as the **solution set** or **truth set**.

Example 6: Determine the solution set of the **P** statement form in the following sets:

(a) $\{y | y \in N \text{ and } 2y = 8\}$.
The solution set is $\{4\}$.

(b) $\{t | t = 2k \text{ where } k = 1,2,3\}$.
The solution set is $\{2,4,6\}$.

(c) $\{z | z \in N \text{ and } 3z = -9\}$.
There are no natural numbers which when multiplied by 3 equal -9. Therefore, there are no solutions or we could say the solution set contains no elements.

A set which contains no elements has its own symbolic representation and is defined as follows:

*For further information regarding infinite sets, see the Suggested Reading List at the end of this chapter.

DEFINITION: The **empty (or null) set,** symbolized $\emptyset$ or $\{ \}$, is the set with no elements.

It is noteworthy in passing to point out that the statement "$x \in \emptyset$" is logically false for all elements x, since there are no elements in $\emptyset$ and the statement "$x \notin \emptyset$" is logically true for all elements x.

An important relation which is defined on sets is the concept of subset.

DEFINITION: Let A and B be any sets, **A is a subset of B,** symbolized $A \subset B$, if and only if every element of set A is an element of set B.
Symbolically, $A \subset B \longleftrightarrow$ for every element x, $x \in A \longrightarrow x \in B$.

If a given set A is not a subset of a set B, then the negation of the definition of subset is as follows:

$A \not\subset B \longleftrightarrow$ there is an element x such that $x \in A$ and $x \notin B$.

Example 7: Let A represent $\{1,2,z,e,a\}$, B represent $\{1,a\}$ and C represent $\{2,b\}$. Then $B \subset A$ since every element of B is an element of A. $C \not\subset A$ since there exists an element, namely, b, in set C that is not an element of A.

As a direct consequence of the definition of subset, we have that for any set A, $A \subset A$, since every element of A is indeed an element of A. Furthermore, the statement $x \in \emptyset$ which is logically false, implies $x \in A$ ($x \in \emptyset \Longrightarrow x \in A$) for any set A. Thus, the empty set is a subset of every set, i.e., $\emptyset \subset A$ for any set A.

Often we are interested in listing all of the subsets of a given set or at least determining the total number of such subsets. If A is the set $\{1,2\}$ then the subsets of A are $\emptyset$, $\{1\}$, $\{2\}$, and $\{1,2\}$. If B is the set $\{a,b,c\}$ then the subsets of B are $\emptyset$, $\{a\}$, $\{b\}$, $\{c\}$, $\{a,b\}$, $\{a,c\}$, $\{b,c\}$, and $\{a,b,c\}$. Set A having two elements has four subsets. Set B having three elements has eight subsets. These examples suggest the following theorem:

THEOREM 1: *If a set A has n elements where $n \in N$, then A has* 2^n subsets.

PROOF: This theorem follows as a direct consequence of the Fundamental Principle of Counting. We can choose to either include the first element of the set in a subset or not choose to include the element in a subset. The choice may be made in two ways. In sequence the same options may be exercised for the second element, the third element, etc., until a choice is made to either include or not to include the nth element. The total number of such choices corresponds to the total number of subsets and may be calculated by the F.P.C. to be

$$\underbrace{2 \times 2 \times 2 \times \cdots \times 2}_{n \text{ factors}} = 2^n.$$

Example 8: Use the F.P.C. and a tree diagram to obtain a listing of the subsets of the set $\{1,2,\{a\}\}$.

***Solution*:** The set contains three elements; namely, 1, 2 and the set containing the element "a" denoted $\{a\}$. By Theorem 1, the total number of subsets is 2^3 or 8.

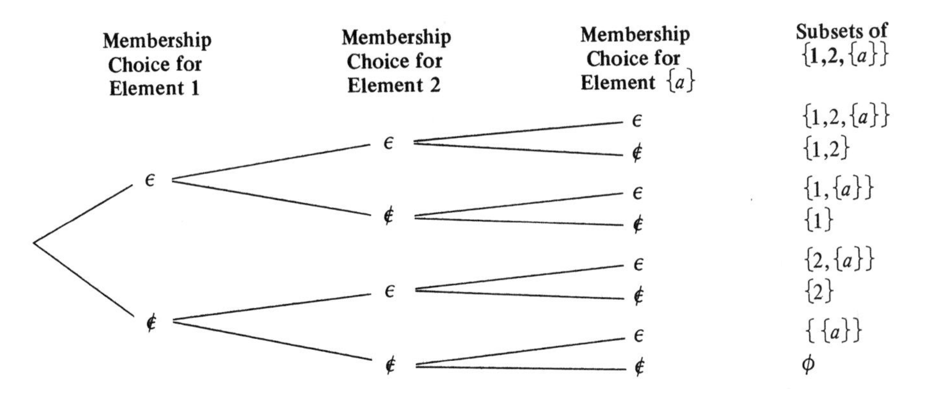

DEFINITION: Two sets **A and B are equal**, denoted $A = B$, if and only if $A \subset B$ and $B \subset A$. Symbolically, $A = B \longleftrightarrow$ for every element x, $x \in A \longrightarrow x \in B$, and $x \in B \longrightarrow x \in A$.

Example 9: The set $A = \{x | x \in N$ and x is less than $10\}$ and the set $B = \{1,2,3,4,5,6,7,8,9\}$ are equal sets.

When we wish to consider the collection of all the subsets of a given set as a set, we do so in accordance with the following definition:

DEFINITION: The set of all subsets of a given set A is called the **power set of A**, symbolized $\mathcal{P}(A)$.

Example 10: If $A = \{a,b,c,d\}$ then

$$\mathcal{P}(A) = \left\{ \begin{matrix} \{a\},\{b\},\{c\},\{d\},\{a,b\},\{a,c\},\{a,d\},\{b,c\},\{b,d\},\{c,d\} \\ \{a,b,c,\},\{a,b,d\},\{a,c,d\},\{b,c,d\},\{a,b,c,d\},\phi \end{matrix} \right\}.$$

(Note: Since A contains four elements, the total number of subsets is $2^4 = 16$.)

Problem Set 2.2

A. Determine whether each of the following is true or false.
1. Any collection of objects is a set.
2. All sets can be represented by the roster method.

3. The empty set may be an element of some set.
4. For the set $A = \{a, \{b\}, c\}$ we have $\{b\} \subset A$.
5. The power set of the empty set is the empty set.

B. 1. Let $A = \{a, 2, 3, b, x\}$. How many elements are in the power set of A?
2. List by the roster method the following sets:
   (a) $\{\alpha | \alpha$ is a digit of the base-ten number system$\}$.
   (b) $\{x | x \in N$ and $(x - 3)(x - 4) = 0\}$.
   (c) $\{y | y \in N$ and $2y$ is less then $11\}$.
   (d) $\{A | A \in \mathcal{P}(x)$, where $x = \{1,2\}\}$.
3. Describe the following sets using set-builder notation. (There may be more than one correct description.)
   (a) $\{1,3,5,7,9\}$.
   (b) $\{a, b, c, d, e\}$.
   (c) $\{2,4,6,8\}$.
   (d) {George Washington, John Adams, Thomas Jefferson}.
4. Let $B = \{\phi, \{1\}, \{1,2\}\}$. Find the power set of B, $\mathcal{P}(B)$.
5. Determine the solution set of the statement from **P** in the following sets:
   (a) $\{x | x \in N$ and $2x = 3\}$.
   (b) $\{y | y \in N$ and $2y$ is less then $19\}$.
   (c) $\{x | x \in N, (x - 2)(x - 4) = 0\}$.
   (d) $\{z | z \in N, (z - 2)(z + 2) = 0\}$.
6. Let $C = \{w, x, \{y, z\}\}$. List all of the subsets of set C.

C. 1. Prove that the empty set is a subset of any set A, using proof by contradiction.
2. Using the definition of subset and logic, show that for the sets N, I, and Q, if $N \subset I$ and $I \subset Q$, then $N \subset Q$.

2.3 Set Operations

In your study of arithmetic you became familiar with such basic operations as addition, subtraction, multiplication, and division which were performed on numbers to obtain other numbers. In our study of logic, the basic connectives of conjunction, disjunction, conditional, and biconditional were used to form compound statements from simple statements. In the same manner, we will define four basic operations for combining sets to form other sets.

DEFINITION: Let A and B be any sets. The **union of sets A and B**, written $A \cup B$, is the set consisting of all the elements in set A or (inclusive) in set B. Symbolically, $A \cup B = \{x | x \in A \vee x \in B\}$.

Example 1: Let $A = \{1,3,5,7\}$ and $B = \{1,2,3,4\}$. Then $A \cup B = \{1,2,3,4,5,7\}$.

DEFINITION: Let A and B be any two sets. The **intersection of sets A and B**, written $A \cap B$, is the set of all elements that are in A and in B. Symbolically, $A \cap B = \{x | x \in A \wedge x \in B\}$.

Example 2: Let A and B be the sets in Example 1, then $A \cap B = \{1,3\}$.

In a given discussion all of the sets we consider will be subsets of some set which we designate by the symbol $\mathcal{U}$, called the universal set, and any element we consider will also be an element of $\mathcal{U}$.

DEFINITION: Let A be a subset of $\mathcal{U}$. The **complement of A**, written $\tilde{A}$, is the set of all elements in $\mathcal{U}$ that are not in A. Symbolically, $\tilde{A} = \{x | x \in \mathcal{U} \wedge x \notin A\}$.

Example 3: Let $\mathcal{U} = \{b,a,d,g,e\}$ and $B = \{g,e\}$, then $\tilde{B} = \{b,a,d\}$.

DEFINITION: Let A and B be any sets, then the **set difference between A and B**, symbolized $A - B$ (read A minus B), is the set of all elements that are in the set A but are not elements of set B. Symbolically, $A - B = \{x | x \in A \wedge x \notin B\}$.

Example 4: Let $A = \{1,3,5,a,b\}$ and $B = \{3,5,b,c\}$. Then

$$A - B = \{1,a\}$$
$$B - A = \{c\}.$$

DEFINITION: If A and B are sets such that $A \cap B = \phi$ then A and B are said to be **disjoint**.

Having defined the basic set operations, we can deal with more complex expressions by repeated applications of the basic definitions as illustrated in the following example:

Example 5: Let $\mathcal{U} = \{1,2,3,4,5,6,7,8,9\}$, $A = \{1,3,5,7\}$, $B = \{2,4,6,8\}$, and $C = \{1,2,4,7,8\}$. Find:
(a) $A \cap (B \cup C)$.
(b) $\widetilde{(C - A)}$.
(c) $(\tilde{A} \cap B) \cap \tilde{C}$.

***Solution*:**
(a) $A \cap (B \cup C) = \{1,3,5,7\} \cap (\{2,4,6,8\} \cup \{1,2,4,7,8\})$
$A \cap (B \cup C) = \{1,3,5,7\} \cap \{2,4,6,8,1,7\}$
$A \cap (B \cup C) = \{1,7\}$.
(b) $\widetilde{(C - A)} = \widetilde{\{1,2,4,7,8\} - \{1,3,5,7\}}$
$\widetilde{(C - A)} = \widetilde{\{2,4,8\}}$
$\widetilde{(C - A)} = \{1,3,5,6,7,9\}$.

(c) $(\tilde{A} \cap B) \cap \tilde{C} = (\widetilde{\{1,3,5,7\}} \cap \{2,4,6,8\}) \cap \widetilde{\{1,2,4,7,8\}}$
$(\tilde{A} \cap B) \cap \tilde{C} = (\{2,4,6,8,9\} \cap \{2,4,6,8\}) \cap \{3,5,6,9\}$
$(\tilde{A} \cap B) \cap \tilde{C} = \{2,4,6,8\} \cap \{3,5,6,9\}$
$(\tilde{A} \cap B) \cap \tilde{C} = \{6\}$.

Problem Set 2.3

A. Determine whether the following statements are true or false.

1. The basic set operations are conjunction, disjunction, and negation.
2. For the sets A and B, the union of A and B consists of the elements of A added to the elements of set B.
3. For any sets C and D, $C - D = D - C$.
4. The sets X and Y are disjoint if and only if X is the empty set or Y is the empty set.
5. The complement of a nonempty set is the empty set.

B. 1. Let $P = \{x | x \in N$, and x is between 2 and 9$\}$, $Q = \{y | y \in N$, and y is between 4 and 8$\}$. Find: (a) $P - Q$, (b) $Q - P$, (c) $Q \cup P$, (d) $P \cap Q$.

2. Let $\mathcal{U} = \{\alpha | \alpha$ is a letter of the alphabet$\}$, $C = \{\beta | \beta$ is a consonant of the alphabet$\}$, and $D = \{\gamma | \gamma$ is a vowel of the alphabet$\}$. Find:
(a) $\tilde{C}$. (b) $D \cap \tilde{C}$. (c) $D - C$. (d) $(\widetilde{C \cup D})$. (e) $\tilde{C} \cap \tilde{D}$.

3. Let $X = \{c | c \in N$ and c is an even number between one and nine$\}$, $Y = \{b | b \in N$ and b is a number between two and six$\}$. Find:
(a) $X \cup Y$. (b) $X \cap Y$. (c) $Y - X$.

C. 1. Use the descriptive set-builder notation to specify the elements (if there are any) in $A - \mathcal{U}$, where $A \subset \mathcal{U}$. Determine what set is equal to the set $A - \mathcal{U}$ and give justification for your answer.

2. By using particular sets $\mathcal{U}$ and A, show that $\mathcal{U} - A = \tilde{A}$ for the sets which you defined. Does $\mathcal{U} - A = \tilde{A}$ for any sets $\mathcal{U}$ and A such that $A \subset \mathcal{U}$? Justify your answer.

3. Let $\mathcal{U} = \{a, b, c, d, e, f, g, h, i\}$, $A = \{a, b, c, d, e\}$, $B = \{b, c, d\}$, $C = \{a, c, e\}$. Find:

(a) $(\widetilde{A - C})$. (b) $\tilde{A} \cap C$. (c) $A - \mathcal{U}$.
(d) $(\widetilde{C \cap B})$. (e) $\mathcal{U} - (C \cap B)$.

4. Let $U = \{1,2,3,4,5,6,7,8,9,a,b,c,d\}$, $A = \{2,4,6,8\}$, $B = \{4,8,b,d\}$, $C = \{2,3,5,b,c,7,1\}$, $D = \{1,3,5,7\}$. Find:

(a) $A \cup B)$. (b) $A \cap B$. (c) $B - A$.
(d) $B \cap D$. (e) $(A \cap B) \cap C$. (f) $A \cap (B \cap C)$.
(g) C. (h) $A - B$.

5. Let $U = \{a, p, \wedge, \vee, \sim, q, s, t, \longrightarrow, \longleftrightarrow, x, y, r\}$, $X = \{a, p, \wedge, \vee, \sim, q, s, t, \longrightarrow, \longleftrightarrow\}$, $W = \{x, y, a, s, \wedge, \longrightarrow\}$, $Y = \{p, q, r, \vee, \sim, \wedge\}$, $Z = \{a, \longrightarrow, \longleftrightarrow\}$. Find:

(a) $X \cup Y$. (b) $X \cap Y$. (c) $Z \cap W$. (d) $\widetilde{Z \cap W}$.
(e) $\tilde{Z}$. (f) $\tilde{W}$. (g) $\tilde{Z} \cup \tilde{W}$. (h) $W - Z$.
(i) $W \cap \tilde{Z}$. (j) $Z \cap Y$.

6. Let $\mathcal{U} = \{u, v, w, x, y, z\}$, $R = \{v, x, z\}$, $S = \{u, v, w, z\}$, $T = \{x, y\}$. Find:

(a) $\tilde{T}$. (b) $S \cup \tilde{T}$. (c) $(S \cap T) \cup R$.

(d) $(S \cup R) \cap (T \cup R)$. (e) $S - T$. (f) $R \cap (S \cap T)$.

(g) $(R \cup S) \cup T$. (h) $\mathcal{U} - T$. (i) $\widetilde{S \cup T}$.

(j) $\tilde{S} \cap \tilde{T}$.

2.4.1 Venn Diagrams

It is convenient and illuminating to represent the universal set and some of its subsets by a diagram. The universal set $\mathcal{U}$ may be represented by a rectangular region and its subsets represented by regions within closed curves which lie entirely within the rectangle. These diagrams are called Venn diagrams after their inventor John Venn.

For example, if we know that a set A is a subset of some set B, we can illustrate this by a Venn diagram (Figure 2.1), where all of region A is drawn inside the region associated with set B.

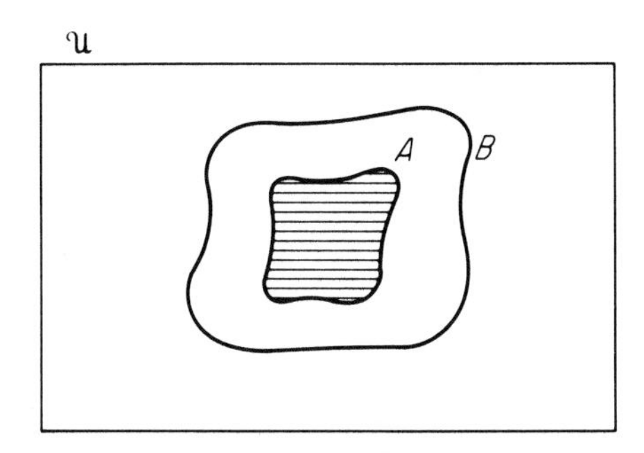

Figure 2.1

To identify which region(s) correspond to a set we may utilize either a shading technique or a procedure in which we associate a number with each region on the diagram.

If a single set A is to be considered, we can shade the region within the closed curve to represent the set A (Figure 2.2).

If two sets A and B are under consideration, we draw the closed curves in an over-

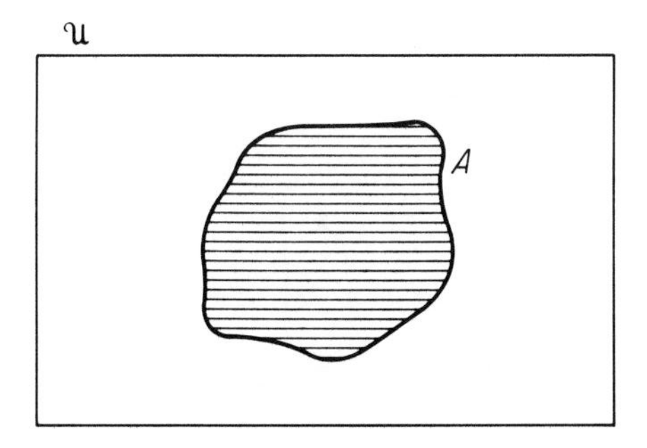

Figure 2.2

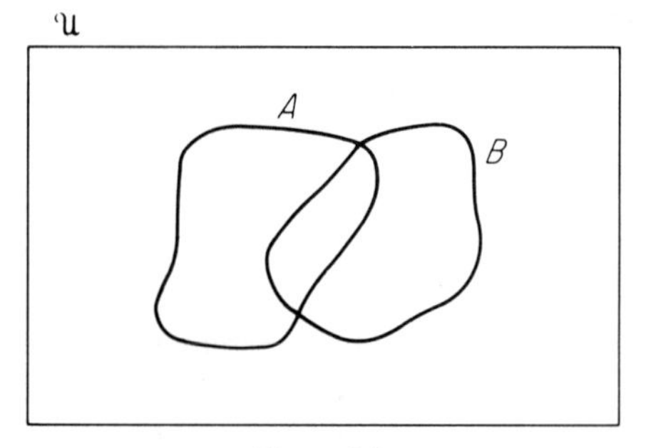

Figure 2.3

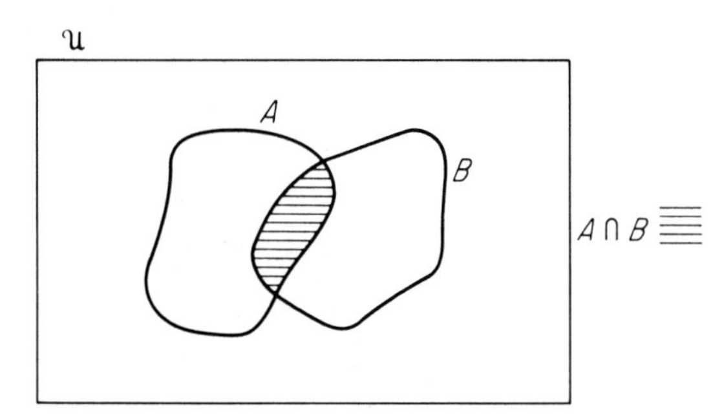

Figure 2.4

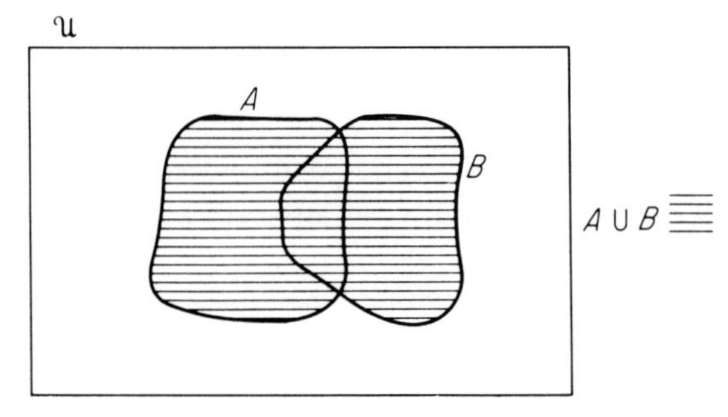

Figure 2.5

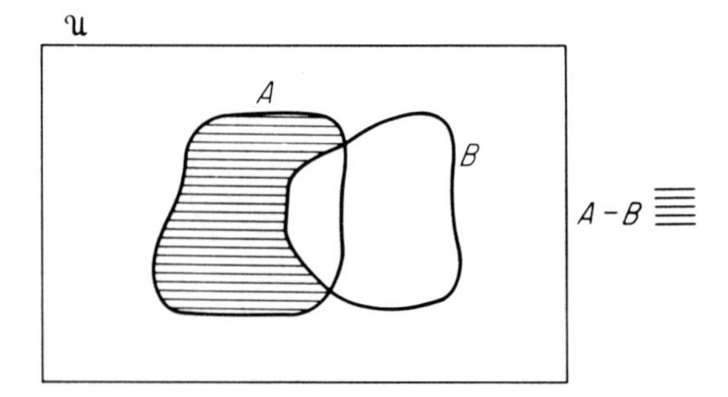

Figure 2.6

lapping manner so as to represent the most general situation (Figure 2.3). If we wish to represent some specific sets such as $A \cap B$, $A \cup B$, and $A - B$ we shade the region corresponding to the intersection (Figure 2.4), the union (Figure 2.5) and the set difference (Figure 2.6) as illustrated.

If we are unable to immediately identify the region(s) associated with a set we can resort to horizontal and vertical shading and then interpret the resulting diagram. For example, if we wish to associate the set $\tilde{A} \cap B$ with the region(s) on a Venn diagram we could use horizontal shading for $\tilde{A}$ and vertical shading for B. The cross hatched region corresponds to the intersection of the two sets (Figure 2.7).

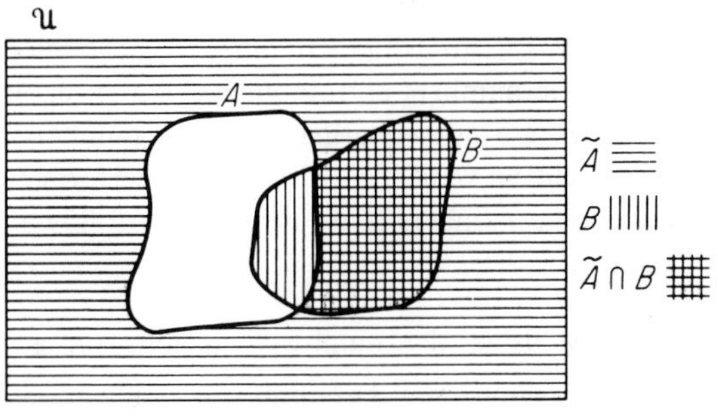

Figure 2.7

Another important use of Venn diagrams is to verify that two sets are equal although represented in a different symbolic form. This method is illustrated in the following example:

Example 1: Verify by means of a Venn diagram that $A \cap (B \cup C) = (A \cap B) \cup (A \cap C)$.

Solution: Treat each side of the equality as a separate shading problem and demonstrate that the regions representing the two expressions are identical. This is illustrated in the accompanying diagram.

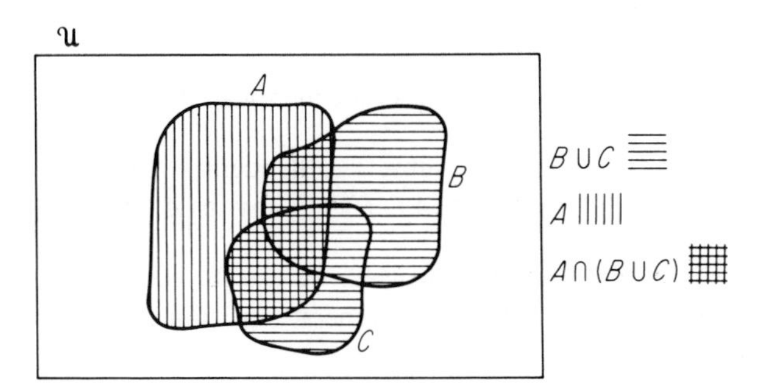

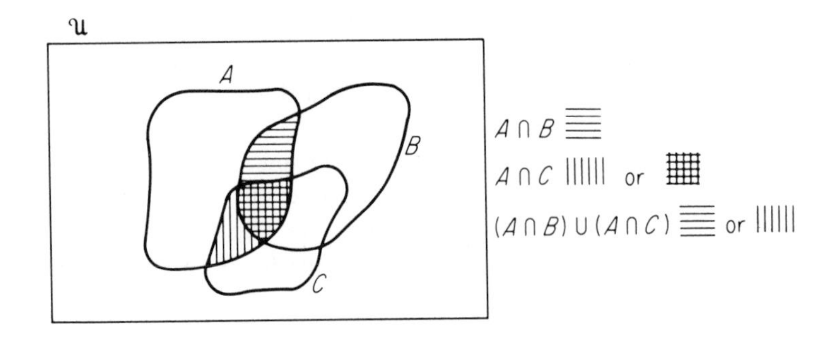

An alternate procedure, which will avoid some of the complications of shading regions for complex expressions, is to assign a number to each region of a Venn diagram. We associate with each set the numbers corresponding to the region(s) which represent the set. In this context we are not equating sets with numbers, but only using the numbered regions as a symbolic convenience. We can then perform routine set operations on the sets which contain these numbers as elements. The following example illustrates this technique.

Example 2: Verify by using the numbered regions on a Venn diagram that $\widetilde{(A \cap B)} = \tilde{A} \cup \tilde{B}$. The regions may be numbered as illustrated (the assignment of the numbers for the regions is arbitrary).

$$A \cap B = \{1\} \qquad \tilde{A} = \{3,4\}, \tilde{B} = \{2,4\}$$
$$\widetilde{(A \cap B)} = \{2,3,4\} \qquad \tilde{A} \cup \tilde{B} = \{2,3,4\}.$$

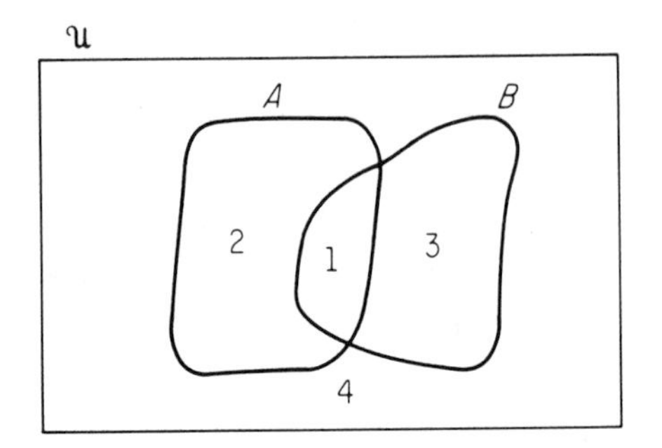

Since the regions obtained for each set are identical, the sets are equal.

Example 3: If A, B, and C are subsets of $\mathfrak{U}$, identify the region(s) on a Venn diagram associated with the sets (a) $(A \cap B) \cap C$, and (b) $\tilde{A} \cap (\tilde{B} \cap C)$.

Solution: Since three sets are involved, we use three closed curves in an overlapping method to illustrate the most general situation and number the regions as illustrated.

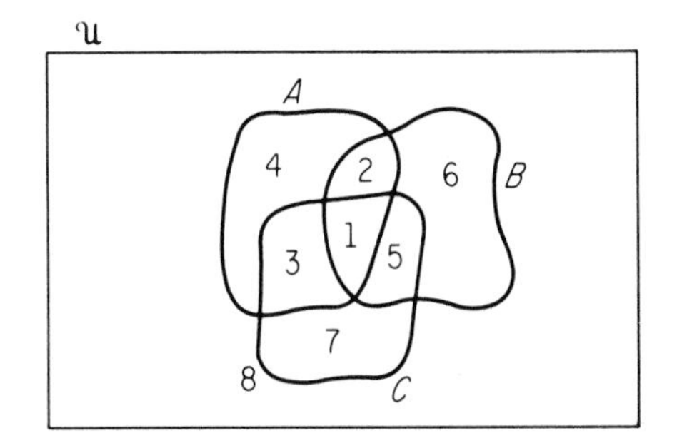

(a) $(A \cap B) \cap C = (\{1,2,3,4\} \cap \{1,2,5,6\}) \cap \{1,3,5,7\}$
$(A \cap B) \cap C = \{1,2\} \cap \{1,3,5,7\}$
$(A \cap B) \cap C = \{1\}$.

(b) $\tilde{A} \cap (\tilde{B} \cap C) = \widetilde{\{1,2,3,4\}} \cap (\widetilde{\{1,2,5,6\}} \cap \{1,3,5,7\})$
$\tilde{A} \cap (\tilde{B} \cap C) = \{5,6,7,8\} \cap (\{3,4,7,8\} \cap \{1,3,5,7\})$
$\tilde{A} \cap (\tilde{B} \cap C) = \{5,6,7,8\} \cap \{3,7\}$
$\tilde{A} \cap (\tilde{B} \cap C) = \{7\}$.

It should be clearly understood that neither the shading method nor the numbered region method constitute a formal proof anymore than a sketch of a geometric figure could be accepted as a proof of any of the properties the figure may possess. Nevertheless, a Venn diagram helps us to visualize set relationships and provides us with many intuitive insights. Later in this chapter we will discuss how we can formally prove these properties when we approach set theory from the point of view of an abstract mathematical system.

Problem Set 2.4.1

A. 1. Let A, B be sets which are subsets of $\mathcal{U}$. Using the shading technique for Venn diagrams, display the region that represents $B - A$.

2. Let C, D be subsets of $\mathcal{U}$ and let $C \subset D$. By use of Venn diagrams and shading, show that (1) $C \cap D = C$ and (2) $C \cup D = D$.

3. Verify by using the numbered regions in a Venn diagram that $A \cup (B \cap C) = (A \cup B) \cap (A \cup C)$.

4. Show by Venn diagram and numbered regions that $(\tilde{\tilde{A}}) = A$ for any set $A \subset \mathcal{U}$.

5. Let A, B, C be subsets of $\mathcal{U}$. By Venn diagram and numbered regions show that $\tilde{A} \cap (B \cup \tilde{C}) = (\tilde{A} \cap B) \cup (\tilde{A} \cap \tilde{C})$.

B. 1. Let A, B be subsets of $\mathcal{U}$. Draw a Venn diagram for these sets and number the regions. Develop a correspondence between the numbered regions and the sets $A \cap B$, $A \cap \tilde{B}$, $\tilde{A} \cap B$, and $\tilde{A} \cap \tilde{B}$.

2. Under what conditions would $B - A = A - B$?

3. Let P, Q, R be subsets of $\mathcal{U}$. Draw a Venn diagram showing that these three sets may divide the Universe into eight disjoint regions. Name each numbered region in the diagram in terms of P, Q, R with the set operations, intersection, and com-

plement. (For example; the region in $\mathcal{U}$ that does not have any element of sets P, Q, R is denoted $\tilde{P} \cap \tilde{Q} \cap \tilde{R}$.)

4. Draw a Venn diagram representing the most general situation for four sets A, B, C and D. (*Hint:* There should be 16 disjoint regions.)

2.4.2 Truth Sets

It is now possible to make explicit some fundamental relations that exist between set theory and symbolic logic. Each of the basic set operations of intersection, union, complementation, and set difference as well as the relations of subset and set equality were defined in symbolic form utilizing symbolic logic. It was also indicated that our set-builder notation, $\{x \mid P(x)\}$, identifies the set of elements for which a given statement P is true. For example, if we consider the set $A \cap \tilde{B}$ as defined in set-builder notation, we would write $A \cap \tilde{B} = \{x \mid x \in A \wedge x \notin B\}$. We have therefore defined a correspondence between a statement "$x \in A \wedge x \notin B$" which is true if and only if x is an element of the set $A \cap \tilde{B}$. The set of all the elements that are members of the set $A \cap \tilde{B}$ constitute the truth set of the statement "$x \in A \wedge x \notin B$."

DEFINITION: The **truth set** for a given statement form P is the set of all the elements for which the given statement form is true.

The criterion for truth or falsity for a statement is associated with membership or nonmembership in a set and may be represented by membership tables for complement, intersection, union, and set difference as shown below:

A	$\tilde{A}$
$\in$	$\notin$
$\notin$	$\in$

A	$\cap$	B
$\in$	$\in$	$\in$
$\in$	$\notin$	$\notin$
$\notin$	$\notin$	$\in$
$\notin$	$\notin$	$\notin$

A	$\cup$	B
$\in$	$\in$	$\in$
$\in$	$\in$	$\notin$
$\notin$	$\in$	$\in$
$\notin$	$\notin$	$\notin$

A	$-$	B
$\in$	$\notin$	$\in$
$\in$	$\in$	$\notin$
$\notin$	$\notin$	$\in$
$\notin$	$\notin$	$\notin$

The similarity between the membership tables and the truth tables discussed in Chapter 1 should be readily apparent. Complementation corresponds to negation, intersection to conjunction, union to disjunction, and set difference to a statement of the form $p \wedge \sim q$. (Why?) Set equality requires that the sets be identical and corresponds to the equivalence relation in logic. In much the same spirit of our previous work with truth tables we can use membership tables to verify relations between sets.

Example 4: Show by membership tables that $\widetilde{(A \cup B)} = \tilde{A} \cap \tilde{B}$.

Solution: Develop a membership table for $\widetilde{(A \cup B)}$ and for $\tilde{A} \cap \tilde{B}$. The two sets are equal since the respective membership tables are the same row by row as follows.

A	B	$\tilde{A}$	$\tilde{B}$	$A \cup B$	$\widetilde{(A \cup B)}$	$\tilde{A} \cap \tilde{B}$
$\in$	$\in$	$\notin$	$\notin$	$\in$	$\notin$	$\notin$
$\in$	$\notin$	$\notin$	$\in$	$\in$	$\notin$	$\notin$
$\notin$	$\in$	$\in$	$\notin$	$\in$	$\notin$	$\notin$
$\notin$	$\notin$	$\in$	$\in$	$\notin$	$\in$	$\in$

The preceding discussion may be generalized into a "dictionary" which summarizes how we can "translate" set form to statement form and vice versa.

Dictionary

Set Form	Statement Form
P	p
Q	q
$\tilde{P}$	$\sim p$
$P \cap Q$	$p \wedge q$
$P \cup Q$	$p \vee q$
$P - Q$ or $P \cap \tilde{Q}$	$p \wedge \sim q$
$\tilde{P} \cup Q$	$p \longrightarrow q \qquad (\sim p \vee q)$
$(P \cup \tilde{Q}) \cap (Q \cup \tilde{P})$	$p \longleftrightarrow q$
$P \subset Q$	$p \Longrightarrow q$
$P = Q$	$p \Longleftrightarrow q$

Additional insight may be gained if we relate the truth sets on a Venn diagram (Figure 2.8) with the dictionary translations. For example, consider the statement forms $p \longrightarrow q$ and $p \Longrightarrow q$. The statement $p \longrightarrow q$ is true in all cases except where p is true and q is false; therefore the corresponding truth set is represented by regions 2, 3, and 4.

In set form we can describe these regions as the union of $\tilde{P}$ (regions 3 and 4) with the set Q (regions 2 and 3). Alternately, the statement $p \longrightarrow q$ can be converted to the equivalent form $\sim p \vee q$ and then translate negation to complement and disjunction to union directly.

The relation of logical implication $p \Longrightarrow q$ exists where there is no case in which p is true and q false. This suggests that in terms of sets the region 1 is empty. The truth set

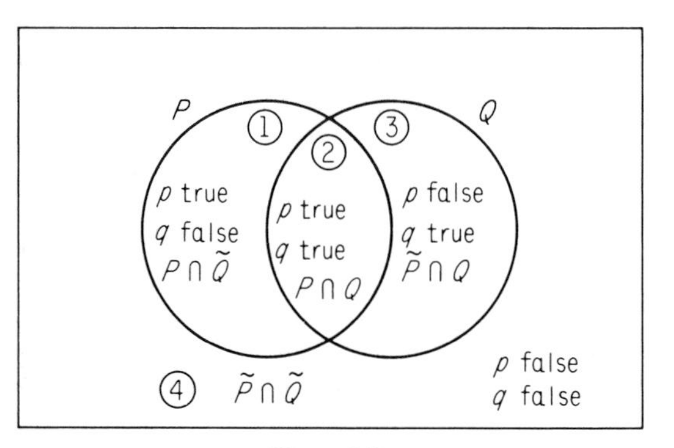

Figure 2.8

of P is therefore restricted to region 2. However, if the truth set of P consists only of region 2 and this region is a part of Q (regions 2 and 3), then P is a subset of Q, $(P \subset Q)$.

Another correspondence between truth sets and symbolic logic can also be made. The designation of a statement p as logically true meant that in every logically possible case the statement is true. Therefore, the truth set P for a logically true statement p is $\mathcal{U}$, since there is no case in which p is false. If a statement p should be logically false, then p is false in every logically possible case and the truth set for P is the empty set ϕ.

As the following examples will illustrate, we can now translate from set form to statement form, and vice versa thus we can display corresponding properties in each system.

Example 5:

1. Translate the following statement forms to set form:
   (a) $\sim(p \vee q) \Longleftrightarrow (\sim p \wedge \sim q)$.

   $\widetilde{(P \cup Q)} = \tilde{P} \cap \tilde{Q}$ (set form).

   (b) $(p \longrightarrow q) \wedge (q \longrightarrow r) \Longrightarrow p \longrightarrow r$.

   $(\tilde{P} \cup Q) \cap (\tilde{Q} \cup R) \subset \tilde{P} \cup R$ (set form).
2. Translate the following set form to statement form:
   (a) $P \cap (Q \cup R) = (P \cap Q) \cup (P \cap R)$.

   $p \wedge (q \vee r) \Longleftrightarrow (p \wedge q) \vee (p \wedge r)$ (statement form).

Example 6: Show by a truth set method that the statement $p \vee \sim(p \wedge \sim q)$ is logically true.

Solution: The corresponding set form of the statement is $P \cup \widetilde{(P \cap Q)}$. Using a numbered region approach for the regions of the Venn diagram as shown we can establish what regions are associated with the set $P \cup \widetilde{(P \cap \tilde{Q})}$.

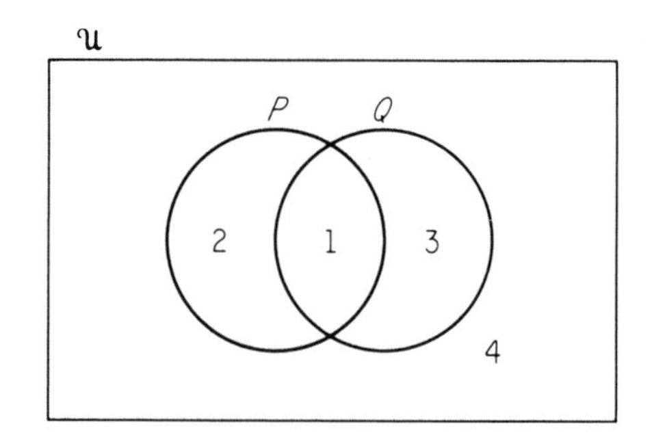

$$P \cup \widetilde{(P \cap \tilde{Q})} = \{1,2\} \cup \widetilde{(\{1,2\} \cap \widetilde{\{1,3\}})}$$
$$P \cup \widetilde{(P \cap \tilde{Q})} = \{1,2\} \cup \widetilde{(\{1,2\} \cap \{2,4\})}$$
$$P \cup \widetilde{(P \cap \tilde{Q})} = \{1,2\} \cup \{\tilde{2}\}$$
$$P \cup \widetilde{(P \cap \tilde{Q})} = \{1,2\} \cup \{1,3,4\}$$
$$P \cup \widetilde{(P \cap \tilde{Q})} = \{1,2,3,4\}.$$

Since the set $P \cup \widetilde{(P \cap \tilde{Q})}$ is the universe $\mathcal{U}$, the statement $p \vee \sim(p \wedge \sim q)$ is logically true.

Problem Set 2.4.2

A. 1. Use Venn diagrams to test whether any of the following statements are logically true, logically false, or neither.

(a) $q \wedge \sim q$. (b) $p \vee \sim p$.
(c) $(p \vee q) \vee (\sim p \wedge \sim q)$. (d) $p \dashrightarrow \sim (p \longrightarrow q)$.
(e) $p \longrightarrow (\sim p \longrightarrow q)$. (f) $p \longrightarrow (p \vee q)$.

2. Use membership tables for the following pairs of sets to test whether one is a subset of the other.

(a) $A; A \cap B$. (b) $B; A \cup B$. (c) $A - B; A$.

3. Translate the following statement forms to set form expressions.

(a) $\sim (p \vee q) \Longleftrightarrow \sim p \wedge \sim q$. (b) $p \Longrightarrow p \vee q$.
(c) $p \wedge q \Longrightarrow q$. (d) $\sim (\sim p) \Longleftrightarrow p$.

4. Show by the truth set method that the statement form $(p \wedge \sim q) \wedge (\sim p \wedge q)$ is logically false.

B. 1. Show by membership tables that:

(a) $\widetilde{(\tilde{A} \cap B)} = A \cup \tilde{B}$. (b) $\widetilde{(A - B)} = \tilde{A} \cup B$.

2. Let C and D be subsets of $\mathcal{U}$. Draw a Venn diagram and number the regions. Name the region (or regions) that geometrically represent the truth set for the following sets:

(a) $C \cap D$. (b) $C \cup D$. (c) $C - D$.
(d) $D - C$. (e) $\widetilde{(D \cup C)}$. (f) $D - \tilde{C}$.

3. Show that $A \cap \tilde{A}$ is the null set by translating the set expression $A \cap \tilde{A}$ into statement form and using properties of logic.

4. The symmetric difference of sets A and B, denoted $A \Delta B$, is defined to be $(A - B) \cup (B - A)$. Draw a Venn diagram and determine the region(s) that represent the symmetric difference of A and B. Does $A \Delta B = B \Delta A$ for any sets A and B?

5. Using the Venn diagram shown, and membership tables, determine which region(s) represent the following sets:

(a) $A \cap B$. (b) $A \cup B$. (c) $A - B$.
(d) $B \cap \tilde{A}$. (e) $\widetilde{(A \cup B)}$.

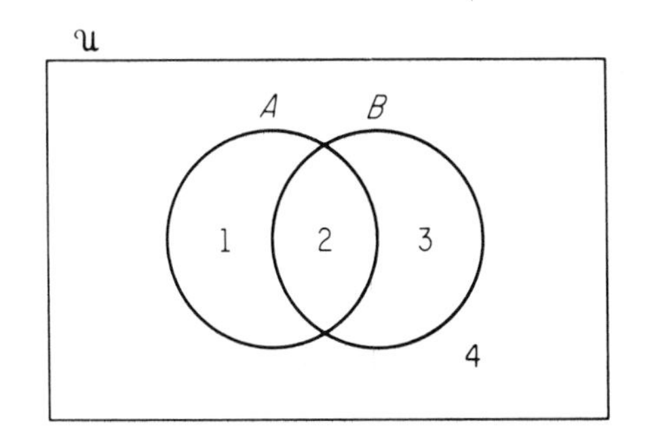

6. Using the Venn diagram shown in Prob. 5, develop an equivalent set form that will represent the following regions:
   (a) 3,4. (b) 1,4. (c) 1,3,4.
   (d) 2,4. (e) 2,3,4.
7. Assume that the Venn diagram in Prob. 5 represents the truth sets for statements "a" and "b."
   (a) What region(s) would be empty if $a \Longrightarrow b$?
   (b) What region(s) would be empty if $a \Longrightarrow b$ and also $b \Longrightarrow a$?

2.5 Properties of Sets

Throughout our discussion of set theory up to this point, we have been careful not to state that we have presented formal proofs of the many properties which we have discussed. Nevertheless, the intuitive approach used and the insights gained by Venn diagrams should have provided sufficient background for the reader not only to understand a set theory proof, but to develop some of his own proofs. In this section we will prove some basic set theorems by using our knowledge of symbolic logic, fundamental definitions, and valid argument forms. Several theorems will be fully discussed in the text and the reader will be asked to supply proofs of other similar theorems in the exercise section.

THEOREM 1: For any sets A and B, $A \cup B = B \cup A$. (Commutative property of union.)

In order to prove this theorem we will have to show that the definition of set equality holds for these two sets; namely $x \in A \cup B \longleftrightarrow x \in B \cup A$.

We can show the set equality holds by showing that $(x \in A \cup B \longrightarrow x \in B \cup A) \wedge (x \in B \cup A \longrightarrow x \in A \cup B)$.

Now let x represent any arbitrary element in the set $A \cup B$; i.e., $x \in A \cup B$. By the definition of union, $x \in A \vee x \in B$. But disjunction is commutative, thus $x \in B \vee x \in A$ and again by definition of union, $x \in B \cup A$. Hence, $x \in A \cup B \longrightarrow x \in B \cup A$.

If we now let $x \in B \cup A$, we can reverse the same line of reasoning and show that $x \in B \cup A \longrightarrow x \in A \cup B$.

Showing each conditional statement holds proves the biconditional statement holds, and proves that $A \cup B = B \cup A$.

The essence of the above proof is to use the definition of set equality and the fact that disjunction is commutative to prove that union is commutative.

THEOREM 2: For any sets A, B, and C, $A \cap (B \cup C) = (A \cap B) \cup (A \cap C)$. (Distributive property of $\cap$ over $\cup$.)

To show the set equality, we must show that $x \in A \cap (B \cup C) \longleftrightarrow x \in (A \cap B) \cup (A \cap C)$.

Let x be an arbitrary element of $A \cap (B \cup C)$ and show that $x \in A \cap (B \cup C) \longrightarrow x \in (A \cap B) \cup (A \cap C)$. By definition of intersection $x \in A \wedge x \in (B \cup C)$. But this simply means that $x \in A \wedge (x \in B \vee x \in C)$ by definition of union. The reader will recall that in Sec. 1-6 we demonstrated that $p \wedge (q \vee r) \Longleftrightarrow (p \wedge q) \vee (p \wedge r)$ which we now use to state $(x \in A \wedge x \in B) \vee (x \in A \wedge x \in C)$. This form is equivalent to stating that $(x \in A \cap B) \vee (x \in A \cap C)$ and this in turn is equivalent to $x \in (A \cap B) \vee (A \cap C)$ which establishes that $x \in (A \cap B) \cup (A \cup C)$. Thus, $x \in A \cap (B \cup C) \longrightarrow x \in (A \cap B) \cup (A \cap C)$.

To prove the converse (which is necessary for the if and only if) let $x \in (A \cup B) \cup (A \cap C)$ and repeat the steps in reverse order to deduce $x \in A \cap (B \cup C)$. Showing both conditionals hold establishes the set equality and proves the theorem.

The reader is cautioned not to jump to the conclusions that every argument is reversible based on the two preceding theorems. Each step should be carefully reviewed and justified.

THEOREM 3: For any sets A and B, $A \subset A \cup B$.

The statement $p \longrightarrow (p \vee q)$ is logically true [Problem Set 2.4.2-A.1.(f)]. If we let x be any arbitrary element of A, i.e., $x \in A$, then $x \in A \longrightarrow x \in A \vee x \in B$ which means $x \in A \longrightarrow x \in A \cup B$. By definition of subset we obtain $A \subset A \cup B$.

THEOREM 4: If A and $\tilde{A}$ are subsets of some $\mathcal{U}$, then $A \cup \tilde{A} = \mathcal{U}$.

We must establish that $x \in \mathcal{U} \longrightarrow x \in A \cup \tilde{A}$ and $x \in A \cup \tilde{A} \longrightarrow x \in \mathcal{U}$.

Let $x \in \mathcal{U}$. Using the tautology $p \vee \sim p$ we have for any $x \in \mathcal{U}$ that $x \in A \vee x \in \tilde{A}$. By definition of union we have $x \in A \cup \tilde{A}$. Since $x \in \mathcal{U}$ and $x \in A \cup \tilde{A}$ are both logically true we obtain the logically true statement, $x \in \mathcal{U} \longrightarrow x \in A \cup \tilde{A}$.

In order to verify that $x \in A \cup \tilde{A} \longrightarrow x \in \mathcal{U}$ we need only observe that since A and $\tilde{A}$ were given as subsets of $\mathcal{U}$ that under the union operation, the set formed will be a subset of $\mathcal{U}$, i.e., $A \cup \tilde{A} \subset \mathcal{U}$. Therefore, $x \in A \cup \tilde{A} \longrightarrow x \in \mathcal{U}$. Since both conditions hold the theorem is proved.

THEOREM 5: If A is any arbitrary set, then $A \cup \emptyset = A$.

Again we must show that $x \in A \cup \emptyset \longrightarrow x \in A$ and $x \in A \longrightarrow x \in A \cup \emptyset$.

Let $x \in A \cup \emptyset$. Then $x \in A \vee x \in \emptyset$. But we know that $x \notin \emptyset$ is logically true and may be used as an additional premise. Our valid argument form $(p \vee q) \wedge \sim q \Longrightarrow p$ permits us to conclude $x \in A$. For the converse argument, let $x \in A$ then $x \in A \vee x \in \emptyset$ by the valid argument form $p \Longrightarrow p \vee q$. Hence, $x \in A \cup \emptyset$. We have shown that the two conditionals hold, thus proving the theorem.

The preceding proofs indicate that we can proceed to develop a mathematical system of sets, called an algebra of sets, using set and element as our undefined terms and the definitions of the set operations and set relations together with the rules of logic to construct theorems. We will ask the reader in the exercises to prove some additional theorems so that the spirit of the activity is fully appreciated. However, we will not attempt to prove all of the fundamental theorems or laws of sets that we wish to employ. Not only would this be a task of considerable magnitude, but it can be shown that an algebra of sets is simply a model of a more abstract mathematical system called Boolean algebra which has the same properties. We will postpone further discussion of Boolean algebra, until the next section, and proceed to list some of the fundamental laws of an algebra of sets.

LAWS OF SET ALGEBRA

1. (a)	$\tilde{\mathcal{U}} = \emptyset$	Complement laws
(b)	$\tilde{\emptyset} = \mathcal{U}$	
2. (a)	$A \cap \tilde{A} = \emptyset$	
(b)	$A \cup \tilde{A} = \mathcal{U}$	
(c)	$(\tilde{\tilde{A}}) = A$	
3. (a)	$A \cap \mathcal{U} = A$	Identity laws
(b)	$A \cap \emptyset = \emptyset$	
(c)	$A \cup \mathcal{U} = \mathcal{U}$	
(d)	$A \cup \emptyset = A$	
4. (a)	$A \cap A = A$	Idempotent laws
(b)	$A \cup A = A$	
5. (a)	$A \cap B = B \cap A$	Commutative laws
(b)	$A \cup B = B \cup A$	
6. (a)	$(A \cap B) \cap C = A \cap (B \cap C)$	Associative laws
(b)	$(A \cup B) \cup C = A \cup (B \cup C)$	
7. (a)	$A \cap (B \cup C) = (A \cap B) \cup (A \cap C)$	Distributive laws
(b)	$A \cup (B \cap C) = (A \cup B) \cap (A \cup C)$	
8. (a)	$(\widetilde{A \cap B}) = \tilde{A} \cup \tilde{B}$	De Morgan's laws
(b)	$(\widetilde{A \cup B}) = \tilde{A} \cap \tilde{B}$	

Example 1: Using the laws of set algebra, prove that $\tilde{A} \cap (\tilde{B} \cap C) = \tilde{A} \cap (\widetilde{\tilde{C} \cup B})$.

Solution:

$\tilde{A} \cap (\widetilde{\tilde{C} \cup B}) = \tilde{A} \cap (\tilde{\tilde{C}} \cap \tilde{B})$	8(b) De Morgan's law
$= \tilde{A} \cap (C \cap \tilde{B})$	2(c) Complement law
$= \tilde{A} \cap (\tilde{B} \cap C)$	5(a) Commutative law

Example 2: Using the laws of set algebra, prove that $(\emptyset \cup \tilde{A}) \cap (\mathcal{U} \cap \tilde{A}) = \tilde{A}$.

Solution:

$(\emptyset \cup \tilde{A}) \cap (\mathcal{U} \cap \tilde{A}) = (\tilde{A} \cup \emptyset) \cap (\tilde{A} \cap \mathcal{U})$	5(a), 5(b)	Commutative law
$= \tilde{A} \cap \tilde{A}$	3(a), 3(d)	Identity law
$= \tilde{A}$	4(a)	Idempotent law

A careful scrutiny of the laws of the algebra of sets discloses another important and useful property. If $\emptyset$ is replaced by $\mathcal{U}$, $\mathcal{U}$ by $\emptyset$, $\cap$ by $\cup$, and $\cup$ by $\cap$, wherever these symbols occur, then the resulting expression is also one of the laws of set algebra. This property is known as the duality principle and permits us to develop a new law from one that has been established by a simple interchange of symbols.

PRINCIPLE OF DUALITY: In the laws for an algebra of sets, if $\emptyset$ is replaced by $\mathcal{U}$, $\mathcal{U}$ by $\emptyset$, $\cap$ by $\cup$, and $\cup$ by $\cap$, the resulting set equality statement is also a law of the algebra of sets. The new law formed is called the "dual" of the original law.

Example 3: Using the principle of duality, develop the second distributive law by forming the dual of the distributive law of intersection over union. The law $A \cap (B \cup C) = (A \cap B) \cup (A \cap C)$ distributes intersection over union.

Replacing $\cap$ by $\cup$, $\cup$ by $\cap$ gives us $A \cup (B \cap C) = (A \cup B) \cap (A \cup C)$ which is the second distributive law of union over intersection.

Example 4: In Example 2 we proved that $(\emptyset \cup \tilde{A}) \cap (\mathcal{U} \cap \tilde{A}) = \tilde{A}$. Write the dual of this statement.

Replacing $\emptyset$ by $\mathcal{U}$, $\mathcal{U}$ by $\emptyset$, $\cup$ by $\cap$, and $\cap$ by $\cup$, we form the dual $(\mathcal{U} \cap \tilde{A}) \cup (\emptyset \cup \tilde{A}) = \tilde{A}$.

Problem Set 2.5

A. State whether true or false.

1. For any sets A and B, $A \cup B \subset A \cap B$.
2. The empty set, $\emptyset$, is the identity element for union in the algebra of sets (i.e., for any set A, $A \cup \emptyset = A$).
3. For sets, intersection is commutative.
4. The complement of the union of set A and set B is equal to the complement of set A union the complement of set B.
5. The union of a nonempty set with its complement results in a set which has no elements.

B. Choose the correct result for each of the following:

1. The set $\tilde{A} \cup (B \cap \tilde{C})$ equals:
   (a) $(\tilde{A} \cap B) \cup (\tilde{A} \cap \tilde{C})$. (b) $(\tilde{A} \cup B) \cap \widetilde{(A \cap C)}$.
   (c) $\mathcal{U}$. (d) $\emptyset$.
2. The complement of the union of sets A and B is a subset of:
   (a) the empty set. (b) $A \cup B$. (c) $A \cap B$.
   (d) $\tilde{A} \cap \tilde{B}$.
3. $\widetilde{(\tilde{A} \cup \tilde{B})} \cup (\tilde{A} \cup \tilde{B})$ equals
   (a) A. (b) $A \cap B$. (c) $\mathcal{U}$.
   (d) $A \cup B$.

4. If $(A \cap B) \subset B$ and $B \subset A \cup B$ then $(A \cap B) \subset$
   (a) $\tilde{B}$. (b) $B \cap \tilde{B}$. (c) $\tilde{A}$.
   (d) $A \cup B$.
5. The complement of $A \cup \tilde{A}$ is
   (a) $\mathcal{U}$. (b) $A \cap B$. (c) $\emptyset$.
   (d) A.

C. 1. Prove: For sets A, B, C, $A \cup (B \cap C) = (A \cup B) \cap (A \cup C)$. Use logic and the definitions of set theory.
2. Using logic and definitions of set theory, show that if $A \subset B$ then $\tilde{B} \subset \tilde{A}$.
3. Show that $A \cap \tilde{A} = \emptyset$ using logic and set definitions.
4. Prove that the complement of the null set is the Universal set.
5. Prove that if $A \subset B$ and $B \subset C$ then $A \subset C$.
6. Show that the following hold using the fundamental laws of the algebra of sets:
   (a) $B \cup (\tilde{B} \cap C) = B \cup C$.
   (b) $A = (A \cap B) \cup (A \cap \tilde{B})$.
   (c) $A \cup B = ((A \cap B) \cup (A \cap \tilde{B})) \cup (\tilde{A} \cap B)$.
   (d) $A \cap (A \cup B) = A$.
7. Show that if $A \subset B$ then $A \cup B = B$.
8. (a) Write the dual of each statement in Prob. C-6.
   (b) Write the dual for the statement:

$$(A \cup \emptyset) \cap (\tilde{A} \cup \mathcal{U}) = (\tilde{A} \cap \emptyset) \cup (\mathcal{U} \cap A).$$

   (c) The laws of set algebra and the principle of duality refer to statements involving set equality. Discuss whether or not the 'duality principle' applies to the set relation subset by considering the two statements $A \cap B \subset A \cup B$; $A \cup B \subset A \cap B$.

2.6.1 Boolean Algebra

In the previous section we pointed out that the algebra of sets is a model of a more abstract mathematical system called Boolean algebra. One should also observe that the close association between symbolic logic and set theory is not accidental. Symbolic logic or, in this context called the algebra of propositions, is also a model of Boolean algebra. In fact, there are many models of Boolean algebra which have no direct reference to either sets or logical propositions. The situation is analogous to the new mathematical insight one gains when one discovers that the algebra studied in secondary school is simply an abstraction of the arithmetic studied on the elementary level. In the abstraction process, the generalizations obtained are applicable not only to familiar numeration systems, but also to a wide variety of nonarithmetic problems.

To develop the general properties of a Boolean algebra, we will use the algebra of sets as a guide in order to smooth the transition to this new system. Any collection of subsets of some set ($\mathcal{U}$) which is:

1. closed under two operations (union and intersection),
2. contains distinct identity elements (universal set and empty set), and
3. includes the complement of each element

is called a **Boolean algebra**.

By "closed" we mean that when the elements of the system are combined under either of the operations, we obtain a unique result which is also in the system. The function of the identity elements were displayed in the algebra of sets. The term *complement* may be thought of as designating the opposite condition, such as negation in logic and complement in sets.

We will now consider the general properties of a Boolean algebra. $A, B, C, D, \ldots$ will represent the abstract elements of our system. Plus (+) and times ($\cdot$) will represent the two operations, having the specific properties designated in the laws of Boolean algebra stated below. Let 0 and 1 be the distinct identity elements, and let $'$ (prime) be the complement of an element. The operations + and $\cdot$ are not to be confused with normal arithmetic operations, nor are the symbols 0 and 1 to be given any interpretation other than specifically denoted in the laws of Boolean algebra. The notation used was chosen simply as a matter of symbolic convenience. The system of Boolean algebra obeys the following laws:

LAWS OF BOOLEAN ALGEBRA

1.	$A + 0 = A$	$A \cdot 0 = 0$	Identity laws
	$A + 1 = 1$	$A \cdot 1 = A$	
2.	$A + A = A$	$A \cdot A = A$	Idempotent laws
3.	$A + A' = 1$	$A \cdot A' = 0$	Complement laws
	$(A')' = A$	$0' = 1$	
		$1' = 0$	
4.	$A + B = B + A$	$A \cdot B = B \cdot A$	Commutative laws
5.	$(A + B) + C = A + (B + C)$	$(A \cdot B) \cdot C = A \cdot (B \cdot C)$	Associative laws
6.	$A + (B \cdot C) = (A + B) \cdot (A + C)$	$A \cdot (B + C) = (A \cdot B) + (A \cdot C)$	Distributive laws
7.	$(A + B)' = A' \cdot B'$	$(A \cdot B)' = A' + B'$	De Morgan's laws

The laws of the algebra of sets which represent one model of Boolean algebra could be obtained by replacing + by $\cup$, $\cdot$ by $\cap$, 0 by $\emptyset$, 1 by $\mathcal{U}$, and $'$ by $\sim$. Rather than reviewing set theory, however, we will study Boolean algebra and other models of this mathematical system.

The simplest and yet one of the most useful Boolean algebras is the one whose variables A, B, $C, \ldots$ can only have the values of 0 or 1. A system in which the variables assume only one of two possible values is referred to as a **binary Boolean algebra**. For any variables A and B that assume only the values 0 and 1 we define the basic operations in the following tabular form:

A	A'
1	0
0	1

A	B	$A + B$	$A \cdot B$
1	1	1	1
1	0	1	0
0	1	1	0
0	0	0	0

The parallel to the truth table method in logic should be obvious. In fact, we can employ the basic tabular definitions to show that expressions are equivalent by showing the tabular forms are the same row by row, exactly as we did in logic.

Example 1: Prove $A \cdot (A + B) = A$ in a binary Boolean algebra.

Solution: Develop in a tabular form the row by row possibilities for $A \cdot (A + B)$. Compare the possibilities of the expression with those appearing under A.

A	B	$A + B$	$A \cdot (A + B)$
1	1	1	1
1	0	1	1
0	1	1	0
0	0	0	0

Since the first column and the last column are the same row by row, $A \cdot (A + B) = A$.

Each of the laws for a binary Boolean algebra could be proved using the basic definitions given in Table 1 and the tabular method illustrated in Example 1. We will not prove these laws, but we will show how we might extend a Boolean algebra by using the basic laws to develop other laws.

Example 2: Prove that $A + [(A' \cdot C) + B] = (A + B) + C$.

PROOF:

$$\begin{aligned}
A + [(A' \cdot C) + B] &= [A + (A' \cdot C)] + B && \text{Associative} \\
&= [(A + A') \cdot (A + C)] + B && \text{Distributive} \\
&= [1 \cdot (A + C)] + B && \text{Complement} \\
&= (A + C) + B && \text{Identity} \\
&= A + (C + B) && \text{Associative} \\
&= A + (B + C) && \text{Commutative} \\
&= (A + B) + C && \text{Associative}
\end{aligned}$$

Example 3: Prove that $(A' \cdot B')' + (A' \cdot B)' = 1$.

PROOF:

$$\begin{aligned}
(A' \cdot B')' + (A' \cdot B)' &= (A + B) + (A + B') && \text{De Morgan's law} \\
&= (A + A) + (B + B') && \text{Associative and commutative} \\
&= A + 1 && \text{Idempotent and complement} \\
&= 1 && \text{Identity}
\end{aligned}$$

Problem Set 2.6.1

A. Determine whether each of the following statements is either true or false.

1. The term "binary" in binary Boolean algebra means that there are only two operations in the system.
2. The algebra of real numbers is an example of a binary Boolean algebra.
3. In a Boolean algebra, elements which act as 0 and 1 are necessary in the algebra.
4. In a Boolean algebra the 0 and 1 are assumed to be distinct.
5. "Prime" and "closed" have the same meaning in Boolean algebra.

B. 1. Prove the following basic laws binary Boolean algebra using the tabular method.
(a) $(A')' = A$. (b) $(A \cdot B)' = A' + B'$. (c) $A + A' = 1$.
(d) $A + A = A$.

2. Using the basic laws of binary Boolean algebra, prove the following:
(a) $A + (A \cdot B') = A$.
(b) $A \cdot (B + B') = A$.
(c) $(A \cdot B) \cdot C = (A \cdot B) \cdot (A \cdot C) \cdot (B \cdot C)$.
(d) $A + (B \cdot C) = (A + B) \cdot (A + C)$.
(e) $A + (A \cdot B) = A \cdot (A + B)$.

3. Show that the following Boolean expressions may be written in terms of times $(\cdot)$ and prime $(')$.
(a) $A + B$. (b) $A \cdot (B + C)$. (c) $(A + B) + C$.
(d) $A' + B'$.

C. 1. Explain why it is perhaps more reasonable to prove some theorems in binary Boolean algebra using the basic laws rather than relying on the use of the tabular method.

2.6.2 Switching Circuits

Binary Boolean algebra is extremely important in the design and construction of electronic computers and telephone communications systems. Although a discussion of electronic circuitry is well beyond the scope of this text, we believe the reader can absorb some simplified concepts of switching circuits and thereby develop some appreciation of the impact applied Boolean algebra has had on the electronics industry.

In order to apply Boolean algebra to a switching circuit, we must give some meaning to the abstract symbols and the basic operations. We will agree to regard the variables $A, B, C, \ldots$ as switches. Each switch will have two positions; namely, the open position designated 0 and the closed position designated 1. Schematically, an open and closed switch will be drawn as follows:

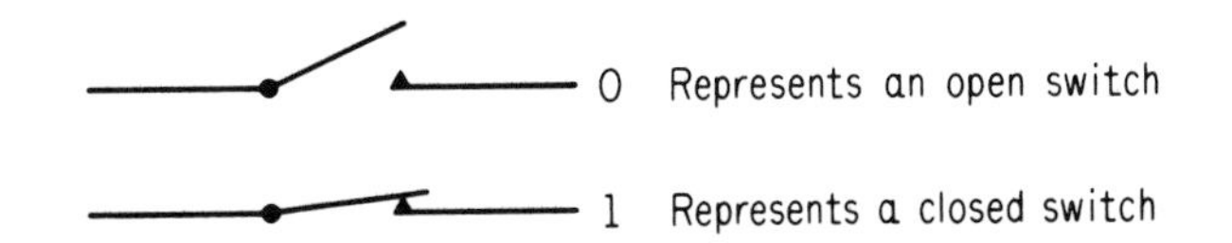

In a switching circuit, A and A' will refer to two switches that operate simultaneously, but have opposite states (i.e., if A is closed, then A' is open and vice versa).

The switches in a network may be interpreted as a binary Boolean expression. Switches may be in series or in parallel. Two basic diagrams are shown below:

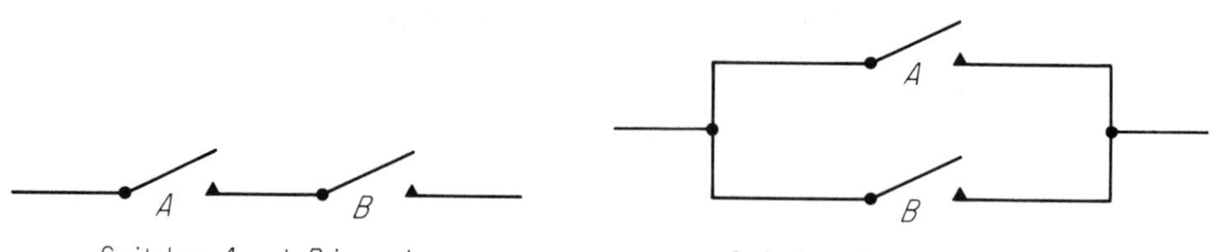

Switches A and B in series

Switches A and B in parallel

In our circuit drawings, we will as a matter of convenience draw A as an open switch and A' as closed. It should be clearly recognized that each switch can assume either the open or closed position. Our diagram will represent only one of the two possible states for each component.

We will use the operation times (·) to indicate the switches are in series and the operation plus (+) to indicate the switches are in parallel. Since each switch has two possible positions (open or closed) a two-switch network will involve four possible conditions. In tabular form we list the possible circuit conditions:

A	B	$A \cdot B$	$A + B$
1	1	1	1
1	0	0	1
0	1	0	1
0	0	0	0

When two switches are in series, the closed circuit state occurs only when both A and B are in the closed position. When A and B are connected in parallel the closed state for the network occurs when either A or B or both are in the closed position.

The physical interpretation to be given to prime (′) in terms of switching circuits is to impose on a circuit or a switch the opposite state. Expressed in tabular form we have

A	A'
1	0
0	1

Often we are interested in comparing two or more networks to determine if they have the same open or closed state under all conditions. Should two networks have the same state for all possible conditions of the switching network, we regard them as **equivalent**.

In practice, to determine equivalence we can show that in a tabular form the networks have the same state row by row or we can use the laws of Boolean algebra.

Example 1: Show by both the tabular method and the laws of Boolean algebra that the networks $A \cdot (A + B)$ and $A + (A \cdot B)$ are equivalent.

Solution: By the tabular method, we determine that column $A \cdot (A + B)$ and $A + (A \cdot B)$ are the same row by row, thereby showing equivalence.

A	B	$A + B$	$A \cdot (A + B)$	$A \cdot B$	$A + (A \cdot B)$
1	1	1	1	1	1
1	0	1	1	0	1
0	1	1	0	0	0
0	0	0	0	0	0

By the laws of Boolean algebra:

$$\begin{aligned} A \cdot (A + B) &= (A \cdot A) + (A \cdot B) && \text{Distributive law} \\ &= A + (A \cdot B) && \text{Idempotent law} \end{aligned}$$

A close inspection of the tabular form for both $A \cdot (A + B)$ and $A + (A \cdot B)$ in Example 1 reveals that they are row-by-row the same as the table for A. In circuit analysis, this would mean that we could replace both networks by a single switch A. The savings that can be realized by eliminating unnecessary components from electronic circuits can be significant and justifies an analysis by Boolean algebra.

In addition to a tabular analysis, a schematic drawing of a network can be helpful in circuit design. Conversely, we can translate a circuit to a Boolean form.

Example 2: Make a schematic drawing of the following switching networks:
(a) $(A \cdot B) + (A \cdot C)$.
(b) $A \cdot (B + C)$.
(c) $A + (A' \cdot B)$.
(d) $(A \cdot B) + (A' \cdot B')$.

Solution:
(a)

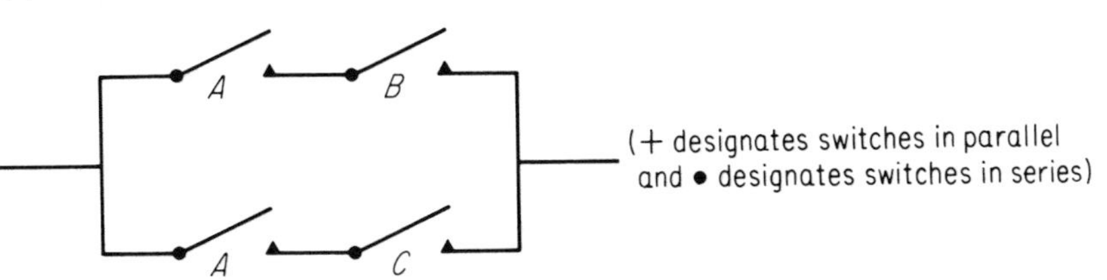

(b)

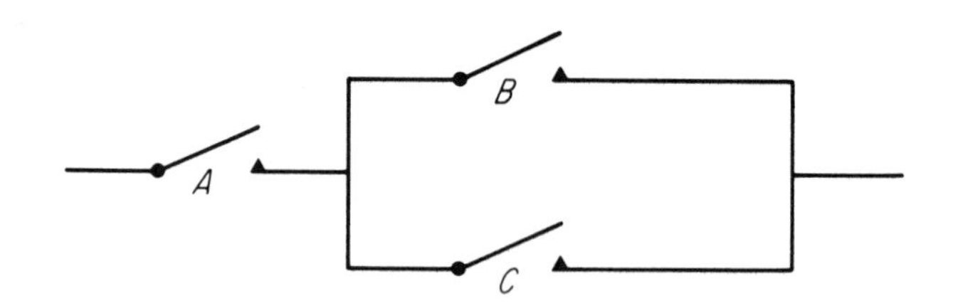

(c)

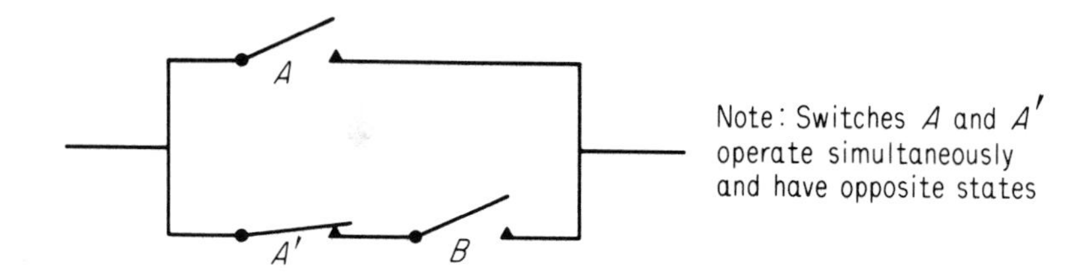

(d)

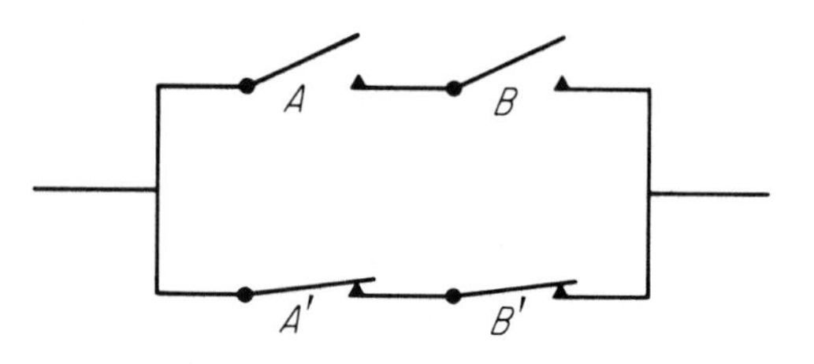

Example 3: Write a network Boolean expression for the following circuit diagrams.
(a)

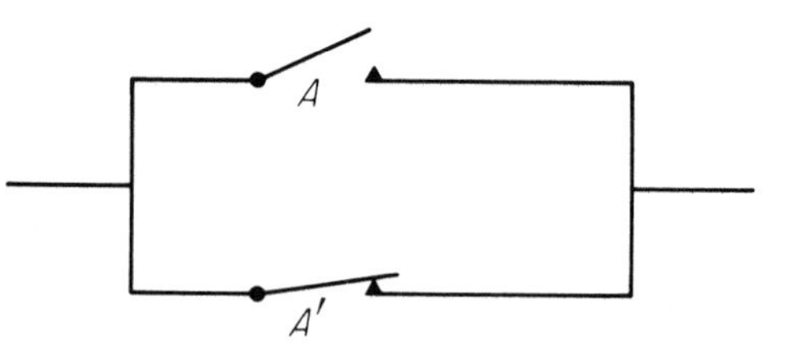

(b)

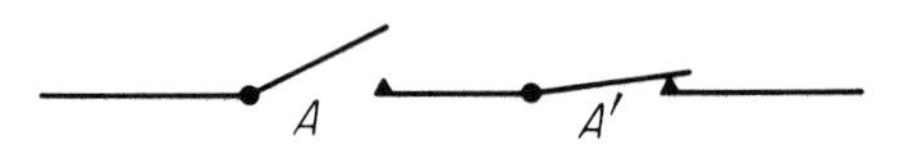

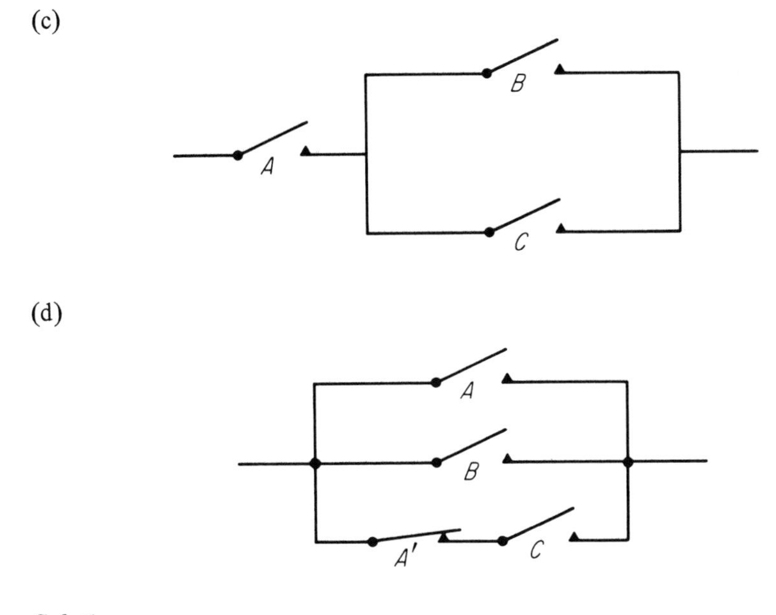

Solution:

(a) $A + A'$

(b) $A \cdot A'$

(c) $A \cdot (B + C)$

(d) $A + B + (A' \cdot C)$

We will conclude our discussion of switching circuits by considering one more example which is common to nearly everyone's experience.

Example 4: Develop a circuit diagram using two switches that will permit a person to operate a hall light from both the bottom and the top of a staircase.

Solution: In order to accomplish the desired result, we will need switches that assume opposite states under normal conditions, but operate independently from the two different locations. Consider the accompanying circuit drawing. Suppose the hall light is off and switch A is at the top of the stairs and B is at the bottom of the stairs. A is open and B' is closed and A' is closed and B is open. By closing switch A the light will go on, since the circuit is then closed. Alternately, close

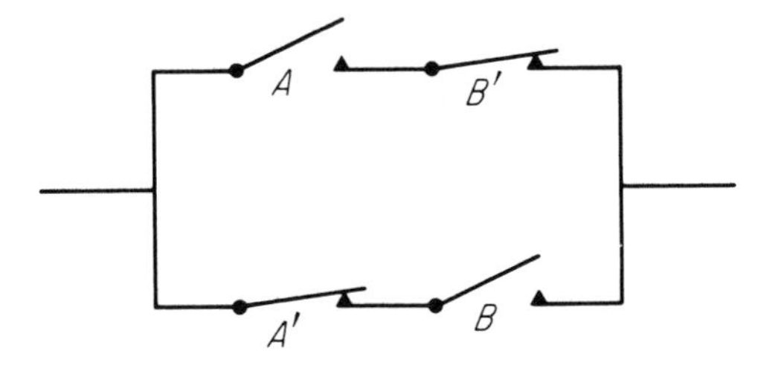

switch B and the light will also go on. The light may be turned off at either location by turning the switch to the opposite position.

Our discussion of Boolean algebra and switching circuits has had a twofold purpose. First, we wished to convey to the reader the process of forming an abstract mathematical system free of a specific model interpretation and limited in its use only by the imagination of the user. Secondly, it was our intent not only to acquaint the reader with a highly utilitarian application of Boolean algebra in terms of switching circuits, but to communicate the spirit of the applied use of mathematics in a technological society.

Problem Set 2.6.2

A. 1. Let A and B be switches. Draw a diagram for the expression $(A + A') \cdot (B + B')$. What can be said about the "state" of the network expression?

2. Draw a circuit diagram for the following switching network expressions. (a) $A + (B \cdot B')$. (b) $(A + B) \cdot (A + C)$. (c) $(A' \cdot B') + (A \cdot B)$. (d) $A + (B \cdot C)$. In this problem, are any of the expressions equivalent? Justify your result.

3. Find a switching network expression for each of the following switching diagrams.

(a)

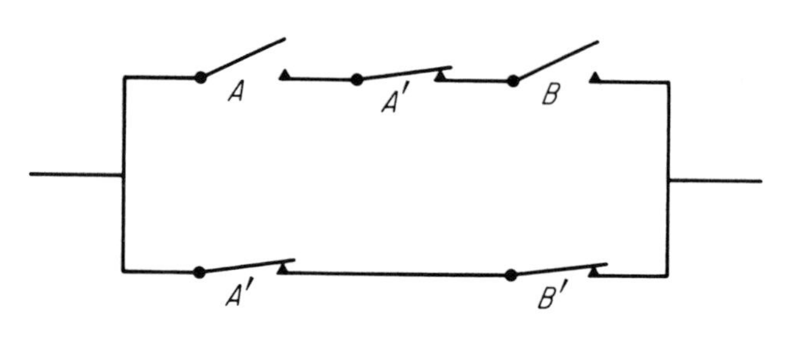

(b)

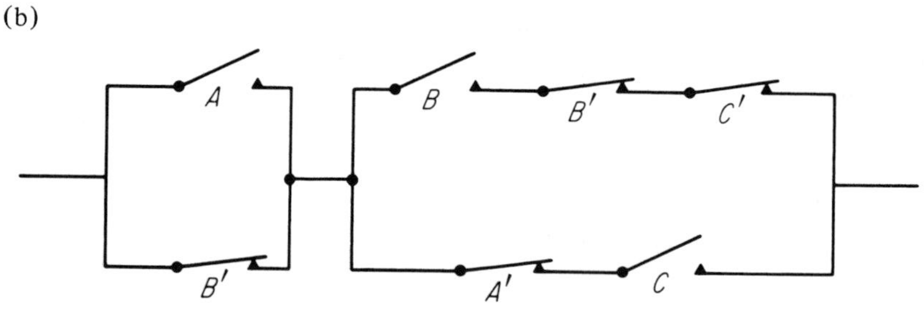

(c)

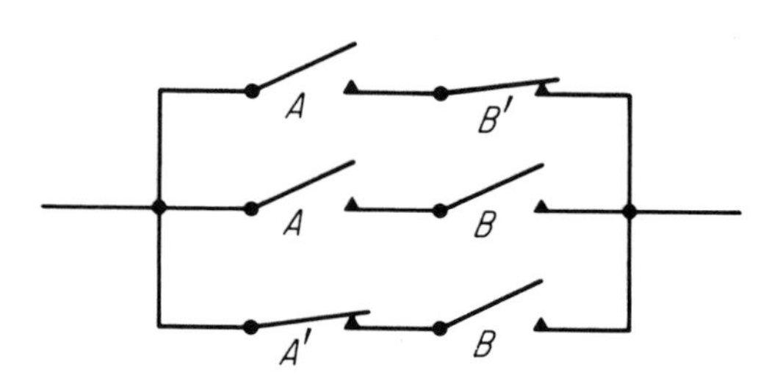

(d)

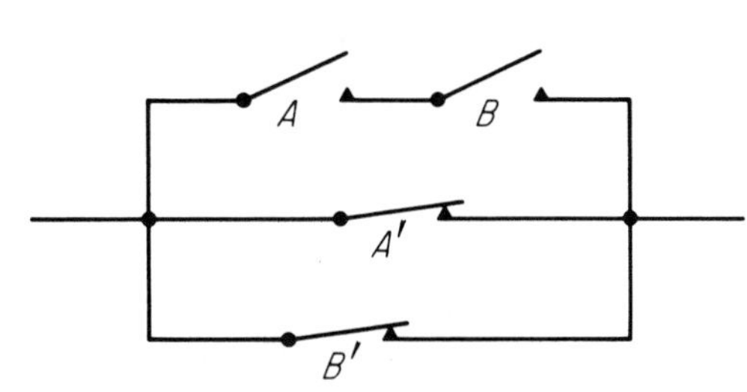

4. For the following network expressions, find the prime of each of them.
   (a) $A + A'$. (b) $A \cdot B$. (c) $A + (B \cdot C)$.
   (d) $A' + B'$.
5. Draw the circuit diagram for the prime of the following expressions.
   (a) $A' + B'$. (b) $A' \cdot (B' + C')$. (c) $A + (B \cdot C)$.
   (d) $A \cdot (A' + A)$. (e) $(A \cdot A') + A$.

B. 1. Determine which of the following network expressions are equivalent.
   (a) A. (b) $A + (A \cdot B)$. (c) $A \cdot (A + B)$.
   (d) $A' \cdot (A' + B')$. (e) $A \cdot B$.
2. Simplify the following switching network expressions, and draw a circuit diagram for the given network and the simplified diagram.
   (a) $(A \cdot B) + (A \cdot C)$. (b) $B \cdot (A + A')$. (c) $(A \cdot B) + (A' \cdot B)$.
   (d) $A + A'$. (e) $A + A$.
3. Which of the following circuit diagrams are equivalent?
   (a)

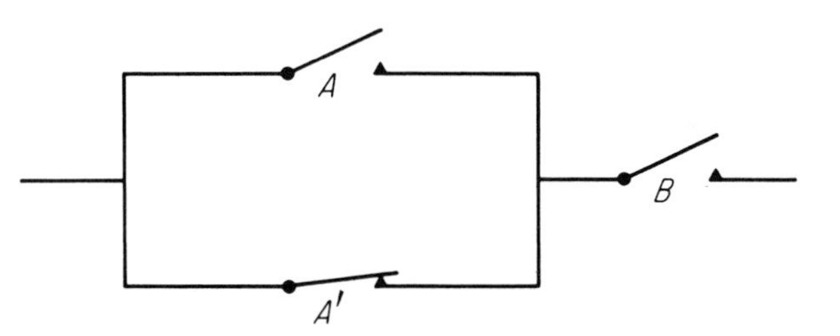

(b)

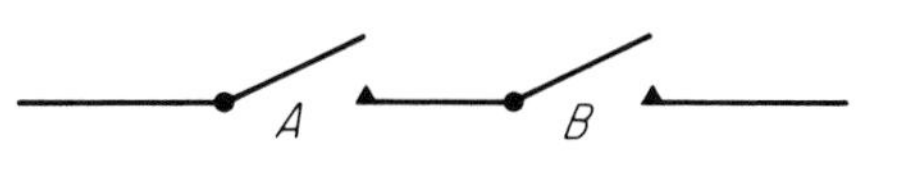

(c)

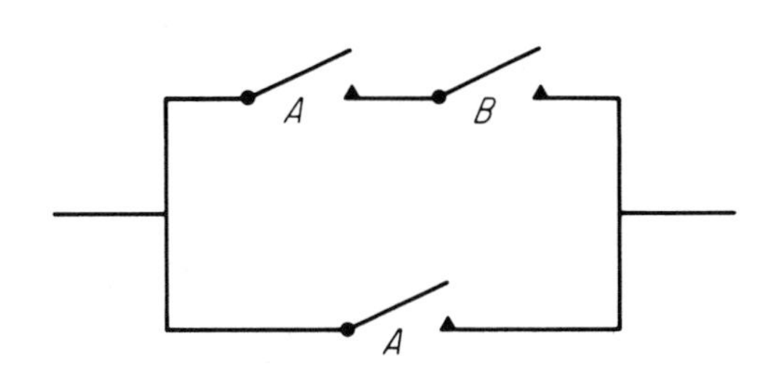

(d)

(e)

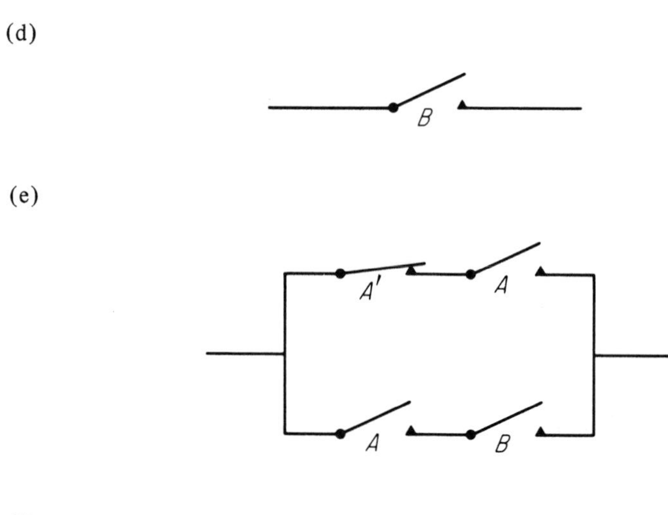

(f)

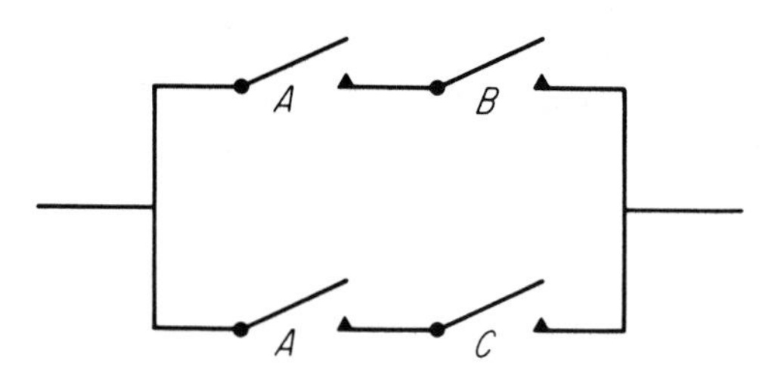

(g)

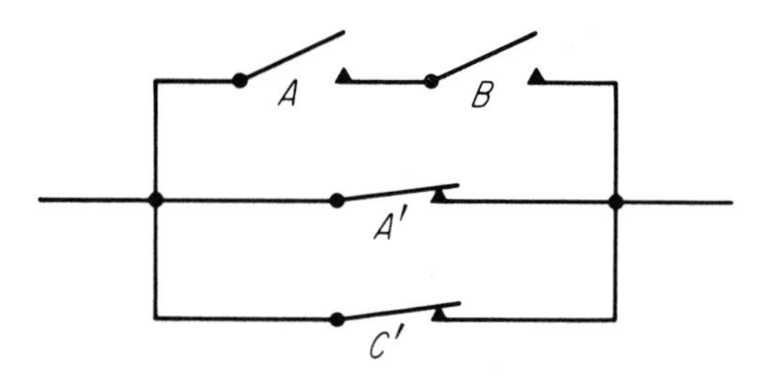

(h)

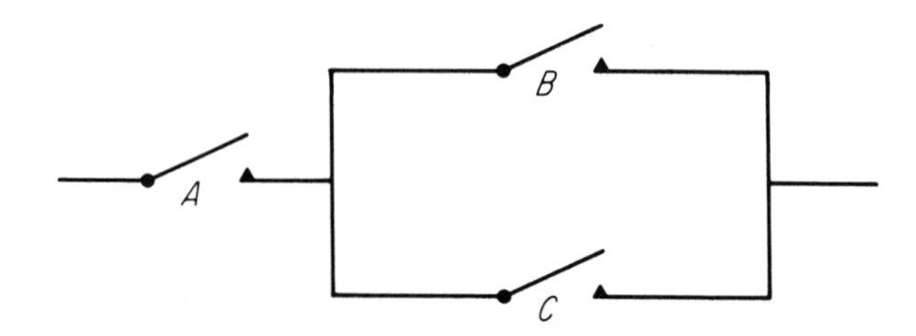

C. 1. Draw a circuit diagram which will enable a person to turn on or turn off a circuit from three different locations, using three switches. (This may be thought of as an extension of the hall-light problem discussed in Example 4.)

2.7 Cartesian Product

In the study of arithmetic you were taught not only such basic operations as addition and multiplication, but also the operations of squaring, square root, forming a reciprocal, etc. Similarly, many operations other than intersection, complement, and union can be defined on sets. Before we complete our discussion in this chapter of set theory we will define one more operation which will be useful to us in the following chapters.

The set operation, **Cartesian product**, differs from the operations we have studied previously in that the elements of the set formed under this operation are ordered pairs. It is common for us to be interested in pairs of information; for example, the height and weight of an individual, the temperature and humidity, rate of speed and gasoline consumption. In organizing the information, it is convenient to introduce an ordering so that, for example, the height would be given first and weight second. We define an ordered pair as follows:

DEFINITION: An **ordered pair** is an element (a,b) formed by taking a from some set and b from some set in such a way that a is designated as the "first" member and b as the "second" member.

Notice the definition does not require that the set from which the element a is selected must be different from the set from which the element b is selected. The concept of order, however, does discriminate between the ordered pair (2,3) and the ordered pair (3,2). Two ordered pairs (a,b) and (c,d) are said to be **equal** if and only if $a = c$ and $b = d$.

Having defined ordered pair, we now define the Cartesian product of two sets.

DEFINITION: The **Cartesian product**, symbolized $A \times B$ (read A cross B) of two sets A and B is the set of all ordered pairs (a,b) with the first member chosen from set A and the second member chosen from set B. Symbolically, $A \times B = \{(a,b) \mid a \in A \wedge b \in B\}$.

Example 1: If $A = \{1,2,3\}$ and $B = \{a,b\}$, write out the Cartesian product $A \times B$.

Solution: A tree diagram provides us with a convenient device by which we can produce all of the ordered pairs. "Reading off" the tree in the order listed we can write $A \times B = \{(1,a),(1,b),(2,a),(2,b),(3,a),(3,b)\}$.

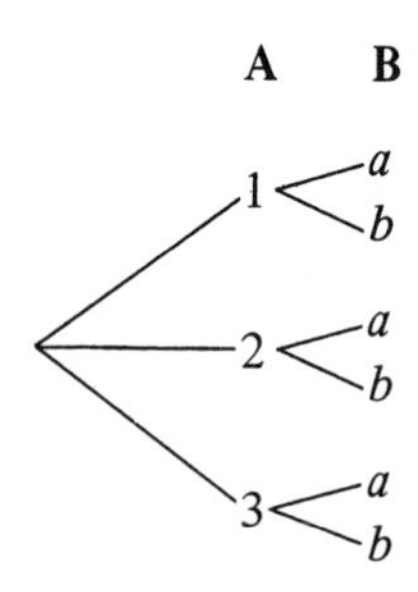

The Cartesian product is not limited to two sets. For example, using ordered triples of the form (a,b,c) and sets A, B, C we can extend the definition as follows:

$$A \times B \times C = \{(a,b,c) \mid a \in A \wedge b \in B \wedge c \in C\}.$$

Example 2: If $A = \{1,2\}$, $B = \{1,a\}$, and $C = \{a,b,c\}$, write out the Cartesian product $A \times B \times C$.

Solution: The tree diagram again provides us with a convenient method for enumerating the ordered triples.

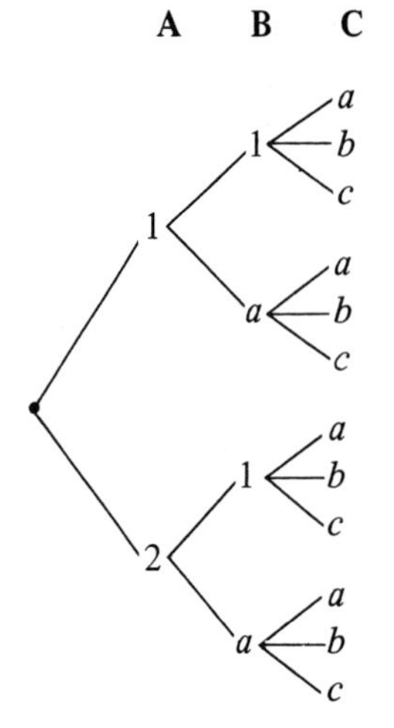

$$A \times B \times C = \{(1,1,a), (1,1,b),(1,1,c),(1,a,a),(1,a,b),\\ (1,a,c),(2,1,a),(2,1,b),(2,1,c),(2,a,a),(2,a,b),(2,a,c)\}$$

As Examples 1 and 2 indicate, we can use a tree diagram to assist us in counting the ordered elements in a Cartesian product. But this is nothing more than an application of the Fundamental Principle of Counting. For finite sets, we can count the number of ordered elements in a Cartesian product by simply multiplying the number of ways we can choose an element from the first set, times the number of ways we can choose an element from the second set, times the number of ways we can choose an element from the third set, continuing in this manner until we have included in the product the number of ways we can choose an element from the last set. Using a notation $n(A)$ which will mean the number of elements in set A, we give a counting formula for the Cartesian product of two sets, and of three sets which can be generalized beyond three sets.

$$n(A \times B) = n(A) \cdot n(B)$$
$$n(A \times B \times C) = n(A) \cdot n(B) \cdot n(C).$$

Example 3: If a set A contains 3 elements, a set B contains 4 elements, and a set C contains 5 elements, determine the number of elements in the following Cartesian products.
(a) $A \times B$.
(b) $A \times B \times C$.
(c) $A \times A$.
(d) $A \times A \times C$.

Solution:
(a) $n(A \times B) = n(A) \cdot n(B) = 3 \cdot 4 = 12$.
(b) $n(A \times B \times C) = n(A) \cdot n(B) \cdot n(C) = 3 \cdot 4 \cdot 5 = 60$.
(c) $n(A \times A) = n(A) \cdot n(A) = 3 \cdot 3 = 9$.
(d) $n(A \times A \times C) = n(A) \cdot n(A) \cdot n(C) = 3 \cdot 3 \cdot 5 = 45$.

An important use of Cartesian products is to associate points in either a two-dimensional or three-dimensional space with the Cartesian product of sets of real numbers. This approach gives rise to an *x-y* coordinate system in two-dimensional space and an *x-y-z* coordinate system in three-dimensional space. In this context we refer to the resulting geometric form as a Cartesian space. Further study along these lines lies properly in a course in algebra or analytic geometry. We will conclude our comments on the Cartesian product by pointing out that the basic concepts of product sets and a Cartesian coordinate system are named after the famous French mathematician René Descartes (1596–1650), who published in 1637 a work entitled *La Géometrie* which for the first time combined the analytical methods of algebra with the concepts of geometry by the use of a coordinate system.

Problem Set 2.7

A. State whether true or false.
1. The Cartesian product of any two sets is another set.
2. The Cartesian product of any sets, say A and B, is commutative, i.e., $A \times B = B \times A$.
3. Each element of the Cartesian product of two nonempty sets is an ordered pair.
4. Let A and B be sets, then $n(A \times B) = n(B \times A)$.
5. The ordered pair (a,b) does not equal the ordered pair (c,d) if and only if $a \neq c$ or $b \neq d$.

B. 1. Let $D = \{1,2,3,4,5,6\}$ a set representing the outcomes on a single die. Find $D \times D$. How many elements are in $D \times D$?
2. Let $A = \{1,2,3\}$, $B = \{1,2\}$. Find (a) $A \times B$, (b) $B \times A$.

3. Let $A = \{x,y\}$. Find (a) $A \times A$, (b) $A \times A \times A$. How many elements are in (a) $A \times A$, (b) $A \times A \times A$? Suppose we form the Cartesian product of n sets of set A, how many elements are there in this Cartesian product of n sets of A?
are there in this Cartesian product of n sets of A?
4. Is the operation of forming a Cartesian product between any two sets commutative? Give an example where the operation is commutative and one where the operation is not commutative.
5. Name two sets such that the Cartesian product between these sets will have exactly seven elements in the Cartesian product. How many elements should each set have?
6. Name two sets such that the Cartesian product of these two sets will have six elements in the set. What is the number of elements of each set?
7. Given the sets $D = \{1,2,3,4,5,6\}$, the outcomes on a roll of a die, and $C = \{H,T\}$, the outcomes on the toss of a coin. Find (a) $C \times D$, (b) $D \times C$.
8. A test is made up of two parts. The first part has 5 true-false questions and the second part has 5 multiple-choice questions, each question having three choices. How many different ways can the test be answered? (Assume each question is answered.)
9. Let $A = \{1,2,3\}$, $B = \{2,4,6\}$, and $C = \{1,2,3,4,6\}$. Find:
   (a) $n(A \times B)$.
   (b) $n(A \times B \times C)$.
   (c) $n(A \times A \times C)$.
   (d) $n[(A \cap B) \times C]$.
   (e) $n[(C - B) \times A]$.

Review Exercises

2.1 State definitions for the following:
   (a) empty set.
   (b) subset.
   (c) solution.
   (d) solution set.
   (e) set union.
   (f) set intersection.
   (g) set difference.
   (h) complement of a set.
   (i) truth set.
   (j) ordered pair.
   (k) Cartesian products of two sets.

2.2 Determine which of the following are sets. For those that are not sets, state the reason.
   (a) The collection of **all** the letters in the word "logic."
   (b) The collection of **all** the symbols in the number, 125,271.
   (c) The group of men's and women's coats in a department store.
   (d) The collection of **all** the letters in the word "algebra."

2.3 Let $A = \{\emptyset, \{x\}, \{y\}, \{x,y\}\}$. List the elements in the power set of set A. How many elements are in the power set of A? Is it possible to have a power set of a power set? Can this be extended indefinitely?

2.4 Let $\mathcal{U} = \{z, y, x, w, v, u, t, s, r, q, p\}$, $A = \{q, s, u, w, y\}$, $B = \{y, x, w, v, u, q, p\}$ and $C = \{r, u, s, t\}$. Find:
   (a) $\tilde{B}$.
   (b) $A - B$.

(c) $B - \tilde{B}$.
(d) $A \cup B \cup C$.
(e) $A \cap B \cap C$.
(f) $A \cap (B \cup C)$.
(g) $(A \cap B) \cup C$.
(h) $C \times C$.
(i) $n(A \times B)$.
(j) $n(B \times A \times C)$.

2.5 Let A, B, and C be subsets of $\mathcal{U}$. Using the Venn diagram with the regions numbered, show that $\tilde{A} \cap (B \cup \tilde{C}) = (\tilde{A} \cap B) \cup (\widetilde{A \cup C})$.

2.6 Let A, B, and C be subsets of $\mathcal{U}$. By membership tables, show that:
(a) $\tilde{A} \cap \tilde{B} = (\widetilde{A \cup B})$.
(b) $\tilde{A} \cup (\tilde{B} \cap \tilde{C}) = (\widetilde{A \cap B}) - (A \cap C)$.

2.7 Let A, B be any sets. Prove that $A \cap B \subset A \cup B$.

2.8 Using the fundamental laws of the algebra of sets given in Sec. 2.5, show that $A \cup (A \cap B) = A$.

2.9 Using the basic laws of Boolean algebra, show that $A + (A \cdot B) = A$.

2.10 Find a switching network expression for the following switching diagram. Can the circuit diagram be simplified and still be functionally equivalent?

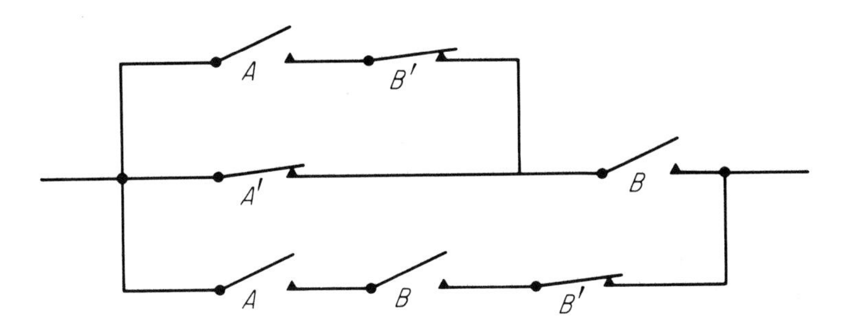

Suggested Reading

Christian, Robert R., *Introduction to Logic and Sets.* Blaisdell, New York, 1965.

Dodge, Clayton W., *Sets, Logic and Numbers.* Prindle, Weber & Schmidt Ind., Mass., 1969.

Kemeny, John G., J. Laurie Snell, and Gerald L. Thompson, *Introduction to Finite Mathematics.* Prentice-Hall, Englewood Cliffs, N.J., 1966.

Lipschutz, Seymour, *Finite Mathematics.* Schaum-McGraw-Hill, New York, 1966.

Lipschutz, Seymour, *Set Theory and Related Topics.* Schaum Outline Series, McGraw-Hill, New York, 1964.

Stoll, Robert R., *Sets, Logic and Axiomatic Methods.* W. H. Freeman, California, 1961.

Selby, Samuel and Leonard Sweet, *Sets–Relations–Functions*, 2d ed. McGraw-Hill, New York, 1969.

God made the integers, all the rest is the work of man.

–L. KRONECKER

CHAPTER 3

Counting Theory

3.1 Introduction

Throughout history mankind has attempted to deal with the question, "How many?" By contemporary standards, some of these efforts seem rather crude. For example, some of our early ancestors when counting their possessions, such as livestock, would collect pebbles of a certain type for each class of objects they wished to enumerate. The pebbles were placed in bags and carried around the waist. At any time, if an individual wished to reaffirm that no enterprising neighbor had relieved him of a prized possession, he would assemble the objects in question and set up a correspondence between each pebble and one of the members of this class of objects. Pebble counting has some rather serious limitations and mankind, out of necessity, invented numeration systems to handle more complex counting problems.

Each numeration system has distinct characteristics. The convenience of the notation and the principle of number base of each numeration system is of primary significance. Consider the multiplication problem in the Roman numeral system of (XVIII) · (CXII) as opposed to the product of (18) · (112) in the Arabic numerations system. Many historians believe that one of the main reasons that the Greeks and Romans did not develop more scientific knowledge was due to the awkwardness of their numeration system.

The goal of this chapter is to acquaint the reader with some fundamental counting methods. We will develop some general procedures for handling such questions as, "How many ways may I rearrange 5 books on a shelf?," and "How many committees of 3 persons can be formed by choosing from 5 individuals?" The knowledge that we gain will aid us in the study of probability in Chapter 4. We begin by first discussing how to count the number of elements in a set. (The reader is reminded that a set may be of a rather simple form such as A or a set may be represented by a more complex form, such as $(A \cap B \cap \tilde{C})$.

3.2 Number of Elements in Sets

Suppose you are given the following information by one of your friends:

30 students have registered for English literature, and 25 students have registered for history.

How would you respond to the question, "How many students are taking English literature or history?" If your response would be that the question cannot be answered because the information is not complete, you have good insight into such problems. What is lacking is some knowledge as to whether or not some students are taking both subjects. If there are no students taking both subjects, then there are 55 students taking one or the other of the two courses. However, if there are some students taking both courses, the total number of students must be less than 55.

Let us look at this situation in a more formal and general way. Let $n(A)$ represent the number of elements in any set A, where $n(A)$ must be greater than or equal to zero, i.e., $n(A) \geqslant 0$. Consider any two sets A and B as shown in the Venn diagram.

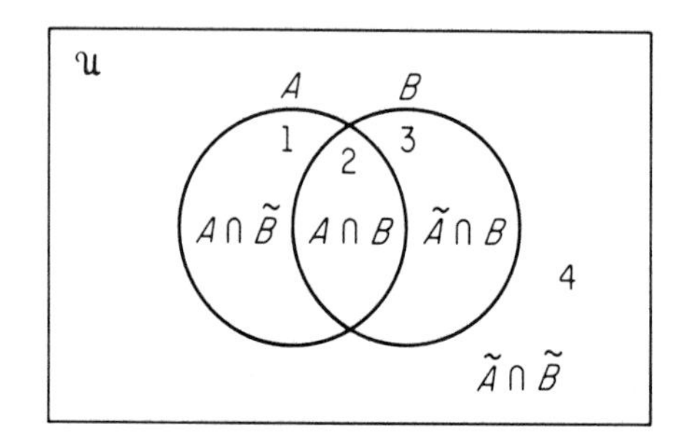

If we wish to determine the number of elements in set A or set B which is simply $n(A \cup B)$ we make use of the fact that regions 1, 2, and 3 are disjoint from each other, and $A \cup B = (A \cap \tilde{B}) \cup (A \cap B) \cup (\tilde{A} \cap B)$. We make the assumption that the total number of objects in the union is the sum of the number of objects in each disjoint region; i.e.,

(1) $\quad \boldsymbol{n(A \cup B) = n(A \cap \tilde{B}) + n(A \cap B) + n(\tilde{A} \cap B).}$

We must now consider two distinct possibilities, $A \cap B = \emptyset$ or $A \cap B \neq \emptyset$.

Case I: Assume that A ∩ B = ∅. The set A is the union of the two disjoint regions $A \cap \tilde{B}$ and $A \cap B$, i.e., $A = (A \cap \tilde{B}) \cup (A \cap B)$. If $A \cap B = \emptyset$ then $A = A \cap \tilde{B}$. Similarly, $B = (A \cap B) \cup (\tilde{A} \cap B)$ and if $A \cap B = \emptyset$, $B = \tilde{A} \cap B$. By definition, the null set is a set without any elements; therefore, $n(\emptyset) = 0$. Replacing in (1) the corresponding equal sets we obtain: $n(A \cup B) = n(A) + n(\emptyset) + n(B)$ which reduces to:

(2) $\quad \boldsymbol{n(A \cup B) = n(A) + n(B).}$

When two sets are disjoint, they are also said to be **mutually exclusive**. Equation (2) is referred to as the **Addition Principle.**

Example 1: Thirty Republicans and forty Democrats attend a charity banquet. How many Republicans or Democrats attend the banquet?

Since
$$R \cap D = \emptyset$$
$$n(R \cup D) = n(R) + n(D)$$

$$n(R \cup D) = 30 + 40$$
$$n(R \cup D) = 70.$$

Case II: Assume that A $\cap$ B $\neq \phi$. Beginning with Eq. (1)

[1] $$n(A \cup B) = n(A \cap \tilde{B}) + n(A \cap B) + n(\tilde{A} \cap B),$$

let us add and subtract a $n(A \cap B)$ term on the right-hand side of the equation to obtain

(3) $$n(A \cup B) = n(A \cap \tilde{B}) + n(A \cap B) + n(\tilde{A} \cap B) + n(A \cap B) - n(A \cap B).$$

This is a permissible operation, since $n(A \cap B) - n(A \cap B) = 0$ and we have in effect added zero to the right-hand side of the equation. Again, we make use of the fact that $A = (A \cap \tilde{B}) \cup (A \cap B)$ and $B = (\tilde{A} \cap B) \cup (A \cap B)$ and as a consequence,

$$n(A) = n(A \cap \tilde{B}) + n(A \cap B)$$

and

$$n(B) = n(\tilde{A} \cap B) + n(A \cap B).$$

Replacing these equivalent forms in Eq. (3) we obtain:

(4) $$n(A \cup B) = n(A) + n(B) - n(A \cap B).$$

The result expressed in Eq. (4) has a simple intuitive explanation. Since we have included $n(A \cap B)$ in both $n(A)$ and $n(B)$, we have included it twice; we must therefore subtract a $n(A \cap B)$ term to preserve the equality.

Example 2: In a given 30-day period, a man watched television 20 days and his wife watched television 25 days. If they watched television together 18 days, determine (a) the number of days one or the other watched television, (b) the number of days which neither one watched television, and (c) the number of days the wife watched television and the husband didn't.

Solution:

(a) Using M for the man watching TV and W for the wife watching TV, we seek $n(M \cup W)$. By Eq. (4),

$$n(M \cup W) = n(M) + n(W) - n(M \cap W)$$
$$n(M \cup W) = 20 + 25 - 18$$
$$n(M \cup W) = 27.$$

(b) Since the total number of days in the universal set under consideration is 30 and the $n(M \cup W) = 27$,

$$n(\widetilde{M \cup W}) = n(\tilde{M} \cap \tilde{W}) = 30 - 27$$
$$n(\widetilde{M \cup W}) = n(\tilde{M} \cap \tilde{W}) = 3.$$

(c) The set corresponding to the wife watching television and the husband not watching is $\tilde{M} \cap W$ and we know that $W = (M \cap W) \cup (\tilde{M} \cap W)$, where

$(M \cap W)$ and $(\tilde{M} \cap W)$ are disjoint. Therefore,

$$n(W) = n(M \cap W) + n(\tilde{M} \cap W)$$
$$25 = 18 + n(\tilde{M} \cap W)$$

and

$$n(\tilde{M} \cap W) = 7.$$

We may gain considerable insight into this problem if we use a Venn diagram. Starting with the condition $n(M \cap W) = 18$, place 18 in the region corresponding to $M \cap W$. Since $n(M) = 20$ and we have 18 in the intersection, there must be 2 in $n(M \cap \tilde{W})$. Similarly, since $n(W) = 25$, there must be 7 in $n(\tilde{M} \cap W)$. To obtain $n(\tilde{M} \cap \tilde{W})$ simply subtract from 30 the sum of 2 + 18 + 7 to obtain 3.

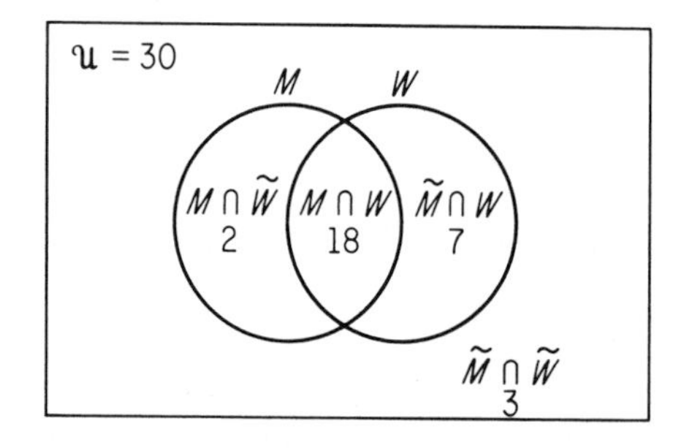

The preceding discussion may be generalized to three or more sets by considering each set as the union of disjoint regions where the number of elements in each set is simply the sum of the elements in each of these disjoint regions.

It is not necessary for us to develop generalized formulas for three or more sets, since we can make repeated applications of the concepts applicable to two sets and we can rely on the use of Venn diagrams to simplify the procedure in finding the number of elements in a specified set. The following example illustrates the general approach.

Example 3: A store specializing in "take home" service keeps a supply of 45 pre-prepared hamburgers available. The manager reports that at a given time he has on hand 20 hamburgers with mustard, 30 with ketchup, 13 with pickles, 12 with mustard and ketchup, 7 with mustard and pickles, 9 with ketchup and pickles, and 5 with mustard, pickles, and ketchup. Determine (a) how many plain hamburgers he has, (b) how many hamburgers have ketchup only, and (c) how many hamburgers have exactly two of the three choices.

Solution: Construct a Venn diagram for the three options. The information that 5 have all three is the initial starting point. In the region corresponding to $M \cap K \cap P$ we place a 5. The statement that 9 items have ketchup and pickles means that $n(K \cap P) = 9$. In set form $K \cap P = (\tilde{M} \cap K \cap P) \cup (M \cap K \cap P)$. Therefore $n(K \cap P) = n(M \cap K \cap P) + n(\tilde{M} \cap K \cap P)$. Using the given information $9 = 5 + n(\tilde{M} \cap K \cap P)$ which tells us that $n(\tilde{M} \cap K \cap P) = 4$. In a similar manner we determine the number to be placed in each of the regions including the region $\tilde{M} \cap \tilde{K} \cap \tilde{P}$ with which we identify the number 5 by taking the $n(M \cup K \cup P)$, namely 40, from the total in the universe 45.

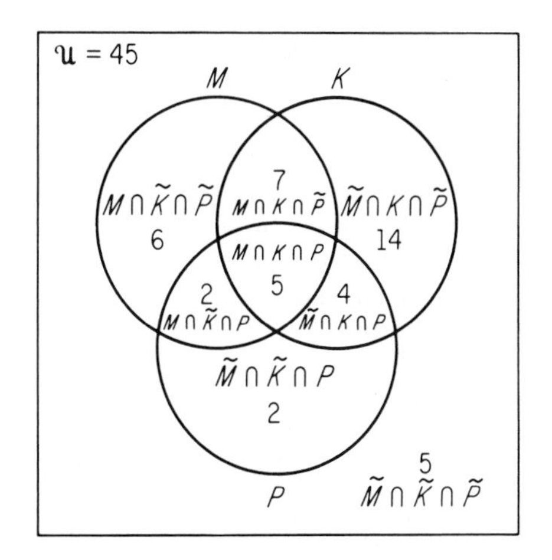

Having constructed the Venn diagram we can determine the requested solutions.

(a) How many plain hamburgers:

$$n(\tilde{M} \cap \tilde{K} \cap \tilde{P}) = 5.$$

(b) How many have ketchup only:

$$n(\tilde{M} \cap K \cap \tilde{P}) = 14.$$

(c) How many have exactly two of the three choices:

$$n(M \cap K \cap \tilde{P}) + n(\tilde{M} \cap K \cap P) + n(M \cap \tilde{K} \cap P) = 7 + 4 + 2 = 13.$$

Problem Set 3.2

A. 1. On a certain day a cafeteria noted the sale of 100 milk shakes and 150 hamburgers. The cashier stated that 175 purchased one or the other of the two items and further noted that 25 bought only milk shakes. Determine how many bought only hamburgers.

2. A survey revealed that 113 people had American made automobiles and 67 had foreign made automobiles and no one in the survey had more than two cars. Also, those that had two autos had an American and a foreign auto; otherwise, they had only one automobile. Determine how many persons were surveyed if
   (a) No person had two cars.
   (b) 27 persons had two cars.
   (c) 89 had only American made automobiles.
   (d) 38 had only foreign made cars.

3. The truth set for the statement p has 11 elements and the truth set for the

statement q has 17 elements. Determine how many elements are in the truth set of $p \vee q$, $(P \cup Q)$ if

(a) $p \wedge q$ holds for exactly three cases.
(b) p and q are inconsistent.
(c) p implies q.

4. There are 21 students in a calculus class and 25 in a physics class. Find the number that take calculus or physics if
   (a) the two classes meet the same hour.
   (b) the two classes meet at different times and 13 students take calculus only
5. A coffee-vending machine received \$10 one day. Each cup of coffee costs ten cents. The machine is rigged so that it will count the following various combinations that people selected in getting their coffee. The readings obtained were:
   (a) 35 had coffee with cream and sugar.
   (b) 55 had coffee with cream.
   (c) 25 had coffee only.

   How many had coffee with sugar only?
6. A car dealer has 200 cars. Some cars have the added accessories of a radio, air conditioner, or electrically operated windows. The dealer has the following cars available:

   15 cars with the three accessories
   10 cars with only radio and electrically operated windows
   5 cars with only electrically operated windows and air conditioners
   40 cars with only radio and air conditioners
   95 cars with only a radio
   15 cars with only an air conditioner
   20 cars with none of the three accessories.

   Determine how many have
   (a) only electrically operated windows.
   (b) only two accessories.
   (c) radio or air conditioner.
   (d) radio, air conditioner, or electrically operated windows.
   (e) air conditioner, electrically operated windows or radio, but not all three accessories.
7. At a small junior college of 500 students, every student is required to select English and then must select mathematics, chemistry, or a foreign language. After registration, the following statistics were available: 300 selected mathematics, chemistry, English, and a foreign language; 390 selected mathematics; 375 selected chemistry; 370 selected a foreign language; 325 selected mathematics, a foreign language, and English; 330 selected English, a foreign language, and chemistry; and 340 selected mathematics, chemistry, and English. Find the number of students that selected
   (a) only English. (b) no foreign language.
   (c) exactly two courses. (d) at least two courses.
   (e) at most two courses.

   Also, how many instructors are needed if no instructor teaches more than two sections in his discipline and each section has no more than 30 students in it?
8. Let A, B be subsets of $\mathcal{U}$, where $n(\mathcal{U}) = 100$, $n(A) = 40$, $n(B) = 57$, and $n(A \cap B) = 19$. Find (a) $n(A \cup B)$, (b) $n(A \cap \tilde{B})$, (c) $n(\tilde{A} \cap \tilde{B})$, (d) $n(\widetilde{A \cup B})$.
9. Given that X, Y are subsets of $\mathcal{U}$ and $n(\mathcal{U}) = 50$, $n(\tilde{X}) = 25$, $n(\tilde{Y}) = 15$,

$n(\tilde{X} \cap \tilde{Y}) = 10$ and $n(X \cap Y) = 20$. Find (a) $n(\tilde{X} \cup \tilde{Y})$, (b) $n(X)$, (c) $n(Y)$, (d) $n(X \cup Y)$, (e) $n(X \cap \tilde{Y})$, (f) $n(\tilde{X} \cap Y)$.

10. Given that R, S are subsets of $\mathcal{U}$ and $n(\mathcal{U}) = 150$, $n(R) = 50$, $n(S) = 90$, $n(\widetilde{R \cup S}) = 40$. Find (a) $n(R \cap S)$, (b) $n(R \cap \tilde{S})$, (c) $n(R \cup S)$, (d) $n(\tilde{R})$, (e) $(\tilde{R} \cup \tilde{S})$.

B. 1. Prove that $n(\tilde{A}) = n(\mathcal{U}) - n(A)$.

2. Let A, B, and C be subsets of a universal set $\mathcal{U}$. Draw a Venn diagram and show that:

   (a) $n(A \cup B \cup C) = n(A) + n(B) + n(C) - n(A \cap B) - n(B \cap C) - n(A \cap C) + n(A \cap B \cap C)$

   (b) also, show by algebra that the property for any two sets A, B; $n(A \cup B) = n(A) + n(B) - n(A \cap B)$ may be extended to three sets

3. Let A, B be subsets of $\mathcal{U}$. Find a formula to calculate $n(\tilde{A} \cap \tilde{B})$. (*Hint*: Use a Venn diagram.)

3.3 Permutations

In the preceding section we dealt with the question, "How many elements are there in a specific set?" Previously the reader will recall we determined that the total number of subsets of a specified set is given by 2^n, where n represents the number of elements in the set. Before we complete our discussion of counting theory, we would like to be able to answer questions which are logical extensions of our earlier work.

For example, if a set has n elements how many subsets are there that have 0, 1, 2, 3, . . . , n elements? Consider the specific set $A = \{a,b,c,d\}$ and the question, "How many three element subsets are there of the set A?" At this point the only techniques available to the reader are the Fundamental Principle of Counting and tree diagrams. Let us try to apply these methods.

If we wish to form 3-element subsets we might think of this as a sequential process in which we fill up the three positions. The first position could be filled by any of the four elements in the set; the second position may be filled by any of the three that are left; and the third position may be filled by either of the remaining two. By the Fundamental Principle of Counting we have:

$$\underline{4} \times \underline{3} \times \underline{2} = 24 \text{ ways.}$$

However, this result is obviously incorrect, since we should recognize that the total number of subsets is equal to 16 (i.e., $16 = 2^4$). We may gain some insight into the error made in using the Fundamental Principle of Counting if we look at the tree-diagram method shown on the following page.

Here we can see that the same subsets have been counted more than once, since simply changing the order of the elements in a set does not give us a new set. For example, there are six cases circled above where "a, b, c" are rearranged, yet they represent only one subset. We see that a clear distinction must be made in any counting problem on the basis of whether or not the order of the elements is or is not to be considered. We will postpone a complete answer to the question of the number of **k** element subsets

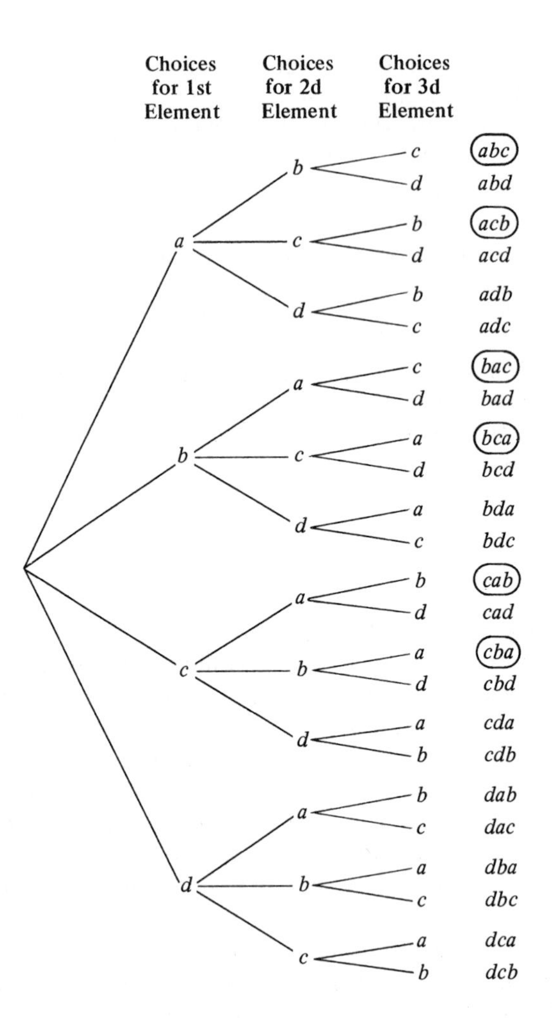

of a set having **n** elements until we can make the aforementioned distinction precise and have developed some additional counting techniques. We will begin with a discussion of those counting problems where the order of the objects must be considered.

Suppose that we are given three slips of paper, one bearing the number 1, a second the number 2, and the third the number 3. How many three-digit numbers may be formed with the three slips of paper? Clearly the number 123 is different than the number 321 and a rearrangement of the elements does give us a new number. The Fundamental Principle of Counting will handle this situation (a tree diagram also) by again thinking of positions to fill and the number of ways each position in sequence may be filled. Since there are three numbers the first position may be filled in three ways, the second position in two ways and the third position in one way. By using the Fundamental Principle of Counting we have

$$\underline{3} \times \underline{2} \times \underline{1} = 6$$

which is the total number of ways of forming three digit numbers. For this simple case it is easy to list them:

123	231	312
132	213	321.

As the number of objects to be ordered increases, it can become tedious to list all of the rearrangements, but we will continue to have an interest in the total number of such arrangements. For example, consider a group of seven students waiting in line for registration. How many ways may this group be arranged in line? The Fundamental Principle of Counting approach gives

$$\underline{7} \times \underline{6} \times \underline{5} \times \underline{4} \times \underline{3} \times \underline{2} \times \underline{1} = 5{,}040 \text{ ways.}$$

We are now ready to introduce some terminology and notation and make generalizations from the preceding examples.

DEFINITION: Any arrangement of **n** distinct objects in a definite order without repetition is called **a permutation.**

Suppose we now consider the case where we have n distinct objects (n is any positive counting number) and we wish to determine the total number of permutations of these n objects. Applying the Fundamental Principle of Counting, we can place any of the n objects in the first position, $n - 1$ in the second, $n - 2$ in the third, continuing in this pattern until the last position can be filled in one way. The total number of permutations of these objects is:

$$\underline{n} \times \underline{(n-1)} \times \underline{(n-2)} \times \underline{(n-3)} \times \cdots \times \underline{1}.$$

This counting procedure occurs so frequently that we utilize a special notation which indicates the desired counting process but does not require us to write out each factor in the product. The symbol "**n!**" (read n factorial) is defined as follows:

$$n! = n \times (n-1) \times (n-2) \times (n-3) \times \cdots \times 1$$

where each factor is a positive integer.

Using this notation, we indicate the total number of permutations of n distinct objects simply as $n!$. The preceding discussion gives rise to a basic permutation formula; namely,

(1) $\quad {}_n\mathbf{P}_n = \mathbf{n!}$

which is read, "The total number of permutations formed by arranging all n of the n distinct objects is n factorial."

Example 1: How many distinct permutations are there formed by arranging all of the letters in the word, JOURNAL?

Solution: The total number of ways in which the seven distinct letters may be arranged is found by using formula (1), thus

$$
\begin{aligned}
{}_7P_7 &= 7! \\
7! &= 7 \times 6 \times 5 \times 4 \times 3 \times 2 \times 1 \\
7! &= 5{,}040.
\end{aligned}
$$

Example 2: A sailboat carries five distinct flags which may be mounted in a line on a mast for signaling other boats. How many distinct "signals" could be made if all of the flags are used for each signal?

Solution: The five flags may be arranged in ${}_5P_5 = 5!$ ways, where

$$
\begin{aligned}
5! &= 5 \cdot 4 \cdot 3 \cdot 2 \cdot 1 \\
5! &= 120 \text{ ways.}
\end{aligned}
$$

Let us now turn our attention to some types of counting problems where we remove some of the basic restrictions which were imposed on the problems we have discussed thus far.

Suppose we consider that we have n distinct objects and we wish to rearrange all n of them, but we remove the "without repetition" requirement and permit repetitions. By the Fundamental Principle of Counting we count the total number of arrangements as follows:

$$\underset{\text{1st}}{\underline{n}} \times \underset{\text{2d}}{\underline{n}} \times \underset{\text{3d}}{\underline{n}} \times \cdots \times \underset{k\text{th}}{n} = n^k.$$

Notice that we have not referred to these arrangements as permutations. (Why?)

Example 3: How many three-digit numbers may be formed from the digits 3, 5, 7, 9 if repetition of the digits is permitted?

Solution: Any of the four digits may occur in any position. Therefore

$$4 \times 4 \times 4 = 4^3$$

$$4^3 = 64 \text{ ways.}$$

Now let us consider the case where we wish to count the number of permutations of n distinct objects where we may choose not to select all of the n objects. Suppose we select r of the objects where r is less than or equal to n. For example, if we wish to form three digit numbers by picking from the digits 1, 2, 3, 4, 5 and we do so without repetition, we can initially resort to the Fundamental Principle of Counting by placing any one of the five digits in the first position, any of the four remaining in the second position, and any of the remaining three in the third position, i.e.,

$$
\begin{aligned}
{}_5P_3 &= 5 \times 4 \times 3 \\
{}_5P_3 &= 60
\end{aligned}
$$

where ${}_5P_3$ is read "the number of permutations of 3 distinct objects chosen from 5 distinct objects."

Before we look at a generalized formula, observe that we could have expressed this particular result in an equivalent form that will more closely parallel the final formula, i.e.,

$$ {}_5P_3 = 5 \times 4 \times 3 \times \frac{2 \times 1}{2 \times 1} $$

$$ {}_5P_3 = \frac{5!}{2!} $$

$$ {}_5P_3 = \frac{5!}{(5-3)!}. $$

If we now consider the number of permutations of r distinct objects (r less than or equal to n) we can once again use the Fundamental Principle of Counting. The first position can be filled in n ways, the second position in $n - 1$ ways, the third in $n - 2$, and the rth position in $n - (r - 1)$ or equivalently $(n - r + 1)$ ways, thus:

$$ {}_nP_r = \frac{n}{\text{1st}} \times \frac{n-1}{\text{2nd}} \times \frac{n-2}{\text{3rd}} \times \cdots \times \frac{(n-r+1)}{r\text{th}}. $$

If the right-hand side of this expression is multiplied by

$$ \frac{(n-r)!}{(n-r)!} $$

we obtain

$$ {}_nP_r = n \times (n-1) \times (n-2) \times \cdots \times (n-r+1) \times \frac{(n-r)!}{(n-r)!} $$

which is equivalent to

$$ (2) \qquad {}_nP_r = \frac{n!}{(n-r)!}, \qquad r \leqslant n. $$

Should $r = n$ we have

$$ {}_nP_n = \frac{n!}{(n-n)!} $$

or

$$ {}_nP_n = \frac{n!}{0!}. $$

For this result to be consistent with formula (1), ${}_nP_n = n!$, we must define $0! = 1$. This is quite reasonable if we keep in mind that we are dealing with the number of ways of arranging objects. The minimum number of ways of arranging any number of objects is one way even if we choose not to rearrange them.

Example 4: Three-digit numbers are formed by selecting without repetition from the digits 1, 2, 3, 4, 5, 6, 7.

(a) How many such numbers are there?
(b) How many are even?
(c) How many are greater than 600?

Solution:

(a) ${}_7P_3 = \frac{7!}{(7-3)!} = \frac{7 \times 6 \times 5 \times 4!}{4!} = 7 \times 6 \times 5 = 210$ ways.

(b) For a number to be even, the last digit must be even. Starting at the last digit and using the Fundamental Principle of Counting we can fill the last position in three ways (2, 4, or 6), the next to the last position in six ways, the first position in five ways. Thus we have

$$\underline{5} \times \underline{6} \times \underline{3} = 90 \text{ ways.}$$

(c) For a number to be greater than 600, the first digit must be at least 6. Starting with the first position, we can fill this in two ways (6 or 7), the second position in six ways, and the last position in five ways. Thus we have

$$\underline{2} \times \underline{6} \times \underline{5} = 60 \text{ ways.}$$

The last restriction we wish to remove for counting purposes is the distinctness of the objects to be arranged. For example, how many distinguishable permutations are there of all of the letters in the word "baa"? One "a" is not distinguishable from the other "a" so that we may not directly apply ${}_3P_3 = 3!$; in fact, if we list the possible arrangements, we find that there are only three distinguishable arrangements:

baa aba aab.

This is one half the total expected if the letters were all distinct. We could reconstruct the total number of permutations of distinct letters if we introduct subscripts to differentiate one "a" from the other "a" and then for each entry permute the subscripts. The distinct permutations are

$$\begin{array}{ccc} ba_1a_2 & a_1ba_2 & a_1a_2b \\ ba_2a_1 & a_2ba_1 & a_2a_1b. \end{array}$$

Extending the concept to the word "baaa" there are only four distinguishable arrangements;

baaa abaa aaba aaab.

Using subscripts and then permuting the subscripts, each entry will give rise to six entries and the total of twenty-four such entries will agree with ${}_4P_4 = 4!$ or 24, which we expect when four distinct letters are permuted. The twenty-four permutations are listed as shown:

$$\begin{array}{cccc} ba_1a_2a_3 & a_1ba_2a_3 & a_1a_2ba_3 & a_1a_2a_3b \\ ba_1a_3a_2 & a_1ba_3a_2 & a_1a_3ba_2 & a_1a_3a_2b \end{array}$$

$ba_2a_1a_3$	$a_2ba_1a_3$	$a_2a_1ba_3$	$a_2a_1a_3b$
$ba_2a_3a_1$	$a_2ba_3a_1$	$a_2a_3ba_1$	$a_2a_3a_1b$
$ba_3a_1a_2$	$a_3ba_1a_2$	$a_3a_1ba_2$	$a_3a_1a_2b$
$ba_3a_2a_1$	$a_3ba_2a_1$	$a_3a_2ba_1$	$a_3a_2a_1b$.

Notice that for "baa" we could have computed the number of distinguishable arrangements by $\frac{3!}{2!} = 3$, for the "baaa" the number of distinguishable arrangements by $\frac{4!}{3!} = 4$. The denominator in each case reflects the number of ways of permuting the indistinguishable objects.

The principle we have illustrated is quite general. When we have a repetition of indistinguishable objects among a group of n objects and we wish to determine the total number of distinguishable permutations, say N, we use the following general counting principle:

> If all n objects are to be arranged where there are n_1 of one indistinguishable type, n_2 of a second indistinguishable type, n_3 of a third indistinguishable type, continuing until we have n_k of the kth indistinguishable type, then the total number of distinguishable permutations N is given by

$$N = \frac{n!}{n_1! \times n_2! \times n_3! \times \cdots \times n_k!} \tag{3}$$

where

$$n_1 + n_2 + n_3 + \cdots + n_k = n.$$

Example 5: Determine the total number of distinguishable permutations that can be formed by rearranging all of the letters in the word "toot" and list all of the permutations.

Solution:

$$N = \frac{4!}{2! \times 2!} = \frac{\overset{2}{\cancel{4}} \times 3 \times \cancel{2} \times 1}{\cancel{2} \times 1 \times \cancel{2} \times 1} = 6.$$

The 4! represents the four letters and the two 2!'s represent the repetition of "t's" and "o's." A tree diagram can be used to enumerate the six possibilities.

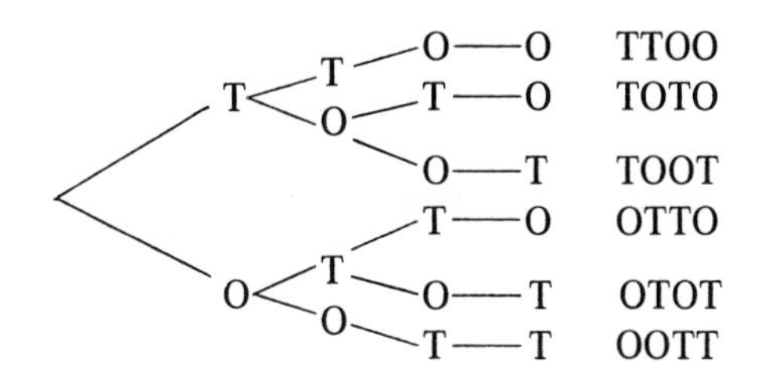

Example 6: How many different permutations are there formed by rearranging all of the letters in the word "arranging?"

Solution: The nine letter word has 2 a's, 2 r's, 2 n's, 2 g's, and 1 i, thus

$$N = \frac{9!}{2! \times 2! \times 2! \times 2! \times 1!}$$

$$= 22{,}680 \text{ permutations.}$$

The three basic permutation formulas (1), (2), and (3) together with the Fundamental Principle of Counting may be used in a variety of combinations to handle more complex counting problems. The following examples illustrate how we may apply and extend the counting concepts we have learned thus far.

Example 7: Six individuals including two brothers are lined up for a race. If the two brothers insist on being next to each other, in how many ways may the six rearrange themselves at the starting line?

Solution: Consider temporarily the two brothers as a unit which will leave five distinct objects to be rearranged. We can compute the number of such permutations by formula (1) to be ${}_5P_5 = 5!$. Now for *each* of these permutations the position of the brothers could be reversed. Since this can be done in two ways, the final result is given by

$${}_5P_5 \times 2 = 5! \times 2$$

$${}_5P_5 \times 2 = 240 \text{ ways.}$$

Example 8: Three boys and four girls are to be seated in a row. If we insist that girls occupy the end positions and we are interested only in the sex of the individuals as regards a seating arrangement, how many boy-girl seating arrangements are there?

Solution: First place a girl at each of the end positions. This can be done in only one way (we are interested only in the sex of the individual). We must fill the remaining five seats with three boys and two girls. This can be handled by formula (3) where we must rearrange five objects, three of one kind and two of another. The desired result may be computed as follows:

$$1 \times \frac{5!}{3! \times 2!} = 1 \times \frac{5 \cdot \overset{2}{\cancel{4}} \cdot \cancel{3} \cdot \cancel{2} \cdot \cancel{1}}{\cancel{3} \cdot \cancel{2} \cdot 1 \cdot \cancel{2} \cdot \cancel{1}}$$

$$= 10 \text{ ways.}$$

Example 9: Five people are seated around a circular table. How many different seating arrangements are there?

Solution: Since the people are in a circular arrangement, a simple rotation of a given seating arrangement does not represent a new seating arrangement. For example, the two diagrams below represent the same seating arrangement of the individuals with respect to each other. In fact, there are five rotations around the table which represent the same seating arrangement of the people. We can, therefore, count the number of different seating arrangements by fixing one person in a seat and rearranging the others with respect to him. We can do this in ${}_4P_4 = 4!$ or 24 ways.

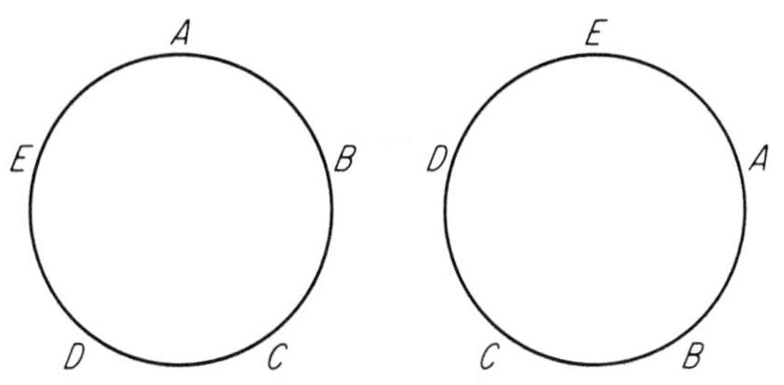

Problem Set 3.3

A.

1. How many permutations are possible using the three letters x, y, and z? List all of the possibilities.
2. Suppose a die is tossed and is then followed by a flip of a coin. List all the possible outcomes for this sequence of events. How many outcomes are possible for this sequence of events?
3. How many four digit numbers can one obtain using the ten digits of our number system, given that the thousands place cannot begin with the number zero?
4. Ten horses are in a race. In how many ways can we have win, place, and show?
5. How many different ways can a true-false test be answered if the test contains
   (a) 3 questions. (b) 5 questions. (c) 10 questions.
   (d) n questions (n a positive integer).
6. Determine how many different outcomes are there when a coin is tossed
   (a) two times. (b) seven times. (c) n times.
7. A man has in his lunch box a ham sandwich, a pickle, a container of soup, and a sweet chocolate candy bar. Determine how many different ways he can eat his lunch if
   (a) the order that he eats his food is not important.
   (b) he chooses to eat the soup first.
   (c) he eats the soup first and then the sandwich.
8. How many distinguishable arrangements, in terms of sex are there if three boys and two girls stand in a line?
9. How many distinguishable permutations may be formed by using all the letters in the word "letter"?
10. A bucket contains golf balls. Three are red, two are yellow, and four are white. State how many line arrangements are possible if
    (a) all the balls are used.
    (b) one yellow is on each end.
    (c) a red is in the middle of the line.

B. 1. Evaluate (a) ${}_7P_3$, (b) ${}_7P_4$, (c) ${}_5P_0$, (d) ${}_4P_4$, (e) ${}_{52}P_5$.

2. In Example 9 in Sec. 3-3, five people were seated around a circular table. In this example we found that there were $(5 - 1)! = {}_4P_4 = 24$ different seating arrangements of the five people. If n people are seated around a circular table, determine an expression which will give the number of seating arrangements for these n people around a circular table.

3. Eight people are seated around a circular table and two people demand to be seated next to each other. How many seating arrangements are possible for these eight people?

4. If we define a word as any arrangement of letters how many word arrangements are possible using all the letters in the word "remember"?

5. How many five-symbol telephone expressions are possible if the first two symbols are letters of the alphabet and cannot be the same, and the last three symbols are the digits of our number system and any one digit may not appear more than once in a given expression?

3.4 Ordered Partitions—Combinations

In the preceding section we pointed out that in problems involving counting, a clear distinction must be made between those where the order of the objects is important and those where order is not considered. Having discussed the application of permutation formulas to the type of counting problems where order is important, we now turn our attention to problems such as counting the number of subsets in a given set and determining the number of ways we may assign objects to different categories where within the category order of the elements is not considered. Our theory will be based on the concept of a partition and a special type of partition called an ordered partition.

DEFINITION: A **partition** of a set $\mathcal{U}$, symbolized $[c_1, c_2, \ldots, c_k]$, is a collection of subsets of $\mathcal{U}$, such that the subsets are pairwise disjoint (the intersection of any subsets taken two at a time is empty) and their union is the set $\mathcal{U}$. The subsets in the partition are called cells or categories.
Note: some cells in the partition may be empty.

Example 1: If $\mathcal{U} = \{1,2,3,4,5,6\}$ determine if any of the collection of subsets (a), (b), (c) is a partition.
(a) $\{1,2\}, \{3\}, \{4,6\}, \emptyset$.
(b) $\{1,2\}, \{2,4\}, \{3,5\}, \{6\}$.
(c) $\{1,3\}, \{2,5\}, \{4,6\}$.

Solution:
(a) is not a partition, since the union of the sets does not equal $\mathcal{U}$.
(b) is not a partition, since the pair of cells $\{1,2\}$ and $\{2,4\}$ are not disjoint, i.e., $\{1,2\} \cap \{2,4\} = \{2\}$.

(c) is a partition, since the cells are pairwise disjoint and their union is $\mathcal{U}$ and could be symbolized $[\{1,3\}, \{2,5\}, \{4,6\}]$.

The definition of a partition makes no distinction between the order of the cells within a partition so that $P_1 = [\{1,3\}, \{2,5\}, \{4,6\}]$ and $P_2 = [\{2,5\}, \{1,3\}, \{4,6\}]$ would be regarded as the same partition of a given set $\mathcal{U}$. If we wish to make a distinction between P_1 and P_2 (and we may), then it must be on the basis of the order of cells in accordance with the following definition.

DEFINITION: A partition of a set $\mathcal{U}$ is called an **ordered partition** of $\mathcal{U}$ if a particular order is specified for the subsets or cells that form the partition.

Our main task in this section will be to develop a procedure for counting the number of ordered partitions of a set $\mathcal{U}$ where we specify how many objects are to go into each cell. Let us consider a few typical examples before we attempt to state a generalized formula.

Suppose we have four individuals (Al, Bill, Charlie, and Dave) and we wish to assign two of them to a Social Committee and two to a Publicity Committee. How many distinct committee assignments are there? We have a two-celled ordered partition with two objects in each cell. The number of possibilities is sufficiently small so that we can list them as follows:

Social Committee	Publicity Committee
AB	*CD*
AC	*BD*
AD	*BC*
CD	*AB*
BD	*AC*
BC	*AD*

The total number of distinct committee assignments is six. Note that we have not included such listings as *BA*/*CD*, *AB*/*DC*, or *BA*/*DC*, since they represent the same committee assignment as the listed entry *AB*/*CD*. We are in effect not counting permutations of *AB* or of *CD* within a cell. Since we know how many ways we could permute the four letters (namely 4!), and we know that in each cell we have 2! ways of rearranging the objects, but that they represent the same committee, we should be able to compute the number of committees as follows:

$$\frac{4!}{2! \times 2!} = \frac{\overset{2}{\cancel{4}} \times 3 \times \cancel{2} \times \cancel{1}}{\cancel{2} \times 1 \times \cancel{2} \times \cancel{1}} = 6.$$

The division by the 2!'s indicates that in each cell the permutation of the two objects must be counted as a single entry and therefore, the total number of permutations of the four objects (4!) must be reduced by the respective 2!'s.

As the number of ordered partitions increases, and the number in each cell increases, we can hardly hope to conveniently list all the possible entries. However, the thought process illustrated in the preceding example will carry over and simplify our task. Consider as another example, that your favorite football team in a given year had four wins, three losses, and two ties. In how many ways could they have compiled this season's record? We can regard this as a three celled ordered partition (wins, losses, and ties) with four in the first cell, three in the second, and two in the third cell. One possible way for this outcome to occur would be as follows:

W	L	T
1,2,3,4	5,6,7	8,9

indicating wins for the first four games, losses for the next three, and ties for the last two. We would make no distinction between this entry and those where we rearranged in the first cell 1,2,3,4 in 4! ways, 5,6,7 in the second cell in 3! ways, and 8,9 in the third cell in 2! ways. Since the nine digits could be rearranged in 9! ways, the total number of ways for the particular season's record is given by

$$\frac{9!}{4! \times 3! \times 2!} = \frac{9 \times \overset{4}{\cancel{8}} \times 7 \times \cancel{6} \times 5 \times \cancel{4!}}{\cancel{4!} \times \cancel{6} \times \cancel{2}} = 1{,}260.$$

The approach we have used can be easily generalized to n objects and ordered partitions of k cells with a specified number in each cell.

The number, N, of ordered partitions of n objects into k cells with n_1 in the first cell, n_2 in the second cell, n_3 in the third cell, and continuing in this manner until we have n_k in the kth cell, is given by

$$(1) \qquad N = \frac{n!}{n_1! \times n_2! \times n_3! \times \ldots n_k!}$$

where

$$n_1 + n_2 + n_3 + \cdots + n_k = n.$$

We will symbolize (1) as follows:

$$N = \binom{n}{n_1, n_2, n_3, \ldots, n_k}.$$

It is not accidental that the formula (3) on page 111 is the same as the above formula (1). The counting technique used is dependent upon the point of view of the reader. In each case we are interested in the total number of distinguishable arrangements whether we are considering some of the objects as indistinguishable or whether we consider the number of objects in particular cells of an ordered partition that are not rearranged.

Example 1: In how many ways may eight waitresses be assigned to four six-hour shifts if two work together on each shift?

Solution: The eight waitresses are to be placed in four cells with two in each cell. The total number is given by

$$N = \binom{8}{2,2,2,2}$$

$$= \frac{8!}{2! \times 2! \times 2! \times 2!} = 2{,}520.$$

Of particular importance is an ordered partition in which only two cells are considered. We call such an ordered partition a **combination** and use a special notation.

DEFINITION: A two-celled ordered partition of n objects with r objects in the first cell and $n - r$ objects in the second cell is called a **combination.** The total number of combinations is given by

$$_nC_r = \binom{n}{r,\, n-r} = \frac{n!}{r!(n-r)!}.$$

In many texts the symbol $\binom{n}{r,\, n-r}$ is abbreviated to $\binom{n}{r}$, where it is understood the second cell contains $n - r$ objects. We will continue to indicate the number in both cells for clarity and for convenience in future work. Let us now consider a problem similar to the one presented at the beginning of Sec. 3.3.

Example 2: If a set $A = \{a,b,c,d,e\}$ how many three element subsets are there of set A?

Solution: We regard this as a combination problem involving two cells (those elements in the subset, and those elements not in the subset) with three in one cell and two in the other there are

$$_5C_3 = \binom{5}{3,2} = \frac{5!}{3! \times 2!} = 10 \text{ subsets.}$$

Example 3: In how many ways may a five-card poker hand be dealt from a deck of 52 cards?

Solution: This combination problem involves two cells, one involving the five cards selected and the other the 47 cards not selected there are

$$_{52}C_5 = \binom{52}{5,47} = \frac{52!}{5! \times 47!} = 2{,}598{,}960 \text{ ways.}$$

Now that we have discussed the basic concepts of counting ordered partitions, we can incorporate more than one of these techniques into a given situation and handle more complex counting problems. The following examples will illustrate these more sophisticated counting techniques.

Example 4: A student must select at registration two history courses and two English courses. If seven history courses and eight English courses are available to choose from, how many different schedules may he obtain?

Solution: We may regard this problem as a sequential counting type of first scheduling the history courses followed by scheduling the English courses (or vice versa) in order to have a complete schedule. By the Fundamental Principle of Counting, the sequential counting would involve the product of the two numbers. In order to obtain two history courses, we must select out of the seven available or ${}_7C_2$ and similarly for the English courses, ${}_8C_2$. The total number of schedules is given by

$${}_7C_2 \times {}_8C_2 = \frac{7!}{2! \times 5!} \times \frac{8!}{2! \times 6!} = 21 \times 28 = 588.$$

Example 5: A large corporation with many divisions has five Production Managers, four Sales Managers, and two Chief Engineers. A management committee is to be formed consisting of two Production Managers, two Sales Managers, and one Chief Engineer. How many different committees may be formed?

Solution: As in Example 4, we can regard this as a sequential counting problem putting first the Production Managers, then the Sales Managers, and finally a Chief Engineer on the committee. The desired result is obtained by:

$${}_5C_2 \times {}_4C_2 \times {}_2C_1 = \frac{5!}{2! \times 3!} \times \frac{4!}{2! \times 2!} \times \frac{2!}{1! \times 1!} = 120 \text{ ways.}$$

Example 6: A given state decides, based on experiments demonstrating ease of recall, to construct their license plates beginning with two digits followed by three letters and then ending with two more digits. The design also required that (1) the first two digits be nonrepeating and nonzero, (2) the three letters be nonrepeating and listed in alphabetical order, and (3) the last two digits be nonzero but with repetition of the digits allowed. How many distinct license plates can be made?

Solution: Each part of the counting process can be treated separately and then since the actions take place in sequence, multiply the results together in accordance with the Fundamental Principle of Counting.

1. The first two digits: since zero is not permitted, we must pick two digits out of nine where order is important, i.e., ${}_9P_2$.
2. The three letters: we must select 3 out of 26 letters, but since they are always listed in one order, namely alphabetically, the letters selected are not permuted. Therefore, we must select the three irrespective of the order which can be done in ${}_{26}C_3$ ways.

3. The last two digits: the digits on the end may not be zero but may repeat. By the Fundamental Principle of Counting we can select each of the last two digits in 9 ways; therefore, the last two positions can be filled in $9 \times 9 = 81$ ways. Forming the product of three separate results, we obtain:

$$_9P_2 \times {}_{26}C_3 \times 81 = \frac{9!}{7!} \times \frac{26!}{3! \times 23!} \times 81$$

$$_9P_2 \times {}_{26}C_3 \times 81 = 15{,}163{,}200 \text{ license plates.}$$

Problem Set 3.4

A.

1. Let S be a set that contains four elements. Determine how many subsets of S have
   (a) no elements of S.
   (b) exactly one element of S.
   (c) exactly two elements of S.
   (d) exactly three elements of S.
   (e) exactly four elements of S.
   (f) How many subsets does S have?
   (g) On the basis of (a) through (e) above, write an expression for the number of subsets of S.
   (h) By the Fundamental Principle of Counting we also know the number of subsets of S is 2^4. Using (g), write an equation for 2^4 in terms of combinations.
2. How many ways can a basketball team be selected from 13 boys
   (a) if 3 specified boys must be included.
   (b) if there are no restrictions.
3. A shipment is received containing nine items, one of which is defective. Determine how many ways can a subset of two items be selected such that
   (a) the defective item is not selected.
   (b) the defective item is selected.
4. In how many ways can one choose four things from nine different things?
5. How many different seven-card poker hands are possible from a deck of 52 cards? (Disregard the order of the cards in each hand.)
6. List all possible ordered partitions of the set $\{a, b, c\}$ into three cells where one or more cells may be empty.
7. How many three element subsets of set $A = \{1,2,3,4\}$ are there that contain the element(s) (a) 1, (b) 1 and 2, (c) 1, 2 and 3?
8. A man has a nickel, dime, quarter, and half-dollar in his pocket. How many different nonzero tips may be given a waitress? How many ways can the waitress be tipped if he gives her at least three of the coins?
9. A bucket contains 8 golf balls. In how many ways may 2 golf balls be selected from the bucket?
10. A man has four colors of paint: red, green, yellow, and blue. How many color combinations are possible if he does not mix more than three of the colors together? In this problem, consider a combination to mean a mixing of two or more paints.

B. 1. Show that ${}_nC_{n-1} = n$ for any positive integer n.

2. Evaluate (a) ${}_7C_3$, (b) ${}_7C_4$, (c) ${}_{52}C_5$, (d) ${}_{52}C_{47}$.

3. Show that ${}_nC_r \times r! = {}_nP_r$.

4. Prove or disprove that ${}_nC_r = {}_nC_{n-r}$ for any positive integer n and $r \leqslant n$.

5. Show by using the definition of a factorial that $\frac{n!}{n} = (n-1)!$ for $n \geqslant 1$, n a positive integer.

6. Show that ${}_nC_{k+1} = \frac{n-k}{k+1} \times {}_nC_k$.

7. Prove that $n \times {}_{n-1}C_{k-1} = k \times {}_nC_k$.

8. Show that ${}_nC_k = {}_{n-1}C_{k-1} + {}_{n-1}C_k$.

3.5 Binomial Expansions

In the previous section and in the exercises the reader was asked to consider many problems which required finding the number of k element subsets of a set having n elements. For example, the number of three-element subsets of a set having 5 elements is 10, found by computing ${}_5C_3 = \binom{5}{3,2}$. An interesting pattern emerges if we compute the number of subsets of many n-element sets and organize the data obtained in a triangular form, called **Pascal's triangle** (illustrated below). Each horizontal row represents a set with the indicated n elements. For example, $n = 0$ represents the set without any elements (the $\emptyset$), $n = 1$ represents a set with one element, etc. Each number in the "k-diagonal" (as shown in the diagram) represents the number of k element subsets of a given set.

Note that each row begins with a 1 indicating there is only one subset with no elements; namely, $\emptyset$, and ends with a 1 indicating there is only one subset of a set

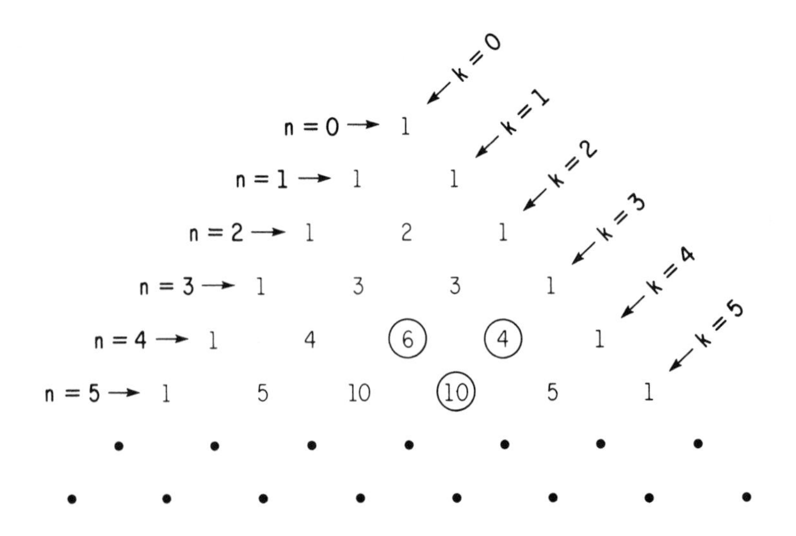

which contains all n elements–namely, the set itself. Other numbers indicate the number of k element subsets of a set with n elements. For example, the circled 10 corresponds to the number of 3-element subsets of a set having 5 elements.

Another interesting pattern is that each entry is the arithmetic sum of the two adjacent numbers in the preceding row. For example, the circled 10 is the sum of the circled 6 and 4 in the row above. Can we prove this conjecture?

Consider a set with five elements, say $A = \{a, b, c, d, e\}$. We know that the number of three-element subsets is 10. Let us focus our attention on a specific element, say a, and consider two separate questions. How many 3-element subsets contain the element a, and how many three-element subsets do not contain the element a?

If we consider those 3-element subsets which contain "a," we have a set $\{a,-,-,\}$ with two positions to fill. We can fill these by picking two letters out of the remaining four without regard to order. This we can do in ${}_4C_2 = \binom{4}{2,2} = \dfrac{4!}{2! \times 2!} = 6$ ways.

Now consider those three element subsets that do not contain a. We have a set $\{-,-,-\}$ with three positions to fill. We can fill these by selecting three out of the four letters (a cannot be chosen) and this can be done in ${}_4C_3 = \binom{4}{3,1} = \dfrac{4!}{3! \times 1!} = 4$ ways.

Since the element a is either in the subset or is not, we have accounted for all the possibilities thereby demonstrating that:

$$\binom{5}{3,2} = \binom{4}{2,2} + \binom{4}{3,1}$$

or

$$10 = 6 + 4.$$

This same argument will hold if we consider a set with n elements and we wish to find the number of k element subsets. We know that the total is ${}_nC_k$ or $\binom{n}{k,n-k}$, but we can also consider this a selection process where we first count the number of subsets that contain a given element ${}_{n-1}C_{k-1}$ or $\binom{n-1}{k-1,n-k}$, and then count the number that do not contain a given element ${}_{n-1}C_k$ or $\binom{n-1}{k,n-1-k}$. The sum of the numbers representing the number that contain the given element and those that do not must be equal to the total number of subsets. In symbolic form, we have

$${}_nC_k = {}_{n-1}C_{k-1} + {}_{n-1}C_k$$

or

$$\binom{n}{k,n-k} = \binom{n-1}{k-1,n-k} + \binom{n-1}{k,n-1-k}.$$

Example 1: Show that formula (1) holds by determining the number of 4-element subsets of a set with 6 elements.

Solution: The total number of such subsets is given by ${}_6C_4$ or $\binom{6}{4,2}$. We must show that $\binom{6}{4,2} = \binom{5}{3,2} + \binom{5}{4,1}$. Computationally ${}_6C_4$ is $\frac{6!}{4!2!}$ or 15 and $\binom{5}{3,2} + \binom{5}{4,1}$ is equivalent to $10 + 5$ or 15.

Interestingly, finding the number of subsets of a given set is only one application of the number pattern displayed by the Pascal triangle. The numbers in each row are used also in probability theory, the study of genetics, and in binomial expansions.

We will discuss at this time the topic of binomial expansions and in Chapter 4 we will relate the material in this section to our study of probability.

Let us begin by looking at a familiar pattern, namely, a tree diagram for a three-variable truth table. However, at the end of each branch we will record not only the entry but the algebraic product of the letters, listing the T's first; for example, TTF is written T^2F.

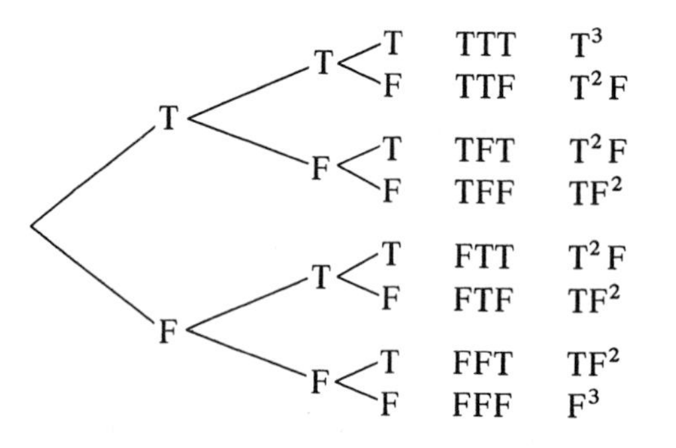

Now if we form the algebraic sum of the products, grouping like terms together, we obtain

$$1 \cdot T^3 + 3T^2F + 3TF^2 + 1 \cdot F^3.$$

In algebra we would normally omit the 1's preceding the T^3 and F^3 but we wish the reader to observe the pattern of these numbers, called the numerical coefficients, that precede the letters, namely, 1, 3, 3, 1. These are exactly the same numbers that appear in the row $n = 3$ of the Pascal triangle. Is this simply a remarkable coincidence or a general property? In order to answer this question we change from a listing of T's and F's by a tree diagram to a consideration of a partial listing of the successive nonnegative integer powers of the binomial $(x + y)$ and its expansion.

$$(x + y)^0 = 1$$
$$(x + y)^1 = 1 \cdot x + 1 \cdot y$$
$$(x + y)^2 = 1 \cdot x^2 + 2xy + 1 \cdot y^2$$
$$(x + y)^3 = 1 \cdot x^3 + 3x^2y + 3xy^2 + 1 \cdot y^3$$
$$(x + y)^4 = 1 \cdot x^4 + 4x^3y + 6x^2y^2 + 4xy^3 + 1 \cdot y^4.$$

The coefficient pattern of the Pascal triangle is preserved and other patterns appear. The exponent of the x factors starts with the power to which the binomial is raised in the first term and decreases one at a time for successive terms. The exponent of the y factors begins with one in the second term increasing one at a time for successive terms, ending when the power of the binomial is reached. The sum of the exponents in any one term is equal to the power to which the binomial is raised. The number of terms is one more than the power.

These observations, however, do not provide us with a generalized formula for the binomial expansion of $(x + y)^n$ where n is a positive integer. But by applying our counting theory we can develop such a formula. Consider the expansion of $(x + y)^3$ by which we mean that $(x + y)$ is to be used as a factor three times; $(x + y)(x + y)(x + y)$. Now let us look at each term in the final expansion beginning with x^3. Out of each of the three binomial factors we must select an x and not select any y's. The number of ways in which we can do this should be the coefficient of the x^3 term and can be found by recognizing that this is simply an ordered partition which we can symbolize as follows:

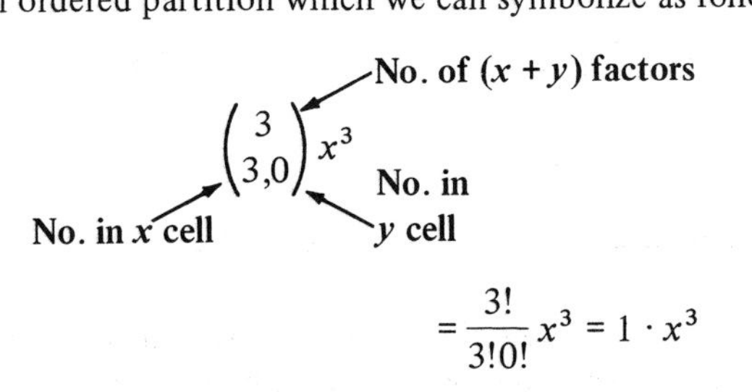

$$= \frac{3!}{3!0!}x^3 = 1 \cdot x^3$$

The same counting process can be used for the next term involving x^2y. Here we must select two x's and one y out of the three factors. We can determine the number of ways to accomplish this as follows:

$$\binom{3}{2,1}x^2y = \frac{3!}{2! \times 1!}x^2y = 3x^2y.$$

The three specific ways are $x \cdot x \cdot y$, $x \cdot y \cdot x$, and $y \cdot x \cdot x$.

In a similar manner, we can obtain the last two terms:

$$\binom{3}{1,2}xy^2 = \frac{3!}{1! \times 2!}xy^2 = 3xy^2$$

$$\binom{3}{0,3}y^3 = \frac{3!}{0! \times 3!}y^3 = 1 \cdot y^3.$$

If we write this out as the algebraic sum of the respective terms we have

$$(x + y)^3 = \binom{3}{3,0}x^3 + \binom{3}{2,1}x^2y + \binom{3}{1,2}xy^2 + \binom{3}{0,3}y^3.$$

Notice that the exponents in each term correspond exactly to the number in each cell of the ordered partition. The argument is general and we can write out a formula for the binomial expansion of $(x + y)^n$:

$$(x + y)^n = \binom{n}{n,0} x^n + \binom{n}{n-1,1} x^{n-1} y + \binom{n}{n-2,2} x^{n-2} y^2 + \cdots + \binom{n}{r,n-r} x^r y^{n-r} + \cdots + \binom{n}{1,n-1} x y^{n-1} + \binom{n}{0,n} y^n$$

where n is a positive integer, is called the **binomial formula.**

Example 1: Use the binomial formula to write out the expansion of $(x + y)^5$.

Solution:

$$\begin{aligned}(x + y)^5 &= \binom{5}{5,0} x^5 + \binom{5}{4,1} x^4 y + \binom{5}{3,2} x^3 y^2 + \binom{5}{2,3} x^2 y^3 + \binom{5}{1,4} x y^4 + \binom{5}{0,5} y^5 \\ &= \frac{5!}{5! \times 0!} x^5 + \frac{5!}{4! \times 1!} x^4 y + \frac{5!}{3! \times 2!} x^3 y^2 + \frac{5!}{2! \times 3!} x^2 y^3 + \frac{5!}{1! \times 4!} x y^4 + \frac{5!}{0! \times 5!} y^5 \\ &= x^5 + 5x^4 y + 10x^3 y^2 + 10x^2 y^3 + 5xy^4 + y^5.\end{aligned}$$

The application of the binomial formula is not limited to the expansions of the binomial $(x + y)^n$. We may expand other binomials by using the expansion of $(x + y)^n$ to the corresponding power as a basic formula.

Example 2: Expand the binomial $(2a - b)^4$.

Solution: First we need to rewrite the expression $(2a - b)^4$ in an equivalent form $[2a + (-b)]^4$, since the binomial formula is expressed as an indicated sum. Next we must be able to write out as a basic formula $(x + y)^4$:

$$(x + y)^4 = x^4 + 4x^3 y + 6x^2 y^2 + 4xy^3 + y^4.$$

Now in the basic formula expansion of $(x + y)^4$ replace each x by $2a$ and each y by $(-b)$ and complete the expansion:

$$\begin{aligned}[2a + (-b)]^4 &= (2a)^4 + 4(2a)^3 (-b) + 6(2a)^2 (-b)^2 + 4(2a)(-b)^3 + (-b)^4 \\ &= 16a^4 + 4(8a^3)(-b) + 6(4a^2)(b^2) + 4(2a)(-b^3) + (b^4) \\ &= 16a^4 - 32a^3 b + 24a^2 b^2 - 8ab^3 + b^4.\end{aligned}$$

Therefore, $(2a - b)^4 = 16a^4 - 32a^3 b + 24a^2 b^2 - 8ab^3 + b^4$.

We may wish to find a particular term in a binomial expansion or the coefficient of a given term and we would like to avoid the necessity of performing a complete expansion in order to obtain this information. The ordered-partition concept provides us with a convenient method for handling this type of problem.

Example 3: In the expansion of $(3a + 2b)^6$, find the coefficient of the a^2b^4 term.

Solution: In our basic formula $(x + y)^6$ the x^2y^4 would have a coefficient of the form $\binom{6}{2,4}$. Replacing x by $3a$ and y by $2b$ we obtain:

$$\binom{6}{2,4}(3a)^2\,(2b)^4 = 15(9a^2)(16b^4)$$
$$= 2160a^2b^4.$$

Thc numerical coefficient of the term is 2160.

Our discussion of binomial expansions may be extended to a consideration of such problems as $(x + y + z)^3$ which are called *multinomial expansions.* A brief introduction to this topic is presented in Appendix A2.

In subsequent sections, we will have occasion to refer to Pascal's triangle and binomial expansions and incorporate these concepts into probability theory.

Problem Set 3.5

A. 1. Use the binomial formula to write out the expansion of (a) $(p + q)^4$, (b) $(x - y)^7$, (c) $(2a + b^2)^5$, (d) $(3r - 4t)^3$.
2. Find the first, second, and third term of $(1 + 0.05)^{20}$.
3. Find the sum of the first three terms of the expansion $(0.1 + 0.9)^5$. If in the expansion of $(0.1 + 0.9)^5$ all the terms are added, what will be the result? Find the sum of the last two terms in the expansion. Is the result a reasonable approximation to the exact sum?
4. How many subsets with 7 elements does a set with 11 elements have?
5. Find 13^4 by using the binomial formula. (*Hint:* $13 = 10 + 3$).
6. In the expansion of $(2x - y)^7$, find the numerical coefficient of the term containing as a factor (a) x^2y^5, (b) xy^6, (c) x^3y^4, (d) x^4y^3.
7. In the expansion of $(x^2 + 2y)^5$ find the numerical coefficient of the term containing x^6y^2 as a factor.

B. 1. For the first six rows of Pascal's triangle, form the triangle in terms of a combination formula, for example, the circled two in the triangle may be written as ${}_2C_1$ and the circled one would be written as ${}_1C_1$.

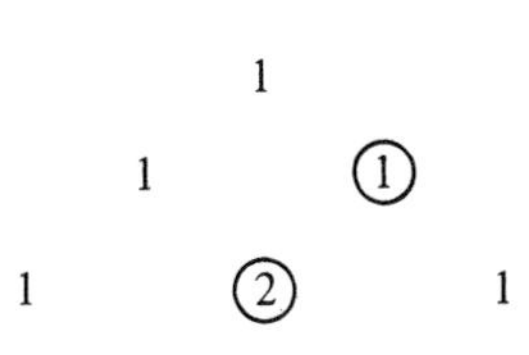

2. Given that 1, 5, 10, 10, 5, 1 is a row of Pascal's triangle, write the next two rows that follow the given row.
3. Solve the following for n(n a positive integer):
   (a) ${}_nC_2 = 45$.
   (b) ${}_nP_2 = 6$.
4. Write out the entries that would be in row $n = 9$ in Pascal's triangle.
5. Use the binomial formula to show that

$$2^n = {}_nC_0 + {}_nC_1 + {}_nC_2 + \cdots + {}_nC_{n-2} + {}_nC_{n-1} + {}_nC_n.$$

Review Exercises

3.1 State definitions for the following:
(a) permutation. (b) partition. (c) ordered partition.
(d) combination. (e) mutually exclusive sets.

3.2 In a poll of 475 people taken by a publishing company with regard to three magazines they publish, the following data was reported:
135 read *See*, *Now*, and *Scientific Report*.
93 read only *Scientific Report*.
79 read only *See*.
57 read only *Now*.
45 read none of the mentioned magazines.
18 read only the magazines *See* and *Scientific Report*.
34 read only *See* and *Now*.
Determine how many read
(a) only *Now* and *Scientific Report*. (b) only one magazine.
(c) *See* or *Now*. (d) *See*, *Now*, or *Scientific Report*.
(e) at most two magazines. (f) at least two magazines.
(g) exactly two magazines.

3.3 How many different permutations are there formed by arranging all the letters in the word "statistics"?

3.4 State how many different ways can a multiple choice test be answered (4 choices to each problem) if the test has
(a) one question. (b) two questions. (c) five questions.
(d) ten questions. (e) n questions.

3.5 A pentagonal pencil has each of its faces numbered one, two, three, four, and five. The pencil is rolled twice. After each roll the number is observed on the face on which the pencil rests. How many different sums are possible and what are the different possible sums?

3.6 A student wishes to take a course in mathematics, literature, social science, and natural science. In mathematics he has a choice of three courses, five choices in the social sciences, four choices in literature, and three choices in natural science. How many different schedule combinations may he select?

3.7 Given a set of 17 elements. Determine how many subsets of this set have
(a) 14 elements. (b) 3 elements.

3.8 Find the numerical coefficient of the term in the expansion of $(x^2 - 3y^3)^9$ which contains the factor
(a) x^{16}. (b) x^6. (c) y^3. (d) y^6. (e) y^{21}.

3.9 Find 7^4 by writing 7^4 as $(10 - 3)^4$ and expand the binomial by using the binomial theorem.

3.10 A group of six faculty members is to be selected from 13 faculty members to do a study. Two of the 13 faculty members will not work together on the study. How many ways can the group be formed to do the study?

3.11 A certain test has five true-false questions and five multiple-choice questions, with four choices to each question. If all questions are answered, how many ways can the test be answered.

Suggested Reading

Bush, Grace A., and John E. Young, *Foundations of Mathematics with Application to the Social and Management Sciences.* McGraw-Hill, New York, 1968.

Fehr, Howard F., Lucas N. H. Bunt, and George Grossman, *An Introduction to Sets, Probability and Hypothesis Testing.* Heath, Boston, Mass., 1964.

Kemeny, John G., J. Laurie Snell, and Gerald L. Thompson, *Introduction to Finite Mathematics.* Prentice-Hall, Englewood Cliffs, N.J., 1966.

Lipschutz, Seymour, *Theory and Problems of Finite Mathematics.* Schaum Outline Series, McGraw-Hill, New York, 1966.

Mosteller, Frederick, Robert Rourke, and George Thomas, *Probability: A First Course.* Addison-Wesley, Mass., 1961.

Richardson, William H., *Finite Mathematics.* Harper & Row, New York, 1968.

It is a truth very certain that, when it is not in our power to determine what is true, we ought to follow what is most probable.

—RENÉ DESCARTES

CHAPTER 4

Probability

4.1 Introduction

Throughout recorded history mankind has been faced with many uncertainties. In his efforts to exert some control over these uncertainties he has tried to obtain some information regarding the likelihood of certain events and to attach some numerical measure to the occurence of these events. Modern weather reporting, for example, not only tells us that "rain is likely" but that "there is a 60 percent probability of rain." Insurance company premiums are all determined by a careful study of the probabilities of accidents, thefts, and deaths. Poll takers predict the probabilities of the public reactions on many issues from voting preference to the effectiveness of an advertising campaign on the latest toothpaste by random-sampling techniques. Games of chance, whether they be horse racing, card playing, or rolling dice, are analyzed on the probabilities of desired outcomes. Even the scientific community, from which we have come to expect exact knowledge, reports more and more of their results, particularly quantum mechanics, in terms of probabilities. For it seems that the basic atomic structure is best described in terms of the probability that a given subatomic particle is in a particular position. Apparently, the microatomic world is one of considerable disorder and it is only the conglomerate probability effects that appear to establish some stability on the world of matter.

Intuitive concepts of probability have been as fundamental a part of the experience of mankind as the parallel experiences of counting or the study of geometric shapes. What is an essential difference, however, is that although man was able at an early date to abstract from his experiences and develop mathematical systems of arithmetic and geometry, it has only been in the last 250 years that the groundwork was laid for an abstract mathematical system of probability. A rapid increase in the mathematical knowledge of probability and its applied counterpart, statistics, has begun to transform such fields as psychology, sociology, economics, and business management into highly quantitative activities.

In the modern world where key decisions must be made and actions undertaken yet where uncertainty is ever present, men increasingly turn to probability analysis for guidelines. The human situation may preclude certainty and exact knowledge about the fundamental issues of mankind, but this need not be cause for despair, particularly if we can continue to refine and improve our ability to act on the basis of highly probable hypotheses.

In this chapter we will attempt to make precise some of our intuitive concepts and develop some of the more important properties of the mathematical system of probability.

4.2 Probability Measure

How shall we impart some precise meaning to the intuitive concept of probability? Consider the familiar question in tossing a coin, "What is the probability of obtaining a head?" Without any formal exposure to the topic, most individuals would respond that the answer is 50 percent or 1/2. The justification for such a response lies in the observation that there are two outcomes, one of which is heads and the probability is the ratio of the number of ways of getting a head to the total number of outcomes. This reasoning process has a long historical tradition and is referred to as the *classical definition of probability*. More formally, the definition is stated as follows:

DEFINITION (classical definition): If a given experiment can produce n equally likely mutually exclusive outcomes, f of which are regarded as favorable, then the **probability of a favorable outcome**, symbolized **Pr(F)**, is given by $\mathbf{Pr(F) = f/n}$.

For the present, the words *mutually exclusive* will mean that none of the outcomes can occur simultaneously. Later we will give a more precise definition of this term.

The expression *equally likely* has raised some theoretical difficulties which are not always easy to overcome. An experiment such as coin tossing or rolling a die seems to satisfy the "equally likely" designation, but many mathematicians have questioned whether we haven't assumed something that only experience could provide. For example, consider an experiment of throwing a dart at a bullseye. There are clearly two outcomes: you either hit the bullseye or you do not. If we regard hitting the bullseye as favorable and if we indiscriminately applied the classical definition, we might state that the probability of getting a bullseye is 1/2. This hardly seems like a reasonable conclusion, however, and the difficulty stems from the inappropriateness of "equally likely" to the outcomes of the experiment. Thus it should be clear that whenever we employ the classical definition we must be aware of the assumption of "equally likely" and exercise caution when we use an abstract mathematical model in an applied sense to problems in the real world.

As a consequence of the possible difficulties in applying the classical definition, mathematicians have considered another definition of probability which involves the concept of a long-run result. This definition is called the *relative frequency definition* and is as follows:

> **DEFINITION** (relative frequency): If an experiment is performed n times with f favorable outcomes and the relative frequency f/n approaches a limiting numerical value as n increases, then the **probability of a favorable outcome, Pr(F),** is defined as that limiting value. **Pr(F) = limit f/n as n increases.**

For example, if we wished to find the probability that a thumbtack would land with the point up, then we might use the relative frequency definition and perform the experiment of throwing the thumbtack perhaps a 100 times, then a 1,000, then 10,000, etc., and observe whether or not the ratio of the number of "points up" to the total number of tosses approaches a limiting numerical value.

However, we still have some difficulties. Shouldn't we really perform the experiment endlessly to determine whether or not there is such a limiting number? But we have only a finite lifetime available! Furthermore, we would not like to base an abstract mathematical system on variable empirical results.

Mathematicians now recognize that the difficulty in making a satisfactory definition of probability is that, if we are going to build an abstract mathematical system free of the real world, we must regard probability as an undefined term just as we regard *point*, *line*, and *space* as undefined terms. We will accept probability as an undefined term, recognizing that the preceding discussion and the two definitions suggested will nevertheless provide us with some basic insight when we attempt to apply a mathematical model to a specific experiment. We will therefore turn our attention to the formulation of a mathematical theory of probability, but without an undue emphasis on rigor and abstractness.

An introductory step in the analysis of a probability experiment will entail a consideration of the possible outcomes. The set of outcomes is called a *sample space* and is defined as follows:

> **DEFINITION:** A set of logical possibilities for a probability experiment is called a **sample space** when each outcome of the experiment corresponds to exactly one element in the set and each element in the set corresponds to an outcome of the experiment.

A given experiment may be represented by more than one sample space and a choice of a sample space will be dictated by the type of information we are seeking, as well as by a matter of convenience. Consider the experiment of picking at random any of the digits 2, 3, 4, 5, or 6. We might quickly select as a sample space the set of digits, namely, $S_1 = \{2,3,4,5,6\}$. However, we could also describe the experiment with a sample space of S_2 = {odd number, even number} or a sample space of S_3 = {prime number, not a prime number}.

DEFINITION: Each element in a sample space is called a **sample point**.

DEFINITION: An **event** is a subset of the sample space for a probability **experiment.**

Example 1: An experiment consists of tossing a fair coin twice. Determine a sample space for the experiment and the events "at least one head" and "the second toss is a head."

Solution: A tree diagram representing the sequential process is helpful. An ordered pair designation where the outcome of the first toss is the first element and the outcome of the second toss is the second element of the ordered pair will give us the sample points. The set of these sample points is the sample space S.

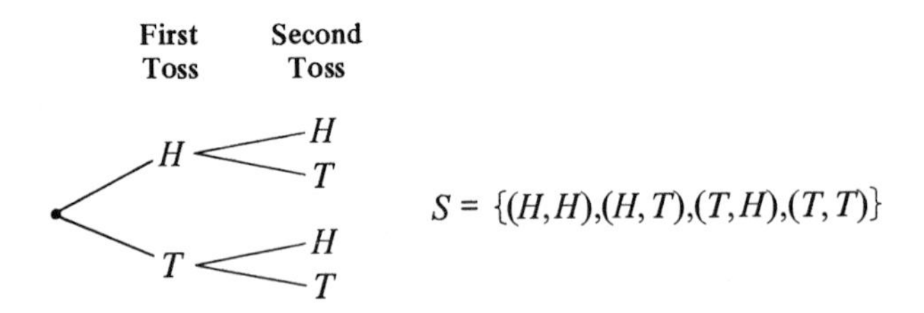

The event "at least one head" which we will designate as E_1 is a subset of S corresponding to $E_1 = \{(H,H),(H,T),(T,H)\}$.

The event "the second toss is a head" is designated as E_2 and corresponds to

$$E_2 = \{(H,H)\ (T,H)\}.$$

Our primary interest is in determining the probability of events. This requires that we assign a probability weight or measure to the sample points of a sample space. For example, in the toss of a single coin to each of the sample points in the sample space $S = \{H,T\}$ we would assign a weight to each sample point of 1/2 so that the $\Pr(H) = 1/2$ and $\Pr(T) = 1/2$.

Unless otherwise indicated, we will assume idealized conditions (fair coins for example) and assign probability weights, using whatever intuitive concepts and counting theory that may be useful. Certain assumptions, which we now make explicit, will underlie the weight assigning process.

A(1) The null set, ϕ, is a subset of any sample space and corresponds to an **impossible event** with $\Pr(\phi) = 0$.

A(2) If a sample space contains n sample points, then the sum of the weights assigned to all the sample points is 1, i.e., if $S = \{S_1, S_2, \ldots, S_n\}$, then $W(S_1) + W(S_2) + \cdots + W(S_n) = 1$. [Note: $W(S_1)$ is read "the weight (measure) of S_1."]

A(3) The weight of any sample point must be a number between zero and one inclusive, i.e.,

$$0 \leqslant W(S_i) \leqslant 1.$$

A(4) If the event E corresponds to a subset of the sample space containing k elements, then the $\Pr(E)$ is the sum of the weights assigned to the sample points of E. Thus if

$$E = \{S_1, S_2 \ldots S_k\}$$

then

$$\Pr(E) = W(S_1) + W(S_2) + \cdots + W(S_k).$$

A(5) If an event E corresponds to the entire sample space, then the event is said to be a **certainty** and the $\Pr(E) = 1$.

Example 2: A coin is tossed followed by the roll of a single die. Determine the probability of the event

(a) E_1, two tails occur.
(b) E_2, a head and an even number occurs.
(c) E_3, an odd or an even number appears.

Solution: A sample space for the experiment can be developed by a tree diagram.

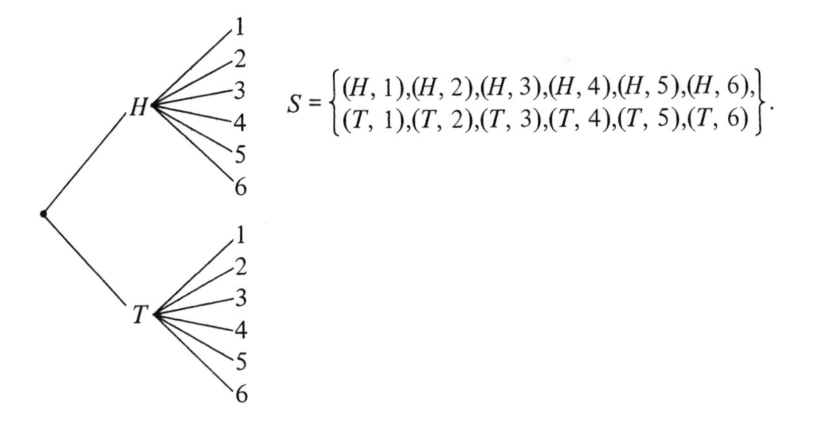

$$S = \left\{ \begin{matrix} (H, 1), (H, 2), (H, 3), (H, 4), (H, 5), (H, 6), \\ (T, 1), (T, 2), (T, 3), (T, 4), (T, 5), (T, 6) \end{matrix} \right\}.$$

Assuming a fair coin and a fair die are used so that the outcomes are equally likely, it is reasonable to assign a weight of 1/12 to each of the twelve sample points.

(a) Event E_1 corresponds to an impossible event or the ϕ, therefore $\Pr(E_1) = 0$.
(b) Event E_2 corresponds to $E_2 = \{(H,2),(H,4),(H,6)\}$. Since each of the points is weighted 1/12, $\Pr(E_2) = 1/12 + 1/12 + 1/12 = 3/12 = 1/4$.
(c) Event E_3 corresponds to the entire sample. Since it is certain that an odd or an even number will appear, $\Pr(E_3) = 1$.

Frequently in probability experiments the sample points are not all equally likely so that the weights assigned to each sample point will take on different numerical values. The next two examples illustrate this type of probability experiment.

Example 3: A record store has on display ten classical (C), eight ballad (B), and twelve rock-and-roll (R) records. If a record is picked at random, determine the probability of the event
(a) E_1, the selection will be a rock-and-roll record.
(b) E_2, the selection will be a ballad or a rock-and-roll record.
(c) E_3, the selection is not a rock-and-roll record.

Solution: A sample space for the experiment where we are interested only in the type of music may be represented as $S = \{C,B,R\}$. Since there are a total of thirty records, we may determine the weight to be assigned to each point by expressing as a ratio the number of each type to the total number (i.e., $\Pr(E) = f/n$) so that $\Pr(C) = 10/30$, $\Pr(B) = 8/30$, $\Pr(R) = 12/30$.
(a) The event E_1 corresponds to $E_1 = \{R\}$. Therefore $\Pr(E_1) = 12/30 = 2/5$.
(b) The event E_2 corresponds to $E_2 = \{B,R\}$. Therefore $\Pr(E_2) = 8/30 + 12/30 = 20/30 = 2/3$.
(c) The event E_3 corresponds to $E_3 = \{C,B\}$. Therefore $\Pr(E_3) = 10/30 + 8/30 = 18/30 = 3/5$.

Example 4: A Republican (R), a Democrat (D), and an Independent (I) are running for a political office. The Republican is twice as likely to win as the Democrat and the Democrat is three times as likely to win as the Independent candidate. What is the probability that each candidate will win?

Solution: A sample space for the experiment representing each possible candidate's winning is given by $S = \{R,D,I\}$. In order to determine the weight to be assigned to each sample point, we make use of the fact that the Democrat's probability of winning is a multiple of the Independent's probability and the Republican's probability is in turn a multiple of the Democrat's probability of winning. If we let w represent the probability of the Independent's winning, then

$$\Pr(I) = w, \Pr(D) = 3w$$
$$\Pr(R) = 6w.$$

Since the sum of the probabilities assigned to all the sample points is 1.

$$w + 3w + 6w = 1$$
$$10w = 1$$
$$w = 1/10.$$

The probability for each candidates winning is

$$\Pr(I) = 1/10, \Pr(D) = 3/10$$
$$\Pr(R) = 6/10.$$

The reader has undoubtedly made note of the fact that in the wording of our examples the familiar words "or," "and," and "not" have reappeared. In the next two sections we will make explicit the relationship between these statement connectives and the probabilities of the union, intersection, and complement of events.

Problem Set 4.2

A. 1. Determine a sample space when the experiment is that of
   (a) tossing a coin three times.
   (b) rolling a die.
   (c) rolling a die followed by tossing a coin.

2. Let the experiment be that of rolling a pair of dice, either simultaneously or one die followed by the other. Determine a sample space if
   (a) the sample points are ordered pairs such that the first component is the number of dots that turns up on the first die and the second component is the number of dots that turns up on the second die.
   (b) the sample points are the sum of the number of dots that turned up on the toss of each die.

3. A bucket contains thirteen golf balls, two have the letter P on them, five have the letter G, and six the letter D. Determine the probability of
   (a) drawing a ball with a letter G on it.
   (b) drawing a ball with the number 5 on it.
   (c) drawing a ball with a P or D on it.

4. A letter is selected at random from the word "contain." Find the probability
   (a) that it is an "n."
   (b) that it is the letter "a."
   (c) that it is a vowel.

5. Let the experiment be that of tossing a coin. Find the probability of at least one head
   (a) if the coin is tossed twice.
   (b) if the coin is tossed three times.
   (c) if the coin is tossed five times.
   (d) if the coin is tossed n times.

6. A married couple desire to have three children. If the probability of having a boy is the same as that of having a girl, determine the probability of having
   (a) three boys.
   (b) no boys.
   (c) three girls.
   (d) at least one boy.
   (e) at most one boy.
   (f) two girls and one boy.

7. The numbers 1, 2, 3, and 4 are written separately on four different slips of paper. The slips of paper are folded so the number cannot be seen, put into a convenient container, and are stirred. A person draws two slips from the container, one after the other, without replacement. He then forms the sum of the two numbers that appeared on the two slips of paper. Determine the sample space and weight each of the sample points. Now suppose that after the first draw, the person notes the number on the slip, folds the paper and returns it to the container and randomly draws a second slip from the container. Again, write out the sample space when there is replacement after the first draw and weight each sample point in the sample space.

8. A bag contains sixteen black or red marbles such that there are three times as many black marbles as red marbles all of which are identical except for color.

One marble is drawn at random. Find the probability that it is
(a) black. (b) yellow. (c) red or black.

9. A card is drawn at random from an ordinary bridge deck. Determine the probability that
   (a) an ace is drawn. (b) a black card is drawn.
   (c) a numbered card is drawn. (d) a queen, jack, or king is drawn.
10. Let $\mathcal{U} = \{x,y,z\}$. Assign weights to the three elements so that no two have the same weight and the weight of each is between zero and one. Also, find the probability for each event of $\mathcal{U}$ with the assigned weights.

B. 1. A die is thrown until a 3 appears. Can a finite sample space represent this experiment?
2. A pyramid has a square base and equilateral triangles for sides. The base has the the letter "a" written on it and the sides of the pyramid have the letters b, c, d, e written one on each side. Let the experiment be that of rolling the pyramid and observing whether the pyramid lands on a side or on its base. Determine a sample space for this experiment. On the basis of your intuition, what is the difficulty in weighting each of the sample points?
3. A coin is tossed repeatedly until a tail first appears, or until heads appear three times in succession. Describe a sample space for this experiment and weight each sample point.
4. A die is "rigged" so that where the four dots appeared, the face is now blank. Let the die be rolled once and observe whether the upturned face is blank, an even number of dots appear, or an odd number of dots appears. Determine a sample space for the experiment and weight each sample point.
5. A three-letter word is formed by randomly choosing three of the letters a, b, e, i, d, o without repetition. Find the probability that the word is
   (a) one which begins with the letter "d."
   (b) one with exactly two vowels in it.
   (c) one which begins and ends with a vowel.
6. If a pair of dice is thrown, what is the probability that each die will land with the same number of dots appearing on each upturned face?

4.3 Probability of the Union and Complement of Events

In the preceding section we determined the probabilities of events by adding together the weights assigned to the sample points that represented the event. We would like to extend our probability theory so that we may take advantage of more powerful techniques and handle more complex problems in an expeditious manner. Fortunately, we can develop useful probability theorems by building directly on our knowledge of logic and set theory.

THEOREM 4.3.1 Let E be an event in a sample space S, then

$$\Pr(E) = 1 - \Pr(\tilde{E}).$$

PROOF: In the sample space S the events E and $\tilde{E}$ are subsets of S such that $E \cup \tilde{E} = S$ and $E \cap \tilde{E} = \emptyset$. Let

$$E = \{s_1, s_2, s_3, \ldots, s_r\}$$

and

$$\tilde{E} = \{s_{r+1}, s_{r+2}, s_{r+3}, \ldots, s_t\}$$

then

$$S = \{s_1, s_2, s_3, \ldots, s_r, s_{r+1}, s_{r+2}, s_{r+3}, \ldots, s_t\}.$$

The probability of E and $\tilde{E}$ is given by

$$\Pr(E) = w(s_1) + w(s_2) + w(s_3) + \cdots + w(s_r)$$

and

$$\Pr(\tilde{E}) = w(s_{r+1}) + w(s_{r+2}) + w(s_{r+3}) + \cdots + w(s_t).$$

Since

$$\Pr(S) = w(s_1) + w(s_2) + w(s_3) + \cdots + w(s_r) + w(s_{r+1}) + w(s_{r+2}) + w(s_{r+3}) + \cdots + w(s_t)$$

then

$$\Pr(S) = \Pr(E) + \Pr(\tilde{E}).$$

But

$$\Pr(S) = 1.$$

Thus

$$1 = \Pr(E) + \Pr(\tilde{E})$$

or

(1) $\mathbf{\Pr(E) = 1 - \Pr(\tilde{E})}$.

Example 1: In tossing a coin three times, what is the probability of the event E, "at least one head?"

Solution: The sample space consists of eight sample points determined by multiplying together the number of outcomes on each toss, namely $\underline{2} \times \underline{2} \times \underline{2}$. The weight for each equally likely sample point is 1/8. We need not write out the entire sample space, however, to determine the probability of the event E, "at least one head," rather we consider the event $\tilde{E}$ which corresponds to "no heads" or the single sample point $(T,T,T,)$. Using Eq. (1) we have

$$\begin{aligned}\Pr(E) &= 1 - \Pr(\tilde{E}) \\ &= 1 - 1/8 \\ &= 7/8.\end{aligned}$$

DEFINITION: If the intersection of two events E_1, E_2 is empty, i.e., $E_1 \cap E_2 = \phi$, then the two events are said to be **mutually exclusive.**

THEOREM 4.3.2: If two events E_1 and E_2 in a sample space S are mutually exclusive, then the $\Pr(E_1 \cup E_2) = \Pr(E_1) + \Pr(E_2)$.

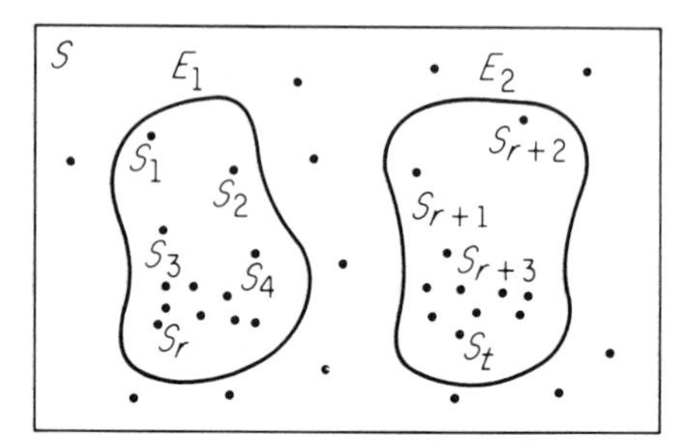

Figure 4.1

PROOF: Let $E_1 = \{s_1, s_2, s_3, \ldots, s_r\}$ and $E_2 = \{s_{r+1}, s_{r+2}, \ldots, s_t\}$ where $E_1 \cap E_2 = \phi$. (See Figure 4.1.)

$$\Pr(E_1) = w(s_1) + w(s_2) + \cdots + w(s_r)$$

and

$$\Pr(E_2) = w(s_{r+1}) + w(s_{r+2}) + \cdots + w(s_t).$$

The event $E_1 \cup E_2$ is a subset of S and since the events are mutually exclusive,

$$E_1 \cup E_2 = \{s_1, s_2, \ldots, s_r, s_{r+1}, s_{r+2}, \ldots, s_t\}.$$

The probability of $E_1 \cup E_2$ is determined by

$$\Pr(E_1 \cup E_2) = w(s_1) + w(s_2) + \cdots + w(s_r) + w(s_{r+1}) + w(s_{r+2}) + \cdots + w(s_t).$$

But the sum of the weights of s_1 through s_r is $\Pr(E_1)$, and the sum of the remaining weights is $\Pr(E_2)$. Hence,

(2) $\mathbf{Pr}(E_1 \cup E_2) = \mathbf{Pr}(E_1) + \mathbf{Pr}(E_2)$.

COROLLARY 4.3.2: If the events $E_1, E_2, E_3, \ldots, E_k$ in sample space S are all pairwise mutually exclusive, then

(3) $\mathbf{Pr}(E_1 \cup E_2 \cup E_3 \cup \cdots \cup E_k) = \mathbf{Pr}(E_1) + \mathbf{Pr}(E_2) + \mathbf{Pr}(E_3) + \cdots + \mathbf{Pr}(E_k)$.

Example 2: In rolling a pair of dice, what is the probability of obtaining a sum of 7, E_1, or a sum of 2, E_2?

Solution: The statement connective "or" is translated as the union of the events, so we are seeking $\Pr(E_1 \cup E_2)$. The event E_1 corresponds to the set of sample points $\{(1,6),(2,5),(3,4),(4,3),(5,2),(6,1)\}$, and the event E_2 corresponds to the set with the sample point $\{(1,1)\}$ (see Exercise A.2, Sec. 4.2) with each point weighted 1/36. Therefore, $\Pr(E_1) = 6/36$ and $\Pr(E_2) = 1/36$. The $\Pr(E_1 \cup E_2)$ where $E_1 \cap E_2 = \phi$ is given by

$$\begin{aligned}\Pr(E_1 \cup E_2) &= \Pr(E_1) + \Pr(E_2)\\ &= 6/36 + 1/36\\ &= 7/36.\end{aligned}$$

THEOREM 4.3.3 (addition theorem): If two events, E_1 and E_2, in sample space S are not mutually exclusive, $E_1 \cap E_2 \neq \phi$, then

$$\Pr(E_1 \cup E_2) = \Pr(E_1) + \Pr(E_2) - \Pr(E_1 \cap E_2).$$

PROOF: Since $E_1 \cap E_2 \neq \phi$ we may represent E_1 and E_2 in the partitioned form

$$E_1 = (E_1 \cap \tilde{E}_2) \cup (E_1 \cap E_2)$$

and

$$E_2 = (\tilde{E}_1 \cap E_2) \cup (E_1 \cap E_2).$$

(See Figure 4.2.)

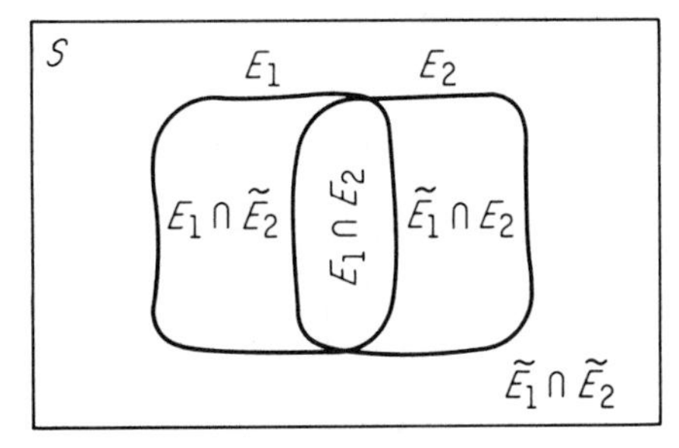

Figure 4.2

By Theorem 4.3.2,

$$\Pr(E_1) = \Pr(E_1 \cap \tilde{E}_2) + \Pr(E_1 \cap E_2)$$

and

$$\Pr(E_2) = \Pr(\tilde{E}_1 \cap E_2) + \Pr(E_1 \cap E_2).$$

Adding the two equations together we obtain

$$\Pr(E_1) + \Pr(E_2) = \Pr(E_1 \cap \tilde{E}_2) + \Pr(E_1 \cap E_2) + \Pr(\tilde{E}_1 \cap E_2) + \Pr(E_1 \cap E_2).$$

However, since the first three terms on the right side of this equation are equivalent to $\Pr(E_1 \cup E_2)$, we have

$$\Pr(E_1) + \Pr(E_2) = \Pr(E_1 \cup E_2) + \Pr(E_1 \cap E_2).$$

Solving for $\Pr(E_1 \cup E_2)$ we obtain

(4) $\mathbf{Pr(E_1 \cup E_2) = Pr(E_1) + Pr(E_2) - Pr(E_1 \cap E_2)}$.

Example 3: In rolling a pair of dice what is the probability we obtain a "match", E_1, or a sum of 6, E_2?

Solution: The event E_1 is the set $\{(1,1),(2,2),(3,3),(4,4),(5,5),(6,6)\}$ and the event E_2 is the set $\{(1,5),(2,4),(3,3),(4,2),(5,1)\}$ with each sample point weighted 1/36. The event $E_1 \cap E_2 = \{(3,3)\}$. Applying Eq. (4), we have

$$\begin{aligned}\Pr(E_1 \cup E_2) &= \Pr(E_1) + \Pr(E_2) - \Pr(E_1 \cap E_2)\\ &= 6/36 + 5/36 - 1/36\\ &= 10/36 = 5/18.\end{aligned}$$

Example 4: If A and B are events in sample space S and $\Pr(A \cap B) = 1/4$, $\Pr(\tilde{A}) = 5/8$, and $\Pr(B) = 3/4$, find (a) $\Pr(A)$, (b) $\Pr(\tilde{B})$, (c) $\Pr(A \cup B)$, (d) $\Pr(A \cap \tilde{B})$, (e) $\Pr(A \cup \tilde{B})$.

Solution: A Venn diagram is a helpful device in visualizing this type of problem. Label each region with the weight assigned to the specific region beginning with the region $A \cap B$ which is weighted 1/4. Since the $\Pr(\tilde{A}) = 5/8$, $\Pr(A) = 1 - 5/8 = 3/8$. Thus region $A \cap \tilde{B}$ must be weighted 1/8, $(3/8 - 1/4 = 1/8)$ and region $\tilde{A} \cap B$ must be weighted 1/2 $(3/4 - 1/4 = 1/2)$.

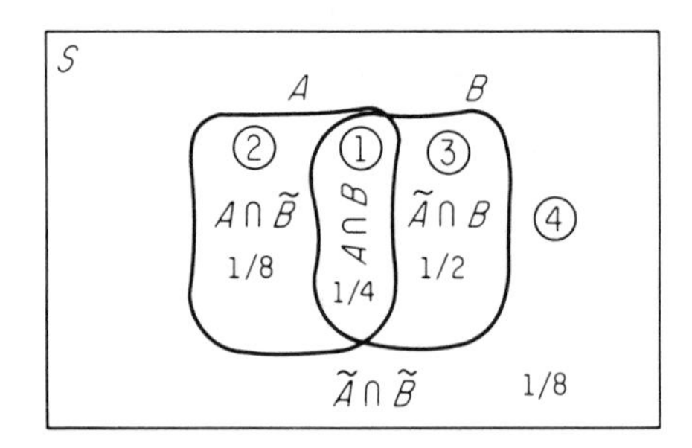

Finally, the sum of the weights assigned to regions ①, ②, and ③ is 7/8 $(1/8 + 1/4 + 1/2 = 7/8)$, leaving 1/8 to be assigned to region ④, $\tilde{A} \cap \tilde{B}$. The desired results can now be obtained quite readily:

(a) $\Pr(A) = 1/8 + 1/4 = 3/8$.
(b) $\Pr(\tilde{B}) = 1/8 + 1/8 = 2/8 = 1/4$.
(c) $\Pr(A \cup B) = 3/8 + 3/4 - 1/4 = 7/8$.
(d) $\Pr(A \cap \tilde{B}) = 1/8$.
(e) $\Pr(A \cup \tilde{B}) = 3/8 + 1/4 - 1/8 = 1/2$.

In games of chance, particularly where betting occurs, it is common to refer to the odds in favor or the odds against a particular outcome. For example, if the odds in favor of a horse winning a particular race are 3 to 2, then the racetrack expects each customer to wager $3 against their $2 for a "fair" bet. Odds are based on probabilities and are defined as follows:

DEFINITION: The **odds in favor** of a particular outcome E are given by the ratio $\frac{\Pr(E)}{\Pr(\tilde{E})}$ and the **odds against** a particular outcome are given by the ratio $\frac{\Pr(\tilde{E})}{\Pr(E)}$.

If we are given the odds in favor of a particular outcome, and we wish to find the probability of that specific outcome, we make reference to the following definition:

DEFINITION: If the odds in favor of a particular outcome E are a/b ("read a to b," sometimes written $a{:}b$), then the **probability of E** is given by

$$\Pr(E) = \frac{a}{a+b}.$$

Example 5: What are the odds in favor of and the odds against getting a sum of 7 when a pair of dice are thrown?

Solution: If E represents getting a sum of 7, then the $\Pr(E) = 6/36$. (See Example 2.) The probability of the complement $\Pr(\tilde{E}) = 1 - 6/36$ or $\Pr(\tilde{E}) = 30/36$.

$$\text{Odds in favor} = \frac{6/36}{30/36} = 6/30 = 1/5(1{:}5).$$

$$\text{Odds against} = \frac{30/36}{6/36} = 30/6 = 5/1(5{:}1).$$

Example 6: If the odds in favor of the stock market going up are quoted as 4 to 3, what is the probability the stock market will go up?

Solution: If E represents the stock market going up, then $\Pr(E) = \frac{4}{4+3} = \frac{4}{7}$.

Problem Set 4.3

A. 1. A fair die is thrown. What is the probability of obtaining a 2 or a 3?

2. Let the experiment be that of rolling a fair die. What is the probability of obtaining an even number or a number divisible by three?

3. A card is drawn randomly from an ordinary bridge deck. Determine the probability that the card drawn was
   (a) a seven or nine. (b) a heart or a diamond.
   (c) neither a seven nor a nine.
4. Let E and F be events of a sample space S such that $\Pr(E \cap F) = 1/5$, $\Pr(\tilde{E}) = 1/3$, and $\Pr(F) = 1/2$. Determine
   (a) $\Pr(E \cup F)$. (b) $\Pr(\tilde{E} \cap \tilde{F})$.
5. Consider an ordinary bridge deck. A card is to be selected at random from the deck. What is the probability of drawing a face card or a diamond?
6. Let A, B, C be events of a sample space S, and $\Pr(A \cap B \cap C) = 1/6$, $\Pr(A \cap B) = 5/18$, $\Pr(A \cap C) = 7/18$, $\Pr(B \cap C) = 5/18$, $\Pr(A) = 5/9$, $\Pr(B) = 1/2$, $\Pr(C) = 2/3$. Find
   (a) $\Pr(A \cap \tilde{B} \cap C)$. (b) $\Pr(\tilde{A} \cap B \cap C)$.
   (c) $\Pr(A \cap B \cap \tilde{C})$. (d) $\Pr(\tilde{A} \cap \tilde{B} \cap \tilde{C})$.
   (e) $\Pr(A \cup B \cup C)$.
7. In a given mathematics course the grades given are A, B, C, D, and F. Furthermore a student in the course has probability .8 of passing the course, and .7 of getting lower than B. Find the probability that a student will get
   (a) a C or D. (b) an A or B.
8. What odds should a person give on his bet that a number divisible by 3 will turn up when he rolls a die?
9. If the probability of an event E is .6, find (a) the odds in favor of E, and (b) the odds against E.
10. A survey of 1,000 electrical items showed that 20 had defective parts, 15 were improperly wired, and 3 had both defects. If a part is picked at random, find the probability
    (a) of the item having a defective part and being improperly wired.
    (b) of the item having a defective part and not being improperly wired.
    (c) of the item having defective parts or being improperly wired.
    (d) of the item not having a defective part and being properly wired.
11. Let the experiment be that of rolling a pair of dice. Find the probability of
    (a) the total number of dots being even or the same number of dots appearing on each die.
    (b) a sum greater than 4 or a sum that is a multiple of 3 appearing.
    (c) a sum greater than 7 is obtained.
    (d) a sum not greater than 7 is obtained.
    (e) a sum of 7 or 11 results.

4.4 Conditional Probability

In determining the probability of an event A we may wish to consider whether or not the occurrence of other events will influence the probability of A. Suppose that we are given as a condition that an event B took place and based on this given information, we wish to determine the probability of A. Symbolically, we are seeking $\Pr(A|B)$, which is read the **conditional probability** of A, given B.

Before we introduce a formal definition of how we can determine $\Pr(A|B)$, let us

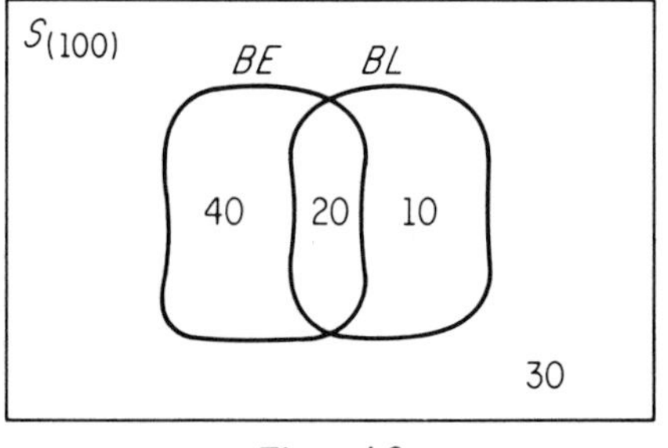

Figure 4.3

consider an example and an intuitive analysis which will help to make the definition seem reasonable.

Suppose 100 people were selected at random and we observe that 60 were blue-eyed (BE), 30 were blonde (BL), and 20 were blue-eyed and blonde ($BE \cap BL$). If we display this information in a Venn diagram form (Figure 4.3), we can determine conveniently the probability that a person picked at random is (a) blue-eyed, (b) blonde, and (c) blue-eyed and blonde. Thus,

(a) $\Pr(BE) = \frac{60}{100} = \frac{6}{10} = \frac{3}{5}.$

(b) $\Pr(BL) = \frac{30}{100} = \frac{3}{10}.$

(c) $\Pr(BE \cap BL) = \frac{20}{100} = \frac{2}{10}.$

Consider now the question, "What is the probability that a person picked at random is blue-eyed, given that the person is blonde?," $\Pr(BE|BL)$. The given condition of "blondeness" reduces the number of persons being considered to 30 of which 20 are blue-eyed. Therefore,

$$\Pr(BE|BL) = \frac{20}{30} = \frac{2}{3}.$$

Alternately, we could refer each group to the entire sample space of 100 and write the initial ratio as

$$\Pr(BE|BL) = \frac{20/100}{30/100} = \frac{2}{3}$$

which the reader will observe that in terms of the regions the result is equivalent to

$$\Pr(BE|BL) = \frac{\Pr(BE \cap BL)}{\Pr(BL)}.$$

In a similar manner we can determine $\Pr(BL|BE)$.

$$\Pr(BL|BE) = \frac{20}{60} = \frac{1}{3}$$

or

$$\Pr(BL|BE) = \frac{20/100}{60/100} = \frac{1}{3}.$$

In terms of the regions considered, the result is

$$\Pr(BL|BE) = \frac{\Pr(BL \cap BE)}{\Pr(BE)}.$$

Notice that the given condition changed the probability of a person being blue-eyed and changed the probability of a person being blonde.

The preceding discussion suggests the following definition of conditional probability.

DEFINITION: If A and B are events in a sample space S, where $\Pr(A) \neq 0$ and $\Pr(B) \neq 0$, then the **conditional probability of A given B** is defined as

(1) $$\mathbf{Pr(A|B) = \frac{Pr(A \cap B)}{Pr(B)}}$$

and the conditional probability of B given A is defined as

(2) $$\mathbf{Pr(B|A) = \frac{Pr(B \cap A)}{Pr(A)}}.$$

Before we apply the conditional probability formulas, we wish to develop from Eq. (2) some very useful results. A formula for the probability that both events occur; namely, the intersection of two events, $\Pr(A \cap B)$ can easily be developed beginning with Eq. (2).

Equation (2) can be rewritten as

$$\Pr(B|A) = \frac{\Pr(A \cap B)}{\Pr(A)}$$

since intersection is a commutative operation. Multiplying both sides of the equation by $\Pr(A)$, we obtain

(3) $$\mathbf{Pr(A \cap B) = Pr(A) \cdot Pr(B|A).}$$

If we wish to consider the probability that three events A, B, and C all take place we can extend Eq. (3) to obtain a formula for the $\Pr(A \cap B \cap C)$ as follows:

$$\Pr[(A \cap B) \cap C] = \Pr(A \cap B) \cdot \Pr(C|A \cap B)$$

thus

(4) $\mathbf{Pr[(A \cap B) \cap C] = Pr(A) \cdot Pr(B|A) \cdot Pr(C|A \cap B).}$

(It is assumed that $\Pr(A \cap B) \neq 0$.)

The generalization of Eqs. (3) and (4) is sufficiently important to be stated as a theorem. The proof of this theorem will, however, be omitted.

THEOREM 4.4.1 (multiplication theorem): If $E_1, E_2, \ldots, E_n$ are events in a sample space S, where $\Pr(E_1 \cap E_2 \cap \ldots \cap E_{n-1}) \neq 0$, then

(5) $$\mathbf{Pr(E_1 \cap E_2 \cap \ldots \cap E_n) = Pr(E_1) \cdot Pr(E_2|E_1) \cdot (Pr(E_3|E_1 \cap E_2) \ldots Pr(E_n|E_1 \cap E_2 \cap E_3 \cap \ldots \cap E_{n-1})}.$$

Equation (5) states that in order to find the probability that a sequence of events all take place, we simply multiply together the probability of each event given all the preceding events took place.

It seems appropriate at this point to take a brief backward glance at our earlier work. We began this text with a discussion of the logical meaning of English statements and developed a symbolic logic for these statements and connectives. In our work with set theory we demonstrated that we may "translate" from logic to set theory, and vice versa. In this chapter we have developed our probability theory by regarding the logical possibilities for an experiment as a set and the statements made about an experiment as a subset or an event. We have thus been able to carry over set-theory concepts into a system where we assign a numerical weight to a set representing the probability of an event. Although there is a hierarchial structure in the presentation of the topics, there is also an underlying unity in both the subject matter and in the reasoning process employed. At this point the reader should have some fuller appreciation of the spirit of mathematical activity as an element of human thought.

Let us now turn to some applications of conditional probability equations and the multiplication theorem.

Example 1: At a golf-driving range a customer receives a bucket containing 15 balls, 9 of which have red markings and 6 green markings. If the balls to be hit are selected at random find the probability that
(a) the first two balls hit have red markings.
(b) the second ball hit has red markings.
(c) given that the second ball hit was red, the first one hit was also red.

Solution: The experiment consists of selecting in sequence two golf balls without replacing the first one in the bucket where we are interested only in the color of the markings on the ball. A tree diagram will be helpful, not only in developing a sample space but in illustrating how we obtain the desired probabilities. Using subscripts to designate the first and second choices, we have as a sample space

$$S = \{(R_1, R_2), (R_1, G_2), (G_1, R_2), (G_1, G_2)\}.$$

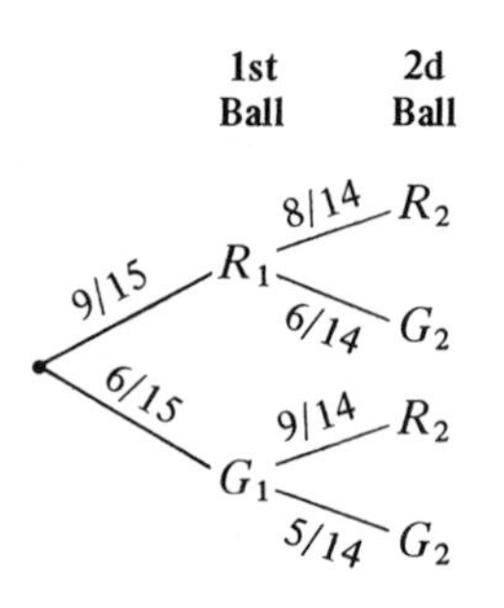

(a) The fact that the first two balls are red corresponds to the sample point (R_1, R_2) which means we have the event red on the first and red on the second, which we symbolize as $R_1 \cap R_2$. By Eq. (3)

$$\Pr(R_1 \cap R_2) = \Pr(R_1) \cdot \Pr(R_2 | R_1).$$

The $\Pr(R_1)$ is simply 9/15 and since one red ball is removed $\Pr(R_2|R_1)$ is 8/14, therefore

$$\Pr(R_1 \cap R_2) = \frac{9}{15} \cdot \frac{8}{14} = \frac{72}{210} = \frac{12}{35}.$$

Notice that this result may be obtained by multiplying together the probability weights on the branches of the tree diagram.

(b) The second ball has red markings, corresponds to the event $\{(R_1, R_2), (G_1, R_2)\}$. We must find the probability weight of each sample point and add them together. From part (a), the $\Pr(R_1 \cap R_2) = 12/35$. The $\Pr(G_1 \cap R_2)$ is obtained by

$$\Pr(G_1 \cap R_2) = \Pr(G_1) \cdot \Pr(R_2 | G_1)$$
$$= \frac{6}{15} \cdot \frac{9}{14} = \frac{9}{35}.$$

Therefore, $P(R_2)$ is given by

$$\Pr(R_2) = \Pr(R_1 \cap R_2) + \Pr(G_1 \cap R_2)$$
$$= \frac{12}{35} + \frac{9}{35} = \frac{21}{35} = \frac{3}{5}.$$

(c) Given the second ball is red, find the probability the first is also red requires that we determine $\Pr(R_1 | R_2)$. By Eq. (1),

$$\Pr(R_1 | R_2) = \frac{\Pr(R_1 \cap R_2)}{\Pr(R_2)}.$$

From (a), $\Pr(R_1 \cap R_2) = 12/35$ and from (b), $\Pr(R_2) = 21/35$; therefore

$$\Pr(R_1 | R_2) = \frac{12/35}{21/35} = \frac{12}{21} = \frac{4}{7}.$$

Example 2: What is the probability that in dealing three cards without replacement from a standard 52-card deck they are all hearts?

Solution: We wish to find the probability of the event $\{(H_1, H_2, H_3)\}$. By Eq. (5) we have

$$\Pr(H_1 \cap H_2 \cap H_3) = \Pr(H_1) \cdot \Pr(H_2|H_1) \cdot \Pr(H_3|H_1 \cap H_2).$$

We are drawing one card each time and we reduce the number of hearts available to be drawn by one each time, therefore

$$\Pr(H_1 \cap H_2 \cap H_3) = \frac{13}{52} \cdot \frac{12}{51} \cdot \frac{11}{50} = \frac{11}{850}.$$

Example 3: Assume that 5 percent of the population are believed to have diabetes. A standard medical test will correctly identify 94 percent of the people that have the disease. However, suppose the test also incorrectly diagnoses 8 percent of those who do not have the disease as having diabetes. A person is picked at random from the population and his test shows he has diabetes. What is the probability he has the disease?

Solution: Let D represent a person who has diabetes and $\tilde{D}$ he does not, T represent a positive test result and $\tilde{T}$ a negative test result. We wish to determine $\Pr(D|T)$. By Eq. (1) this can be found by

$$\Pr(D|T) = \frac{\Pr(D \cap T)}{\Pr(T)}.$$

Let us use a tree diagram to represent the possibilities. Thus,

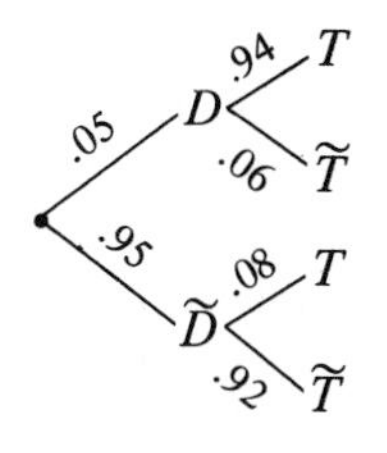

$$S = \{(D,T),(D,\tilde{T}),(\tilde{D},T),(\tilde{D},\tilde{T})\}$$
$$\Pr(T) = \Pr(D \cap T) + \Pr(\tilde{D} \cap T)$$
$$= (.05)\,(.94) + (.95)\,(.08) = .123$$

and the $\Pr(D \cap T) = (.05)\,(.94) = .047$. The conditional probability $\Pr(D|T)$ is determined by $\Pr(D|T) = \dfrac{.047}{.123} \doteq .382$. (The symbol $\doteq$ means "approximately equal to.")

The result indicates that even though the medical test is positive, there is a better than 50–50 chance he really does not have the disease.

Probability problems may be handled by more than one method or by a combination of methods. We will complete this section by considering one more example with two methods of solution. The alternate solution involves counting methods and a ratio of the number of successes to the total number of ways different actions may be taken is typical of what is referred to as the **combinatorial method.**

Example 4: Three prizes are to be awarded to three of ten individuals by drawing slips of paper numbered 1 through 10 without replacement out of a hat where each number corresponds to one person in the group. What is the probability that 2 of the 3 winning numbers are odd?

Solution: The problem does not stipulate the order in which the two odd, 0, numbers or the one even, E, number were drawn. Let us consider one possible sequence, namely $0_1 0_2 E_3$, i.e., the first two numbers were odd and third was even. The multiplication theorem provides us with the following approach:

$$\Pr(0_1 \cap 0_2 \cap E_3) = \Pr(0_1) \cdot \Pr(0_2 | 0_1) \cdot \Pr(E_3 | 0_1 \cap 0_2)$$

$$= \frac{5}{10} \cdot \frac{4}{9} \cdot \frac{5}{8} = \frac{5}{36}.$$

However, this is not the only possible sequence. The 0 0 E series of letters can be arranged in three ways $\left(\frac{3!}{2!}\right)$; namely, OOE, OEO and EOO. Each of these sequences occur with the same probability; therefore,

$$\Pr(2 \text{ odd and } 1 \text{ even}) = 3 \cdot \frac{5}{36} = \frac{5}{12}.$$

Alternate Solution: We will solve the problem using the f/n approach where f will correspond to the number of ways we can select 2 odd and 1 even number out of the 10 numbers and the n will correspond to the total number of ways of selecting 3 numbers out of the 10 numbers. Let us first consider the numerator f. We can consider the selection of 2 odd and 1 even as a sequential counting process where order is not important. The 2 odd must be selected out of the 5 that are odd and followed by selecting one even out of the five that are even:

$${}_5C_2 \times {}_5C_1 = \frac{5!}{2! \times 3!} \times \frac{5!}{1! \times 4!} = 50 \text{ ways}.$$

The denominator n represents how many ways we can select 3 numbers out of 10, again where order is not important. Thus,

$${}_{10}C_3 = \frac{10!}{3! \times 7!} = 120 \text{ ways}.$$

Therefore,

$$\Pr(2 \text{ odd and } 1 \text{ even}) = \frac{50}{120} = \frac{5}{12}.$$

Problem Set 4.4

A. 1. Of 100 people polled, 55 watch football on television, 35 watch basketball on television, and 25 watch both sports. What is the probability that a person chosen at random from the people polled
   (a) watches only basketball.
   (b) watches basketball or football.
   (c) watches neither basketball nor football.
   (d) watches basketball, given he watches football.
   (e) watches basketball given that he does not watch football.

2. Let the experiment be that of rolling a pair of dice.
   (a) What is the probability that a 1 appears on one die or the other, given that the sum is less than five?
   (b) What is the probability that a 3 or 4 appears on either die, given that the sum is greater than six?

3. Suppose a thumbtack is tossed twice and we are given the probability of the thumbtack landing up on a single roll is $1/4$, $\Pr(U) = 1/4$.
   (a) Find the probabilities of all possible outcomes.
   (b) Find the probability that the second toss fell up (U), given the first toss fell down (D).

4. A child has 24 marbles in a bag. Five are blue, nine yellow, six red, and four white. The child is asked to draw three marbles, one at a time without replacement out of the bag. Find the probability that
   (a) all three marbles are blue; red; yellow; white.
   (b) all three marbles are not the same color.

5. Let E and F be events in a sample space S with nonzero probabilities. Find the $\Pr(E|F)$ if
   (a) F is a subset of E.
   (b) E and F are mutually exclusive.

6. Let E, F be events in a sample space S such that $\Pr(E \cap F) = \frac{1}{6}$, $\Pr(E) = \frac{5}{12}$, $\Pr(F) = \frac{1}{2}$. Find (a) $\Pr(E \cup F)$, (b) $\Pr(\tilde{E})$, (c) $\Pr(E|F)$, (d) $\Pr(F|E)$, (e) $\Pr(\tilde{E}|\tilde{F})$, (f) $\Pr(\tilde{F}|\tilde{E})$.

7. If five cards are randomly selected from a bridge deck without replacement, find the probability that (symbolize but do not compute an answer)
   (a) all five are spades.
   (b) all five cards are red.
   (c) exactly three are red.
   (d) all cards are of the same unit.

8. In a drawer there are four brown socks, five red and three green. If two socks are drawn in succession, without replacement, what is the probability of getting a

color match? How many draws are necessary to guarantee a matched pair of socks?

9. On a shelf there are five different mathematics books, three different literature books, and two different philosophy books. If these books are placed randomly on a shelf, what is the probability that they remain grouped together by subject?

4.5 Independent Events

The conditional probability theory developed in the previous section gave us a method for finding the probability of an event given the condition that a second event had occurred. In many situations the occurrence of the second event may have no influence on the first event and will leave the probability of the first event unchanged. For example, if we are given the condition that "it is a cloudy day" and then are asked, "What is the probability of obtaining a head on the toss of a fair coin?," we do not expect the $\Pr(H) = 1/2$ to change. An event whose probability does not change when a second event occurs is said to be independent of the second event. More precisely we define the independence of two events as follows:

DEFINITION: Let A and B be two distinct events in sample space S where $\Pr(A) \neq 0$ and $\Pr(B) \neq 0$. The events A and B are said to be **independent** if and only if

$$\mathbf{Pr(A|B) = Pr(A)} \quad \textbf{and} \quad \mathbf{Pr(B|A) = Pr(B).}$$

An immediate consequence of the definition of independence is the following theorem:

THEOREM 4.5.1: If A and B are events in sample space S with $\Pr(A) \neq 0$ and $\Pr(B) \neq 0$ then A and B are independent if and only if

$$\text{(1)} \quad \mathbf{Pr(A \cap B) = Pr(A) \cdot Pr(B).}$$

PROOF: The biconditional statement given $\Pr(A) \neq 0$ and $\Pr(B) \neq 0$ requires a proof of (a) if A and B are independent events then $\Pr(A \cap B) = \Pr(A) \cdot \Pr(B)$, and (b) if $\Pr(A \cap B) = \Pr(A) \cdot \Pr(B)$ then the events are independent.

Consider (a) first. If A and B are independent events, then $\Pr(A|B) = \Pr(A)$ and $\Pr(B|A) = \Pr(B)$.

The basic definition of conditional probability for the event A given B is

$$\Pr(A|B) = \frac{\Pr(A \cap B)}{\Pr(B)}$$

Replacing $\Pr(A|B)$ by $\Pr(A)$ we have

$$\Pr(A) = \frac{\Pr(A \cap B)}{\Pr(B)}.$$

Multiplying both sides by $\Pr(B)$ gives the conclusion

$$\Pr(A \cap B) = \Pr(A) \cdot \Pr(B).$$

Thus we have deduced, if A and B are independent then $\Pr(A \cap B) = \Pr(A) \cdot \Pr(B)$. Next we will consider (b). We know that $\Pr(A \cap B) = \Pr(A) \cdot \Pr(B|A)$ and the hypothesis is that $\Pr(A \cap B) = \Pr(A) \cdot \Pr(B)$. We can equate the two expressions for $\Pr(A \cap B)$ giving

$$\Pr(A) \cdot \Pr(B|A) = \Pr(A) \cdot \Pr(B).$$

Dividing both sides by $\Pr(A)$ we obtain

$$\Pr(B|A) = \Pr(B).$$

We could also start with $\Pr(B \cap A) = \Pr(B) \cdot \Pr(A|B)$ and $\Pr(B \cap A) = \Pr(A \cap B) = \Pr(A) \cdot \Pr(B)$ and equate the expressions for $\Pr(B \cap A)$ to obtain $\Pr(B) \cdot \Pr(A|B) = \Pr(B) \cdot \Pr(A)$. Dividing both sides by $\Pr(B)$ we obtain the second part of the definition of independence; namely,

$$\Pr(A|B) = \Pr(A).$$

Since $\Pr(B|A) = \Pr(B)$ and $\Pr(A|B) = \Pr(A)$ the events are independent.

Example 1: A fair coin is tossed twice. Let A represent the event "head on second toss" and B represent "tail on first toss." Show that the events A and B are independent.

Solution: Our familiar sample space for the experiment is $S = \{(H,H), (H,T), (T,H), (T,T)\}$ with each sample point weighted 1/4. The event $A = \{(H,H), (T,H)\}$ with $\Pr(A) = 1/2$ and the event $B = \{(T,H), (T,T)\}$ with $\Pr(B) = 1/2$. We must show that $\Pr(A \cap B) = \Pr(A) \cdot \Pr(B)$. The event $A \cap B$ corresponds to $\{(T,H)\}$. Therefore, $\Pr(A \cap B) = 1/4$. Since the equality $\Pr(A \cap B) = \Pr(A) \cdot \Pr(B)$ does hold, i.e., $1/4 = 1/2 \cdot 1/2$ the events are independent.

It is not surprising that such experiments as coin tossing or rolling a die frequently involve a sequence of independent events. For example, at each toss of a fair coin, the $\Pr(H) = 1/2$ no matter what outcomes precede each toss. There may be a psychological impact on you if you have tossed, let us say, 5 heads in a row that the probability of a head on the next toss will not be 1/2, but the coin has no memory of previous outcomes and the $\Pr(H)$ on each toss remains constant.

To obtain the probability of the sequence of three or more independent events, we may of course generalize (1) and obtain a theorem which is really a special case of the multiplication theorem (4.4.1). However, we must insist that if we have a sequence of n events that the events be independent two at a time, three at a time $\cdots$ n at a time for all possible combinations. We can then designate the events as **completely independent**.

The generalized theorem, the proof of which is omitted, is as follows:

THEOREM 4.5.2: If the events $E_1, E_2, \ldots, E_n$ are completely independent then

$$\Pr(E_1 \cap E_2 \cap \cdots \cap E_n) = \Pr(E_1) \cdot \Pr(E_2) \cdots \Pr(E_n).$$

From this point on we will assume independence means complete independence.

Example 2: A die is rolled four times. What is the probability that three 6's and any number other than a 6 is obtained in (a) that specific order, and (b) any order?

Solution: Assuming independence we must determine in (a) $\Pr(6_1 \cap 6_2 \cap 6_3 \cap \tilde{6}_4)$; by Theorem 4.5.2 we have:

$$\Pr(6_1 \cap 6_2 \cap 6_3 \cap \tilde{6}_4) = 1/6 \cdot 1/6 \cdot 1/6 \cdot 5/6 = \frac{5}{1296}.$$

For part (b) we must first consider how many ways we can arrange three 6's and one nonsix. But this is simply a permutation of four symbols where three are indistinguishable. There are $\frac{4!}{3! \times 1!} = 4$ distinct ways, $(6\,6\,6\,\tilde{6}, 6\,6\,\tilde{6}\,6, 6\,\tilde{6}\,6\,6$, and $\tilde{6}\,6\,6\,6)$.

Each of the events has the same probability, therefore the final probability of the event three 6's and one nonsix in any order is Pr (three 6's and one nonsix) = $4 \cdot \frac{5}{1296} = \frac{5}{324}$.

Example 3: A company has two machines that independently produce electric switches. Five percent of the switches produced by machine A are defective and four percent of the switches made by machine B are defective. An inspector picks at random a switch made by A followed by the random selection of a switch made by B.

(a) What is the probability that both switches are defective?
(b) What is the probability one or two of the switches are defective?
(c) What is the probability they are both "good" switches?

Solution: Let A represent the event that a defective part is made by machine A and B represent the event that a defective part is made by machine B.

(a) $\Pr(A \cap B) = \Pr(A) \cdot \Pr(B)$
$\Pr(A \cap B) = (.05) \cdot (.04) = .002.$

(b) $\Pr(A \cup B) = \Pr(A) + \Pr(B) - \Pr(A \cap B)$
$\Pr(A \cup B) = (.05) + (.04) - (.002) = .088.$

(c) $\Pr(\widetilde{A \cup B})$ represents the probability, they are both good since $\Pr(\tilde{A} \cap \tilde{B}) = \Pr(\widetilde{A \cup B})$. Thus,

$$
\begin{aligned}
\Pr(\widetilde{A \cup B}) &= 1 - \Pr(A \cup B) \\
&= 1 - 0.088 \\
&= 0.912.
\end{aligned}
$$

Problem Set 4.5

A. 1. Let A, B be events in a sample space S. If $\Pr(A) = 5/8$, $\Pr(B) = 3/8$, and $\Pr(A \cap B) = 1/8$, are A and B independent? Verify your result.

2. A green die and a white die are rolled. Let E be the event that "an even number of dots appears on the white die" and let F be the event that "the number of dots appearing on the green die is a prime number." Find
   (a) $\Pr(E)$. (b) $\Pr(F)$.
   (c) $\Pr(E \cap F)$. (d) $\Pr(E \cup F)$.
   Are the events E and F independent?
3. A pair of dice are rolled one after the other. Let E be the event "an even number of dots appears on the first die" and F is the event "the sum of the dots appearing is even." Determine whether the events E and F are independent.
4. The probability that a man hits a target is $2/5$ and the probability that his son hits the target is $1/3$. What is the probability that the target will be hit if the man and his son shoot at the target?
5. A red die and a green die are rolled. If A is the event, "an even number of dots on the red die" and B is the event, "an even number of dots on the green die" and C is the event "each die has the same number of dots," determine if
   (a) A and B are independent. (b) A and C are independent.
   (c) B and C are independent. (d) A, B, C are independent.
6. A basketball player hits 60 percent of his foul shots. If he shoots two fouls and E represents that he makes the first shot and F represents he makes the second shot and E and F are independent, determine if
   (a) E and $\tilde{F}$ are independent. (b) $\tilde{E}$ and F are independent.
   (c) $\tilde{E}$ and $\tilde{F}$ are independent.

B. 1. Show that if two events, each with nonzero probability, are independent, then they are not mutually exclusive.

2. Prove or disprove: If $\Pr(A \mid B) = \Pr(A)$ and $\Pr(A) \neq 0$ and $\Pr(B) \neq 0$, then $\Pr(B \mid A) = \Pr(B)$.
3. Let A and B be independent events. Prove the following pairs of events are also independent. (a) A and $\tilde{B}$, (b) $\tilde{A}$ and B

4.6 Binominal Probability Model

Many different probability experiments have characteristics in common and may be thought of as examples of a more general mathematical probability model. In this section we will consider one particularly useful probability model, the binomial model. This model is widely used in genetics, quality-control work, and any situation where certain basic assumptions are satisfied. For our purposes it will be instructive to observe that its development flows naturally out of our previous work.

The binomial probability model has four distinguishing characteristics.

1. As the experiment is performed repetitively from trial to trial, the trials are independent.
2. At each trial one of two outcomes is possible.
3. The probability of either outcome remains fixed from trial to trial.
4. The experiment is performed a specified number of trials.

From our earlier work, such examples as tossing a coin where we are interested in a head or a tail, rolling a die where we are interested in a six or nonsix, and selecting numbers with replacement where we categorize the results as odd or even, come naturally to mind as typical binomial experiments.

In order to provide some intuitive motivation for the abstract model, let us first consider a specific example.

Example 1: What is the probability that in five rolls of a fair die we obtain three 6's?

Solution: So that we may generalize this example, let us call a six appearing a "success" and a nonsix appearing a "failure." We wish to find the probability that we obtain three S's and two F's in any order. Consider one possible sequence; namely $SSSFF$. The probability of this sequence is $\Pr(S \cap S \cap S \cap F \cap F) = 1/6 \cdot 1/6 \cdot 1/6 \cdot 5/6 \cdot 5/6 = (1/6)^3\ (5/6)^2$. However, this is not the only way of obtaining the three S's and two F's. In fact the number of such rearrangements is $\dfrac{5!}{3! \times 2!}$ which may also be written $\binom{5}{3,2}$. Each sequence has the same probability. Therefore, the final result can be written in the form

$$\Pr(\text{three 6's in 5 rolls}) = \binom{5}{3,2}(1/6)^3\ (5/6)^2 = \frac{125}{3888}.$$

The form of the solution to Example 1 may appear familiar to the reader as a typical term in a binomial expansion.

Suppose we go one more step and consider a binomial experiment which involves three trials and we call the probability of a success on each trial p; $\Pr(S) = p$, and the probability of a failure q, $\Pr(F) = q$. A tree diagram (shown on the opposite page) will provide us with a list of logical possibilities and the probability of each outcome.

As we did previously in our binomial expansions, we can group like terms together and add to obtain

$$p^3 + 3p^2q + 3pq^2 + q^3$$

which is the expansion of $(p+q)^3$ namely,

$$(p+q)^3 = \binom{3}{3,0}p^3 + \binom{3}{2,1}p^2q + \binom{3}{1,2}pq^2 + \binom{3}{0,3}q^3.$$

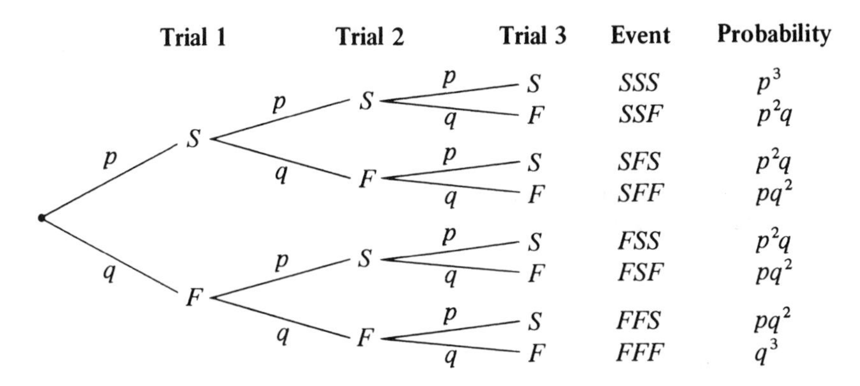

We can now, however, impart a probability meaning to each term in the expansion. The first term is the probability of obtaining three successes in three trials, the second term the probability of two successes and one failure in three trials, the third term the probability of one success and two failures in three trials, and the last term the probability of three failures in three trials. Notice that since $p + q = 1$, $(p + q)^3 = (1)^3 = 1$ and the sum of the terms on the right-hand side of the equation is also 1, a reasonable result, since all possible outcomes are represented.

If we extend the discussion to a binomial experiment performed n times where $\Pr(S) = p$ and $\Pr(F) = q$, we would apply the binomial formula directly to obtain

$$(p + q)^n = \binom{n}{n,0}p^n + \binom{n}{n-1,1}p^{n-1}q + \cdots + \binom{n}{1,n-1}pq^{n-1} + \binom{n}{0,n}q^n.$$

The general term for this expansion where we have "k" successes and "n-k" failures is given by

$$\binom{n}{k,n-k}p^k q^{n-k}.$$

We summarize the previous discussion and introduce an abbreviated notation in the following definition.

DEFINITION: If a binomial experiment is performed n times with the probability of one outcome p (success) and the probability of the other outcome q (failure), then the **probability of obtaining k successes and n-k failures** is given by

$$\mathbf{b(k;n,p)} = \binom{\mathbf{n}}{\mathbf{k,n-k}}\mathbf{p^k q^{n-k}}, \ \mathbf{k = 0,1,2, \ldots, n.}$$

The notation $b(k;n,p)$ is read "the binomial probability of k successes in n trials where the probability of success is p." The binomial probability model has widespread application. For extended work in probability, detailed binomial probability tables have been prepared and are readily available. However, we have a limited objective and will restrict our attention to applications where either the computation is not too onerous or where we may symbolize but not compute a final numerical result.

Example 2: A student guesses on each question of a 10-question multiple choice test where there are four choices for each question. What is the probability he will obtain (a) exactly seven correct and (b) a passing grade, if seven or more is regarded as passing.

Solution: For each question $\Pr(S) = 1/4$, $P(F) = 3/4$.
(a) The probability of exactly seven correct is

$$b(7;10,1/4) = \binom{10}{7,3}(1/4)^7(3/4)^3 = 0.0031.$$

(b) 7 or more represents the possibility for 7 or 8 or 9 or 10 correct answers. The probability is the sum of

$$b(7;10,1/4) + b(8;10,1/4) + b(9;10,1/4) + b(10;10,1/4)$$

or

$$\binom{10}{7,3}(1/4)^7(3/4)^3 + \binom{10}{8,2}(1/4)^8(3/4)^2$$
$$+ \binom{10}{9,1}(1/4)^9(3/4)^1 + \binom{10}{10,0}(1/4)^{10}(3/4)^0.$$

We will leave the result in the symbolized form. (From tables the result is approximately .0035.)

Example 3: Suppose that 90 percent of the launchings of satellites that are fired into orbit occur without any mishap. What is the probability that between two and four inclusive of the next five launchings will occur without a mishap?

Solution: We must determine the probability of 2, 3, or 4 successive launchings We can sum the exact probabilities for each:

$$b(2;5,.9) + b(3;5,.9) + b(4;5,.9)$$

or

$$\binom{5}{2,3}(.9)^2(.1)^3 + \binom{5}{3,2}(.9)^3(.1)^2 + \binom{5}{4,1}(.9)^4(.1)^1.$$

Again we will leave the result in the symbolized form. (From tables, the result is given as .409.)

Example 4: A binomial experiment is performed four times. It is observed that the probability for three successes and one failure is exactly twice the probability of two successes and two failures. Find the probability of a success, p, and the probability of a failure, q.

Solution: $\binom{4}{3,1}p^3q$ represents 3 S's and 1 F and $\binom{4}{2,2}p^2q^2$ represents 2 S's and 2 F's. From the given information,

$$\binom{4}{3,1}p^3q = 2 \cdot \binom{4}{2,2}p^2q^2$$

or

$$\frac{4!}{3!1!}p^3q = 2 \cdot \frac{4!}{2!2!}p^2q^2$$

which is equivalent to $4p^3q = 12\,p^2q^2$. Dividing both sides by $4p^2q$, we obtain $p = 3q$. Combining this result with the fact that $p + q = 1$ and replacing p with $3q$ we have

$$3q + q = 1$$

$$4q = 1 \quad \text{and} \quad q = 1/4.$$

Since $p + q = 1$, p must equal 3/4.

Problem Set 4.6

A. 1. Determine whether each of the following experiments is binomial, and if not, why not?
   (a) Tossing a dart until one gets a bullseye.
   (b) Shooting a rifle ten times at a target with a variable crosswind.
   (c) Flipping a coin, followed by rolling a die, each done ten times, where a favorable result is that of getting a head on the coin and an even number of dots on the die.
   (d) Performing five trials of flipping a fair coin and if a head appears we roll a die, otherwise the coin is flipped.

2. Find the probability that in four rolls of a die we obtain a prime number on
   (a) exactly two rolls. (b) exactly three rolls. (c) at least two rolls.

3. A basketball player's foul snooting average is 80 percent. What is the probability that he makes exactly three foul shots in his next five shots?

4. Of a dozen eggs, three have been hard-boiled and then placed back in the container with the other fresh eggs. If an egg is randomly selected and this is done four times with the egg being replaced after each draw, find the probability that
   (a) one egg is hard boiled. (b) two eggs are hard boiled.
   (c) three eggs are hard boiled. (d) four eggs are hard boiled.
   (e) at least one egg is hard boiled.

5. Assuming that, on the average, eight out of ten airplanes arrive on schedule, determine the probability that out of six planes chosen at random
   (a) all arrive on schedule. (b) three arrive on schedule.
   (c) at least four arrive on schedule. (d) at most four arrive on schedule.
   (e) none arrive on schedule.

B. 1. Evaluate
   (a) $b(2;3,1/2)$. (b) $b(3;4,1/4)$. (c) $b(1;4,3/4)$.

2. Show that $b(k;n,p) = b(n - k;n, 1 - p)$.

3. A binomial experiment is performed five times. It is noted that the probability

for four successes is three times the probability of three successes. Find
(a) the probability of a success. (b) the probability of a failure.

4.7 Finite Stochastic Processes

The binomial probability experiments discussed in the previous section were characterized by a finite number of trials where only independent events having two outcomes with fixed probabilities were considered. In this section we wish to analyze experiments where we remove the restrictions that the events must be independent and that we must limit ourselves to two outcomes on each trial. If a sequence of events is not independent, then of course, the probabilities for each outcome need not be fixed. Experiments of this type are referred to as **stochastic** processes. The word stochastic is derived from the Greek word "stochos" which may be translated as "guess" or "chance" and is employed in this sense to describe the chance element involved in determining probabilities. At each trial the probability of a specific outcome may take on different values, based on what events took place before the trial under consideration.

Fortunately we do not have to develop any new probability theory, but simply apply our basic probability concepts, particularly the conditional probability formula and the multiplication theorem as required in the context of a given problem. Before we look at some specific examples, let us illustrate by a tree diagram some typical characteristics of a stochastic process by considering an experiment with three possible outcomes on the the first trial, two outcomes on the second trial, and one outcome on the third trial.

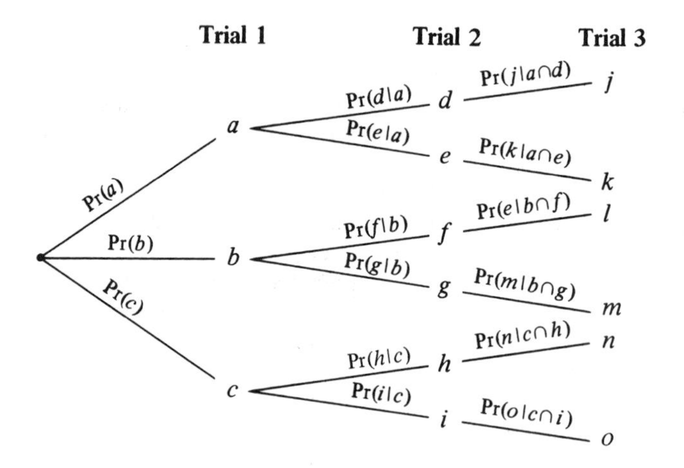

At the first trial, since either a or b or c take place, $\Pr(a) + \Pr(b) + \Pr(c) = 1$ and at each successive trial given a particular outcome preceded that trial, the sum of the probabilities assigned to the branches must be one. For example, if "a" took place on the first trial, then $\Pr(d|a) + \Pr(e|a) = 1$. If we wish to determine the probability of a specific sequence, such as a on the first trial, e on the second trial, and k on the third trial, we use the multiplication theorem and multiply together the respective conditional probabilities, so that $\Pr(a \cap e \cap k) = \Pr(a) \cdot \Pr(e|a) \cdot \Pr(k|a \cap e)$. Each of the three step sequences may be regarded as a sample point in a sample space and assigned a probability weight. The probability of an event may be found, as usual, by adding the weights assigned to the sample points corresponding to the event. Let us now consider some specific examples of a stochastic process.

Example 1: In a certain community 50 percent of the voting residents are registered Republican, 30 percent are registered Democrat, and the remaining 20 percent are registered Independent. A bond issue is under consideration and it is known that 60 percent of the Republicans, 40 percent of the Democrats, and 40 percent of the Independents favor the issue. What is the probability that a voting resident picked at random will favor the bond issue?

Solution: Let us sketch a tree diagram and list on the branches the respective probabilities.

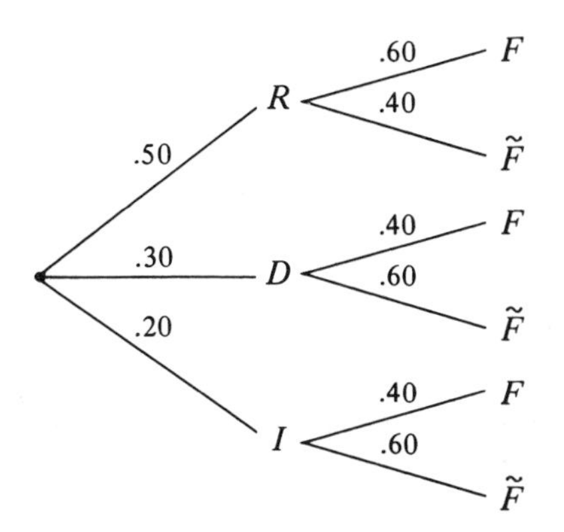

The event, E, a person favors the bond issue, corresponds to $\{(R,F),(D,F),(I,F)\}$:

$$\begin{aligned}\Pr(E) &= \Pr(R \cap F) + \Pr(D \cap F) \mathrm{E} \Pr(I \cap F)\\ &= \Pr(R) \cdot P(F|R) + \Pr(D) \cdot \Pr(F|D) + \Pr(I) \cdot \Pr(F|I)\\ &= (.50)(.60) + (.30)(.40) + (.20)(.40)\\ &= .50.\end{aligned}$$

Example 2: A coin weighted so that $\Pr(H) = 2/3$ is tossed twice. If two heads appear, a die with three faces painted red and three faces painted green is rolled. If other than two heads appear a die with two faces painted red, two faces painted green, and two painted blue is rolled.

(a) What is the probability a red face will appear?

(b) Given that a red face appeared, what is the probability two heads in a row occurred?

Solution: Consider a tree diagram for the stochastic process.

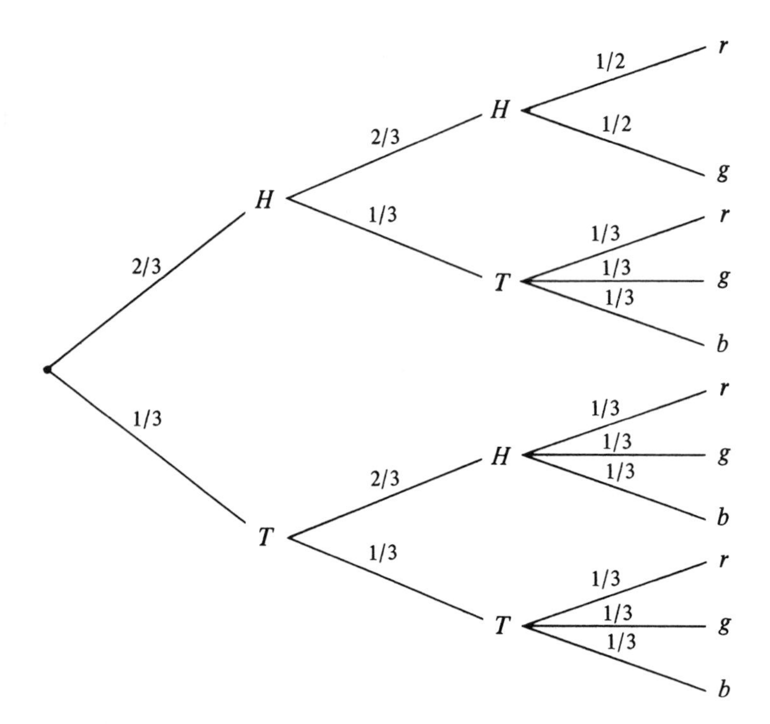

(a) The event, E, that a red face appears, corresponds to $\{(H,H,r),(H,T,r),(T,H,r),(T,T,r)\}$. The probability of E is

$$\begin{aligned}\Pr(E) &= 2/3 \cdot 2/3 \cdot 1/2 + 2/3 \cdot 1/3 \cdot 1/3 + 1/3 \cdot 2/3 \cdot 1/3 + 1/3 \cdot 1/3 \cdot 1/3 \\ &= 4/18 + 2/27 + 2/27 + 1/27 \\ &= 22/54 = 11/27.\end{aligned}$$

(b) For the given condition we wish to determine $\Pr(HH|r)$. The probability of $\Pr(HH|r)$ is given by

$$\Pr(HH|r) = \frac{\Pr(HH \cap r)}{\Pr(r)}$$

and using the result in (a)

$$\Pr(HH|r) = \frac{4/18}{11/27} = \frac{12/54}{22/54} = 12/22 = 6/11.$$

Example 3: A baseball manager prior to an important three game series rates his chances based on his pitching choices as 70 percent for game 1, 50 percent for game 2, and 40 percent for game 3. Using the manager's probability estimates, what is the probability his team wins the series (2 or 3 wins)?

Solution: Using W for win and L for loss, we sketch a tree diagram for the stochastic process:

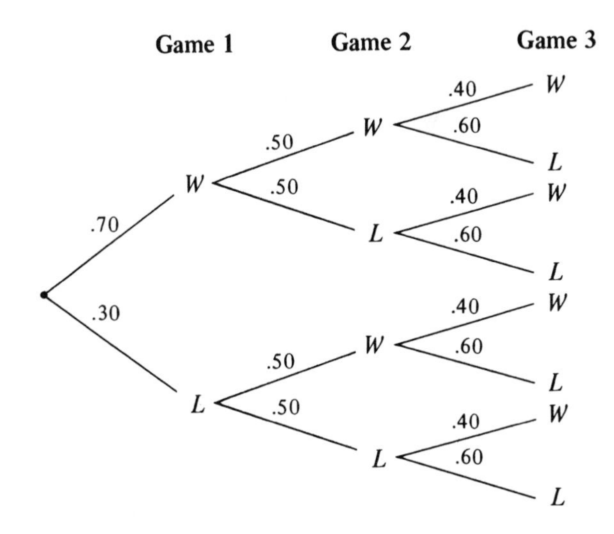

The event, E, corresponding to two or three wins, is $\{(W,W,W),(W,W,L),(W,L,W),$ $(L,W,W)\}$. The probability of E is

$$\begin{aligned}\Pr(E) &= (.70)(.50)(.40) + (.70)(.50)(.60) + (.70)(.50)(.40) + (.30)(.50)(.40)\\ &= .140 + .210 + .140 + .060\\ &= .550.\end{aligned}$$

Example 4: A weighted coin, $\Pr(H) = 1/4$, is tossed repeatedly until a tail first appears or until a head appears three times in succession. Find the probability of obtaining a tail.

Solution: Using H for head and T for tail, we sketch a tree diagram for the stochastic process:

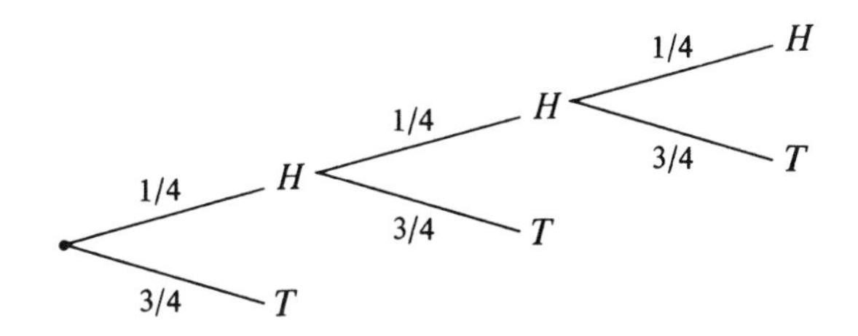

The event, E, corresponding to obtaining a tail is $\{(T),(H,T),(H,H,T)\}$:

$$\begin{aligned}\Pr(E) &= 3/4 + 1/4 \cdot 3/4 + 1/4 \cdot 1/4 \cdot 3/4\\ &= 3/4 + 3/16 + 3/64\\ &= 63/64.\end{aligned}$$

Alternately, $\Pr(E)$, could be found by determining the $\Pr(\tilde{E})$, i.e., $\Pr(H,H,H) = 1/4 \cdot 1/4 \cdot 1/4 = 1/64$ therefore $\Pr(E) = 1 - 1/4 = 63/64$.

Problem Set 4.7

A. 1. A fair penny and a weighted dime, $\Pr(H) = 2/7$, are randomly chosen from a container. If the fair penny is chosen then a die is rolled, and if the weighted dime is chosen then the dime is flipped.
   (a) What is the probability of a head on the dime in the experiment?
   (b) What is the probability a prime number is obtained on the roll of the die?
   (c) What is the probability that the weighted dime is selected?
2. A cafeteria has five pieces of lemon pie, four pieces of apple pie, and seven pieces of blueberry pie. Three people one after the other make a random selection of pie. Draw a tree diagram and list the probability of selecting a pie on each branch.
3. A food store has three bags of oranges. In one bag there are ten oranges with three that are bad, in the second bag there are fifteen with six bad ones, and in the third bag there are twelve with three bad oranges. If a bag is selected at random and an orange drawn, what is the probability of selecting a bad orange?
4. Three printing machines (P_1, P_2, and P_3) produce respectively 45 percent, 30 percent and 25 percent of the printed material for a publishing concern. The percentages of defective output are 2 percent, 3 percent, and 5 percent for the respective machines. If a piece of printed material is selected at random, find the probability
   (a) that the printed piece of material is defective.
   (b) that printing machine P_2 produced the material given that the material was defective.
5. A college faculty consists of 60 percent liberals, 30 percent conservatives, and 10 percent malcontents. Assume each individual is in one grouping only. A policy on academic freedom, proposed by the administration, is under consideration and and it is known that 30 percent of the liberals favor the proposal, 75 percent of the conservatives favor the proposal, and 20 percent of the malcontents favor the proposal. What is the probability that a faculty member chosen at random favors the proposal?
6. An urn has three red balls, four green, and two yellow. A ball is randomly chosen from the urn. If a red ball is drawn, another ball is drawn from the urn without replacing the first ball. If a green ball is obtained on the first draw, then a die is rolled. If a yellow ball is drawn on the first draw, the game ends. Draw a tree diagram for this process. What is the probability of obtaining a yellow ball in the experiment?

4.8 Expected Value

In many probability experiments that are performed repetitively we are interested in the expected number of each possible outcome. For example, if we were to toss a fair coin 100 times we intuitively expect to obtain on the average 50 heads and 50 tails. By "on the average" we mean that if the experiment of tossing a coin one hundred times were repeated over and over again, we would average 50 heads and 50 tails for each experiment. A single experiment might quite reasonably produce 55 heads and 45 tails or any one of many other outcomes, but the long run average should produce the same number of each outcome. Similarly, if we rolled a fair die 180 times, we would expect on the average 30 of each of the six faces to appear.

Often a statistician will attempt to fit a collection of empirical data with a probability model and will accept or reject the model based on whether or not there is close agreement between the expected number of each outcome predicted by the model and the observed number of each outcome actually obtained.

Another important application of the concept of expected value occurs in the field of game theory where we may wish to determine whether or not a game is fair and if betting is involved, what is the expected earning or loss per game on the average.

In this section we will present some of the basic mathematical concepts of expected value and illustrate how we may apply these concepts. Before we make any precise definitions, let us look at an example from an intuitive point of view.

Example 1: A box contains nine slips of paper; two bearing the number 2; three bearing the number 3; and four slips bearing the number 4. A slip of paper is drawn at random, the number recorded and the slip of paper is returned to the box. The experiment is repeated again and again. What is the expected value or numerical average of all the numbers recorded.

Solution: Since we are not given a precise number of trials, nor the outcome on each trial, let us consider an analysis based on probability. There are a total of nine slips of paper in the box and by f/n, $\Pr(2) = 2/9$, $\Pr(3) = 3/9$, and $\Pr(4) = 4/9$. In other words, 2/9ths of the time we expect to record a "2", 3/9ths of the time we expect to record a "3," and 4/9th's of the time we expect to record a "4". The expected value or numerical average can be found by either finding the average of the nine slips of paper, i.e.,

$$\frac{2+2+3+3+3+4+4+4+4}{9} = \frac{29}{9}$$

or by forming a weighted average using the probability of each outcome as follows:

$$\begin{aligned}\text{Expected value} &= (2)(2/9) + 3(3/9) + 4(4/9)\\ &= 4/9 + 9/9 + 16/9\\ &= \frac{29}{9}.\end{aligned}$$

Observe that the expected value is not necessarily an outcome of the experiment.

If in Example 1 we had literally thousands of slips of paper, but only one of three numbers on each slip of paper, we might be reluctant to find the average of the numbers by finding the sum and then dividing by the total number of slips. However, the weighted average approach using the probability of each outcome is easy to apply since we are only concerned with the probability of each outcome. We will now give a precise definition of the expected value or sometimes referred to as the mathematical expectation of a probability experiment.

DEFINITION: If the outcomes of a probability experiment are denoted by a variable x that may randomly assume the values $a_1, a_2, \ldots, a_k$ where the probabilities of each corresponding outcome are $p_1, p_2, \ldots, p_k$ then the **expected value**, symbolized $E(x)$, is given by

$$E(x) = a_1 p_1 + a_2 p_2 + \cdots + a_k p_k.$$

Example 2: An experiment consists of tossing four coins and recording the number of heads that appear. If the experiment is repeated many times, what is the $E(x)$ of the number of heads?

Solution: The outcomes of the experiment are 0, 1, 2, 3, or 4 heads. The probability of each outcome can be determined by the binomial probability model where we regard a head as a success. For example, the probability of two heads is $b(2;4,1/2) = \binom{4}{2,2}(1/2)^2\ (1/2) = 6/16$. If we organize the information in a tabular form we have

x = No. of heads	0	1	2	3	4
Probability of x Pr(x)	$\frac{1}{16}$	$\frac{4}{16}$	$\frac{6}{16}$	$\frac{4}{16}$	$\frac{1}{16}$

Then $E(x)$ is given by

$$\begin{aligned} E(x) &= (0)(1/16) + (1)(4/16) + (2)(6/16) + (3)(4/16) + 4(1/16) \\ &= 2. \end{aligned}$$

DEFINITION: The expected value, $E(x)$, of a binomial probability experiment is given by

$$E(x) = np$$

where n represents the number of trials and p represents the probability of a "success" on each trial.

Example 3: Twelve light bulbs are in a box and it is known that three are defective. A random sample of two bulbs are selected and tested. The bulbs are replaced and another random sample of two selected. The experiment is repeated many times. What is the $E(x)$ of the number of defectives?

Solution: Of the 12 bulbs, three are defective, therefore $\Pr(D) = 3/12 = 1/4$. The selection of two bulbs at random with replacement is a binomial experiment. Then $E(x)$ is given by

$$\begin{aligned} E(x) &= np \\ &= (2)(1/4) = 1/2. \end{aligned}$$

A word of caution is in order regarding the result. On a given trial we would have either 0, 1, or 2 defectives and not 1/2 defective. However, over a long run, the average of the number of defectives is 1/2.

When we apply expected value to a game of chance where the expected outcome is the money that is being earned or lost, we are interested in whether the game is fair, favors the player, or is unfavorable to the player. For example, if you and another player each wager $1 on the toss of a coin and you pick heads and he picks tails, the outcomes from your point of view are either +$1 or −$1, each with the probability of 1/2. Then $E(x) = (+1)(1/2) + (-1)(1/2) = 0$. *Whenever the expected value of game is zero then the game is regarded as fair.* Should the expected value be a positive number, the game is said to favor the player and if the expected value is negative, the game is said to be unfavorable to or against the player.

Example 4: A man is paid $2 if he draws a jack or queen from a deck of 52 cards, and is paid $5 if he draws an ace from the deck. If any other card appears he loses $1. What is the expected value of this game from the player's point of view? Is the game fair?

***Solution*:** Representing the experiment in tabular form, we have the following outcomes with their respective probabilities

x = $ earned or lost	2	5	−1
Probability of x $\Pr(x)$	$\frac{8}{52}$	$\frac{4}{52}$	$\frac{40}{52}$

$$E(x) = (2)(8/52) + (5)(4/52) + (-1)(40/52)$$

$$= -\frac{4}{52} = -\frac{1}{13} \doteq -7.7 \text{ cents.}$$

Since the $E(x)$ is a negative number, the game is not fair and is unfavorable to the player.

Example 5: An insurance company mortality tables indicate that the probability that a 40-year-old person will die in the next year is .005. What premium should the company charge for a $5,000 life insurance policy if they wish to make an average profit on each policy holder of $10?

***Solution*:** If we let "p" represent the premium rate then the outcomes are either "p" indicating the person did not die or "$-\$5000 + p$" representing their loss if the person did die. Note that the company does not lose a total of $5,000, since they retain whatever premium they charged for the year. In tabular form we have

x = $ earned or lost	$-5000 + p$	p
$\Pr(x)$	.005	.995

$$E(x) = (.005)(-5000 + p) + .995p.$$

However, the company wishes to make a \$10. profit therefore $E(x) = \$10$. Thus,

$$\begin{aligned}10 &= (.005)(-\$5000 + p) + .995p \\ &= -25 + .005p + .995p \\ &= -25 + p \\ 35 &= p.\end{aligned}$$

The company should charge a premium of \$35.

With this discussion of expected value, we complete our presentation of the theoretical aspects of probability. The exercises and examples should have provided the reader with an appreciation of the wide range of probability applications. Most modern curriculum revisions in the mathematics area are recognizing the increasing importance of probability and in the future we can anticipate the introduction of some basic probability concepts in the elementary grades and then a spiraling of the topic through the secondary level. Beyond the secondary level it is expected that there will be an increasing emphasis on probability theory and its applied counterpart, statistics, in the areas of business management and the behavioral and social sciences. In the next chapter we will introduce some of the fundamental concepts of statistics so that the reader may gain some additional perspective on the material we have now completed as well as an appreciation of its extension to one of the most dynamic areas of applied mathematics.

Problem Set 4.8

A. 1. In a class of 30 students it was found that 2 received A's, 5 received B's, 15 received C's, 3 received D's, 5 received F's. The grades are weighted as follows: $w(A) = 4$, $w(B) = 3$, $w(C) = 2$, $w(D) = 1$, and $w(F) = 0$. If a student is chosen at random, what is the expected value of the grade he received? (Round off your answer to 0, 1, 2, 3, or 4.)

2. In a group of ten people, seven are men and three are women. If four people from the ten are selected at random for a committee, what is the expected number of men on the committee?

3. On a gambling wheel the numbers, 1, 2, 3, 4, and 5 are sectioned in such a way that $\Pr(1) = 1/10$, $\Pr(2) = 1/10$, $\Pr(3) = 1/2$, $\Pr(4) = 1/10$, and $\Pr(5) = 1/5$. To try his luck on the wheel, a player must pay five dollars and is paid \$20 if a "1" comes up, \$5 if a "2" comes up, \$1 if a "3" comes up, \$3 if a "4" comes up, and \$4 if a "5" comes up. Is this a fair game? If not, suggest a way to make it a fair game.

4. A private college sells 1,000 tickets at \$5. each. The single prize offered is worth \$2,500. Are the tickets overpriced? Why?

5. A game consists of flipping a coin followed by tossing a die. If the player gets a head on the coin and an even number on the die, he gets \$5, otherwise he pays \$3. Is this a fair game?

6. Let the game be that of tossing two fair coins. If two heads occur, the player wins \$6; if one head occurs, the player wins \$4; and if no heads occur, the player loses \$10.

   (a) What is the expected value of the game?

(b) If \$2 is subtracted from each amount given for the occurence of an outcome, what is the expected value of the game?

(c) If \$3 is added to each amount given for the occurence of an outcome, what is the expected value of the game?

(d) If k dollars (k any real number) is algebraically added to the given amount of each outcome, what is the expected value of the game?

(e) If each given amount is multiplied by 4, what is the expected value of the game?

(f) If each given amount is multiplied by (-3), what is the expected value of the game?

(g) If each given amount is multiplied by some nonzero constant k, what is the expected value of the game?

7. Let a pair of dice be tossed. What is the expected sum of the dots appearing on the upturned faces?

8. Let a die be rolled ten times and let x be the number of dots which appear on an upturned face. Find the expected value of x, $E(x)$.

B. 1. Show that if each outcome is multiplied by a nonzero constant k, then the expected value, $E(x)$, is multiplied by the constant k, i.e., $E(kx) = kE(x)$.

2. Prove that if a constant k is added to each outcome of a game, then this constant is also added to the expected value of this game, i.e., $E(k + x) = k + E(x)$.

Review Exercises

4.1 Define the following:

(a) probability of an outcome.
(b) sample space.
(c) event.
(d) mutually exclusive events.
(e) odds in favor of an event.
(f) conditional probability.
(g) independent events.
(h) stochastic process.
(i) expected value.
(j) fair game.

4.2 The probability of a tack landing with the point up, $\Pr(U)$, is $1/10$ of the probability of the tack landing with its point in contact with the landing surface. Find the probability of the tack landing with its point up.

4.3 A letter is selected at random from the word "radar." Find the probability that the letter

(a) is a vowel.
(b) is a consonant.
(c) is not a vowel.
(d) "e" is selected.
(e) "r" or "a" is chosen.

4.4 A bag contains five lemons and nine limes. If a person picks two objects from the bag, find the probability that they are both the same kind of fruit.

4.5 Let E, F be events in S with the $\Pr(E \cap \tilde{F}) = 1/3$ and $\Pr(\tilde{E} \cap \tilde{F}) = 1/2$. Find (a) $\Pr(F)$, (b) $\Pr(E \cup F)$.

4.6 A class consists of seven girls and five boys. Find the probability that a pair chosen at random from this class will be of the same sex.

4.7 Let A and B be events in S. If $\Pr(\tilde{A} \cap \tilde{B}) = 7/12$, $\Pr(A \cap \tilde{B}) = 1/6$, and $\Pr(\tilde{A} \cap B) = 1/6$, are A and B independent events? Find (a) $\Pr(A|B)$, (b) $\Pr(B|A)$, (c) $\Pr(\tilde{A}|\tilde{B})$, (d) $\Pr(\tilde{B}|A)$, and (e) $\Pr(A \cup B)$.

4.8 Assume that 90 percent of the moon launchings occur without any mishap. What is the probability that the first and second launching in the next four launchings both occur without a mishap?

4.9 A letter is chosen at random from the word "chance." If a vowel is chosen, a die is rolled; otherwise a card is drawn from a standard bridge deck. Find the probability

(a) of getting an even number on the die.
(b) of drawing an ace of spades given that a vowel was chosen.
(c) of selecting a vowel.
(d) of drawing a club given a consonant was chosen.

4.10 A card is drawn from a standard bridge deck. The card is replaced after each draw. How many times can we expect to draw a face card from the deck in 20 draws?

Suggested Reading

Crouch, Ralph, *Finite Mathematics with Statistics for Business.* McGraw-Hill, New York, 1968.

Feller, W., *An Introduction to Probability Theory and Its Applications.* Wiley, New York, 1957.

Goldberg, Samuel, *Probability: An Introduction.* Prentice-Hall, Englewood Cliffs, N.J., 1960.

Hoyt, John P., *A Brief Introduction to Probability Theory.* International Textbook, Scranton, Pa., 1967.

Lipschutz, Seymour, *Probability.* Schaum Outline Series, McGraw-Hill, New York, 1968.

Mendenhall, William, *Introduction to Probability and Statistics.* Wadsworth, California, 1967.

Mosteller, Frederick, Robert Rourke, and George Thomas, *Probability: A First Course.* Addison-Wesley, Mass., 1961.

Munroe, M. E., *Theory of Probability.* McGraw-Hill, New York, 1951.

Statistical thinking will one day be as necessary for efficient citizenship as the ability to read and write.

–H. G. WELLS

CHAPTER 5

Statistics

5.1 Statistics

Statistics is that field of knowledge which is concerned with the collecting, organizing, and interpreting of numerical data. The name "statistics" was applied historically to the collection of data important to the state, such as population, income, trade, etc. Although the recording of such data has always been important, statistics now are used for interpretive purposes. This use of data for statistical inference is based almost entirely on the theory of probability.

Basic statistical methodology is the same no matter what the field of application may be. Statistical procedures now not only form an important part of the physical sciences but also play a significant role in such areas as psychology, sociology, and economics.

We are living in a time of unprecedented knowledge expansion, population growth, and changing economic conditions. The raw data produced by modern society seem at times to be almost overwhelming. Decisions are extremely critical and actions frequently cannot be postponed. If we must make decisions in a time of uncertainty, then it is wise to collect what information we can and choose the course of action that seems most probable to produce the desired results. This is the aim of statistical analysis.

Recognition of the increasingly important role that probability and statistics have assumed in our society has prompted leading educators to recommend an early introduction of these topics into our educational system. Some experimental text material is now available that introduces some of the basic concepts of probability and statistics at the fifth-grade level. Undoubtedly, the 1970's and 1980's will see an increasing emphasis being placed on these topics by advocates of educational reform.

In this chapter we will discuss methods by which data can be organized, measurements that may be made to characterize the central tendency or variability of a set of data, and how we may apply probability theory to predict or infer conclusions from a set of data.

5.2 Organizing Data

When we begin to observe and record data for the purposes of a statistical analysis, it becomes necessary to relate the observations to some numerical scale. However, in making numerical measurements we are not required to refer our findings to any absolute standards. It is sufficient to obtain a scale which permits us to compare and rank the observations in some order. For example, there exist many intelligence tests that enable us to rank the individuals taking the test by their test scores and use the results as a measure of the relative intelligence of one individual in comparison to others taking the same test. No claim is made that the test scores measure intelligence in other than a relative sense.

The observations which we make in collecting data are of two general types, **continuous** or **discrete.** By **continuous** we mean that the results obtained may take on, within the limits of our measuring device, all possible values in a given interval. For example, the daily recorded temperature in a given area may take on a value such as 65.7°F or 73.21°F as well as a whole range of different temperatures. A **discrete variable,** however, may assume only certain specified values. For example, the number of runs scored by a baseball team in a given game could be 0, 1, 2, 3, . . . , but no team could score 2.75 runs.

After observed data is collected more than likely we will wish to organize the data in some way that will more concisely describe the results obtained. For example, we may develop a graph or visual picture of the information, calculate an average and measure the spread of the data. Reporting of this type of information falls under the heading of **descriptive statistics.**

If the results obtained are used to test a particular hypothesis or infer some conclusion based on the results obtained, then we are using the statistical data for the purpose of **statistical inference.**

In subsequent sections we will discuss in more detail the descriptive measures of central tendency and variability and the basic concepts of statistical inference. In this section we wish to consider a fundamental question, how do we organize or display the data which we obtain? It will not be possible for us to examine all of the possible methods that are used, but a review of some typical approaches will be sufficient for our purposes.

Assume that the data in Table 5.1 represents the grades obtained on the final examination of a history course.

Table 5.1 Grades of 40 Students on a History Final Examination

96	86	72	75	73	75	78	83
61	82	77	89	62	98	80	71
77	71	92	74	85	76	72	75
91	65	70	54	81	73	87	82
79	70	88	77	74	82	67	99

The raw data collected must now be organized. One convenient method is to prepare a **frequency distribution** which will separate the data into classes and will indicate the number in each class. For example, a typical grading distribution pattern is to use a class interval of 10 starting with all grades 90 to 99 inclusive. Organizing the results in Table 5.1 by a frequency distribution of this type will give us Table 5.2.

Table 5.2 Frequency Distribution of Grades

Class Interval	Frequency
90–99	5
80–89	11
70–79	19
60–69	4
50–59	1
	Total...... 40

Table 5.2 might be useful for assigning letter grades or for making some general observations. For example, more individuals obtained scores of 70–79 than any other group. If 70 is regarded as the minimum passing grade, 35 out of the 40 passed the final examination.

In addition to a tabular form of presenting the data, we may wish to display the data graphically. One common method for constructing a graph is to use a uniform class interval as a base for a rectangle and the frequency as the height of a rectangle. The figure formed is called a **histogram**. The histogram for the frequency distribution in Table 5.2 is shown in Figure 5.1.

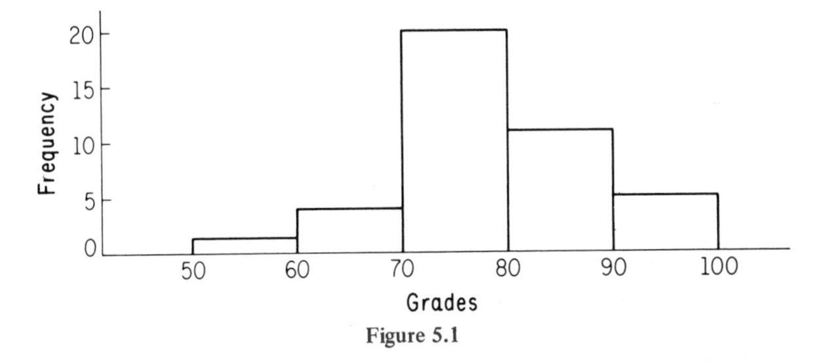

Figure 5.1

Another common method for graphically displaying the data is by means of a **frequency polygon**. In this method, instead of using rectangles, the midpoint of each class interval is associated with the frequency and straight lines are used to connect the points. For example, the midpoint of the class interval 90–99 is 94.5 and this is associated with the frequency 5. Figure 5.2 illustrates the frequency polygon for the data in Table 5.2.

The histogram and frequency polygon are particularly useful in providing us with a visualization of the distribution of the data. If the final examination which we have been considering was given to a large number of students and the data handled similarly,

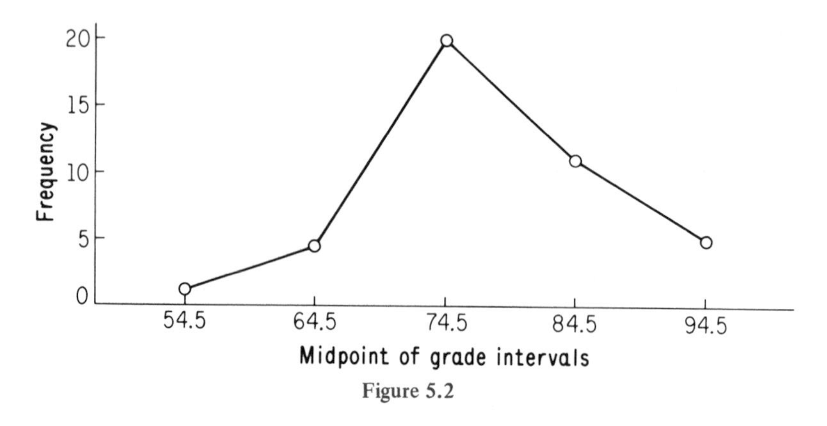

Figure 5.2

we could make smaller and smaller class intervals so that the histogram and frequency polygon would take on the same general shape. The shape of the distribution obtained could then be analyzed to see if it conformed closely to any known probability distribution. If there is a general agreement between the data and a probability model, we might wish to use the mathematics of probability theory to further analyze the test scores, state the probability of different test results, and predict with specified probabilities the performance level of other groups taking this final examination after completing a similar course.

The most common probability models which are used in statistical analysis depend on two variables: the **arithmetic mean** and the **variability** or dispersion of the data. Therefore, before we can apply probability model theory we must discuss (and we will do so in the following sections) descriptive measures of **central tendency** and variability.

Problem Set 5.2

A. 1. Identify which of the following collections of data would be regarded as discrete observations and which would be regarded as continuous observations:

(a) The number of leaves on apple trees.
(b) The weights of male athletes.
(c) The daily relative humidity in the summer months.
(d) The number of sixes obtained when a die is rolled ten times.
(e) The number of bacteria left on a glass plate following the use of an antiseptic.

B. 1. The weights of 30 people are recorded below.

114	124	135	101	105	149
107	115	97	152	114	132
123	133	160	125	137	126
157	147	135	116	112	119
136	102	151	119	102	151

(a) Construct a frequency distribution table of the above weights with weight interval of 5 beginning with the interval 95–99 inclusive.

(b) Draw a histogram for the weights given with a class interval of 5 using the distribution table from part (a).
(c) Superimpose a frequency polygon on the histogram prepared in part (b) of this problem.

2. The lengths of some book titles (in words) chosen at random from a library book list are:

7	4	4	3	1	2	3	3	5	5	5	4	5	1	2	3
3	4	4	3	2	7	1	5	4	3	2	4	1	3	4	4

Make a frequency distribution, histogram and frequency polygon for (a) an interval of 2, (b) an interval of 1. Superimpose the frequency polygon on the histogram.

3. A person's temperature was measured in four-hour intervals for three days. The readings recorded were as follows:

First day	97.0	97.5	97.8	98.0	98.4	99.0
Second day	97.3	97.7	98.3	98.6	98.8	99.2
Third day	97.9	98.2	98.6	98.7	98.8	98.6

(a) Make a frequency distribution of the temperatures using an interval of 0.5.
(b) Construct a histogram for the frequency distribution of the temperatures.
(c) Draw a frequency polygon for the set of given temperatures.

4. A housewife recorded over the period of two months the amount of money she spent on groceries each time she shopped at a particular store. The amounts she spent on each trip were:

$ 2.73	$ 3.37	$ 2.39	$ 4.07	$2.51
10.31	11.92	13.27	6.59	3.90
5.46	1.79	3.61	7.13	4.39
8.02	4.87	1.81	14.63	7.21

Make a frequency distribution, histogram and frequency polygon for (a) an interval of 3, (b) an interval of 1.

5. A set of books is weighed one at a time. The weights are 6 oz, 10 oz, 1 lb, 1-1/2 lb, 1-1/4 lb 8 oz, 14 oz, 2 lb, 7 oz, 1-1/4 lb, 1-3/4 lb, 6 oz, 10 oz, 1 lb,

Figure 5.3

11 oz, 1-1/4 lb, 2-1/4 lb, 7 oz, 5 oz, 3 oz. Make a frequency distribution, histogram, and frequency polygon for (a) an interval of 1 oz, (b) an interval of 2 oz, and (c) an interval of 4 oz.

6. A frequency polygon representing the weights of a group of ninth-grade students is shown in Figure 5.3. Reconstruct a frequency distribution from the given information.

C. 1. A fair coin is tossed four times. Draw a histogram where the horizontal axis corresponds to the number of heads that could appear and the vertical axis corresponding to the probability of each outcome. (*Hint:* The probability of each outcome is given by the binomial probability model $b(k;n,p)$.)

2. Table 5.3 lists the number of days of vacation taken by a group of employees of a given company. Also tabulated is the cumulative frequency which is the total number below the upper limit of that interval. Draw a frequency polybon and on the same axes draw a cumulative frequency polygon. (Have cumulative frequency on the vertical axis and class interval on the horizontal axis.)

Table 5.3

Number of Vacation Days–Class Interval	Frequency	Cumulative Frequency
0–2	8	8
3–5	2	10
6–8	12	22
9–11	15	37
12–14	10	47
15–17	5	52
18–20	4	56

5.3 Measures of Central Tendency

Although the frequency distribution of a group of data and a histogram or frequency polygon are useful, they do not provide us with numerical measurements which will independently characterize the distribution. In this section we will present certain measurements that are used to describe the central tendency of a set of data.

One of the more common measures of the center of a group of data is the arithmetic mean defined as follows:

DEFINITION: The **arithmetic mean** of a set of data, symbolized $\overline{x}$, is the sum of all the numerical observations divided by the total number of observations. If we represent the observations by $x_1, x_2, x_3, \ldots, x_n$ then

$$\overline{x} = \frac{x_1 + x_2 + x_3 + \cdots + x_n}{n}.$$

Example 1: If the ages of five people picked at random in a class are 18, 24, 23, 17, and 21, what is the arithmetic mean age of the five individuals chosen?

Solution:

$$\bar{x} = \frac{18 + 24 + 23 + 17 + 21}{5} = 20.6 \text{ years.}$$

The arithmetic mean need not be one of the entries in a list of data, yet a change in any single entry will affect the arithmetic mean. Although there are other measurements of central tendency referred to as "mean," such as the geometric mean, we will in context, hereafter, refer to the arithmetic mean simply as the mean. The ease with which we can calculate the mean and the sensitivity of this measure to changes in the data make the mean the most widely used measure of central tendency.

If the data collected consist of many repeated entries, the computation of the mean can be shortened by grouping the data in a frequency table. Then instead of adding up the entire list we can form the product of each entry and the frequency with which it occurs. The sum of these products divided by the total number of entries will give us the mean. The computation procedure for this method is illustrated in the following example.

Example 2: Prizes in a lottery are to be awarded by drawing numbered tickets out of a large box. The box contains 300 tickets imprinted with the number 1, 200 tickets imprinted with the number 2, 100 imprinted with the number 3, 50 imprinted with the number 4, and 25 imprinted with the number 5. What is the arithmetic mean of all of the numbers on the tickets in the box?

Solution: The data can be organized into a frequency table including a third column representing the product of the frequency and entry number.

Entry	Frequency	Product
1	300	300
2	200	400
3	100	300
4	50	200
5	25	125
Total	675	1325

The arithmetic mean, $\bar{x}$, is given by

$$\bar{x} = \frac{1325}{675} \doteq 1.96.$$

In general, if we let $f_1, f_2, f_3, \ldots, f_n$ represent the corresponding frequencies then $\bar{x}$ can be obtained as follows:

$$\bar{x} = \frac{x_1 f_1 + x_2 f_2 + x_3 f_3 + \cdots + x_n f_n}{N}$$

where $N = f_1 + f_2 + f_3 + \cdots + f_n$.

If a frequency table should be organized on the basis of class intervals, then a close approximation of the mean can be found by using the midpoint of each interval to represent all the entries in the class interval. The products of each midpoint times the number in each interval are summed and divided by the total number to obtain the mean.

Although the mean is the most commonly used measure of central tendency, it is adversely affected by the extreme values. For example, if five individuals chosen at random in a certain town have yearly incomes of \$5,000, \$6,000, \$8,000, \$9,000, and \$42,000 it could be very misleading to report that the average income of the residents in this town is \$14,000 per year. When the data collected contains extreme values it is desirable to not only calculate the mean but also to determine the median.

DEFINITION: The **median**, symbolized $\tilde{x}$, of a set of data arranged in order of magnitude is the middle entry or if there is no single middle entry the arithmetic mean of the two middle entries.

The median of the five yearly incomes \$5,000, \$6,000, \$8,000, \$9,000, and \$42,000 is \$8,000 per year which more accurately reflects the center of this set of data than the arithmetic mean.

Example 3: Find the median of the set of data 12, 7, 5, 10, 18, 13, 6, 21, 15, 7, 11, 10, 11, 8.

Solution: Arranging the data in order of magnitude, we obtain 5, 6, 7, 7, 8, 10, 10, 11, 11, 12, 13, 15, 18, 21. Since there is not a unique middle entry of the fourteen observations we find the mean of the two middle entries.

$$\tilde{x} = \frac{10 + 11}{2} = 10.5.$$

When a set of data is not homogeneous the median is a useful statistic. However, the median is not as sensitive to changes in the data as is the arithmetic mean. For example, in the list of data presented for Example 3 changing the largest entry 21, to 25 or 100 or any other larger number would leave the median unchanged. If the data collected is a set of sample information drawn from a larger group which is referred to as the **population**, where the intent is to use the sample data to characterize the population, then the sample arithmetic mean is likely to be a better estimate of the population mean than is the sample median.

The median divides the group into two equal parts and is also referred to as the 50th percentile. The divisions of a group of data arranged in order of magnitude into four equal parts will give us **quartile** scores, into five equal parts will give us **quintile** scores, and into ten equal parts will give us **decile scores** and into one hundred parts will give us **percentile** scores. Subdivisions of this type are common on standardized educational tests and in reporting the class ranking of high school or college graduates.

The third and last measure of central tendency which we will discuss is the mode.

DEFINITION: **A mode** of a set of data is that entry which occurs with the greatest frequency. The mode is not necessarily a unique value.

If each entry in a group of data should occur only once, as in Example 1 in this section, then in a technical sense the mode is each entry and the information provided by the mode is not very useful. However, as the amount of data increases and certain entries appear to occur with more frequency than others, then the mode can provide us with significant information. However, changes in the less frequent entries can be made without affecting the mode and thus, as a measurement of central tendency it is not as sensitive to data changes as the mean. The sample mode is also not as reliable as measure of the central tendency of a population as is the sample mean.

Example 4: The grading distribution of 20 students on a given mathematics test arranged in order of magnitude is as follows:

$$53, 62, 65, 70, 71, 71, 71, 74, 74, 76, 76, 76, 78, 80, 82, 82, 85, 87, 92, 94.$$

Find the (a) mean, (b) median, and (c) mode.

Solution:

(a) The mean can be found by adding all the grades together and dividing by 20, i.e.,

$$\bar{x} = \frac{53 + 62 + 65 + \cdots + 92 + 94}{20} = 75.95 \ .$$

(b) The median will be the average of the two middle entries i.e.,

$$\tilde{x} = \frac{76 + 76}{2} = 76 \ .$$

(c) The distribution is bimodal, since 71 and 76 each occur three times and are the most frequent entries.

Each of the measures of central tendency have their own relative merits. The arithmetic mean is the most commonly used, since it is the most sensitive measurement of central tendency and in a sampling from a population provides us with the best estimator for the center of a population. Whenever practicable, however, it is desirable to have available the mean, median, and mode of a set of data so that appropriate comparisons can be made.

Problem Set 5.3

A.

1. The heights of ten people was recorded as 60 in., 66 in., 68 in., 63 in., 70 in., 75 in., 78 in., 68 in., 67 in., 65 in. What is the mean height of the ten people?
2. A student received a C for a four-credit mathematics course, a B for a three-credit psychology course, an A for a two-credit health course, and a D for a five-credit science course. Let the grade A be worth 4 quality points, B 3 quality points, C 2 quality points, D 1 quality point, and F 0 quality points. Find the mean when the grades are weighted according to the credits given in each course. Also find the mean of the grades when each grade is weighted the same irrespective of the number of credits.

3. The vowels in a given paragraph were counted and the data recorded in the following table.

Vowel	Frequency
a	23
e	23
i	18
o	8
u	3

Find the mode(s) for the given data.

4. A set of numbers consists of two 2's, four 3's, five 4's, four 5's, three 6's, two 7's, three 8's, four 9's, five 10's, four 11's, and two 12's. Find the (a) mean, (b) mode(s), and (c) median.
5. The yearly incomes of people chosen at random were as follows: ten earned \$10,000, seven earned \$8,000, six earned \$13,000, nine earned \$7,000, three earned \$18,000. Find the (a) mean, (b) mode(s), and (c) median.
6. Find the mean and median of the sets of numbers: (a) 5, 9, 7, 3, 6, 4, (b) 5, 13, 6, 1, 7, 2.
7. Find the mean and median of the following numbers: 4, 9, 3, 10, 5, 2, 6, 8, 4, 5, 7, 6, 1, 11.
8. Let the numbers 5, 9, 7, 3, 6, 4 be given.
   (a) Find the mean and median of the given numbers.
   (b) Add 3 to each of the numbers and find the mean and median.
   (c) Multiply each of the numbers by 4 and find the mean and median.
   (d) How does the mean obtained in parts (b) and (c) compare to the mean of the given numbers in part (a).
9. A box contains four slips of paper numbered 2, 4, 6, and 8. Two slips of paper are drawn without replacement which will be called a sample. List all possible samples and find the mean of each sample. Find the mean of all the sample means. How does the mean of the sample means compare with the mean of the numbers 2, 4, 6, 8?
10. If the mean, mode and median are approximately the same, what can be said about the distribution of the data?

B.

1. Using the data given in Table 5-1, find the exact mean of the 40 entries. Using the frequency distribution given by Table 5-2, find the mean by first forming the product of the midpoint of each interval times the frequency, then divide the sum of the products by the total number, 40. Compare the results obtained by each method.
2. Show that if $X_1, X_2, X_3, \ldots, X_n$ represents a set of data with mean of $\overline{X}$ then a new set of data which is formed by adding a constant K to each of the original entries has a mean of $\overline{X} + K$.
3. Show that if $X_1, X_2, X_3, \ldots, X_n$ represent a set of data with a mean of $\overline{X}$ then a new set a data which is formed by multiplying each of the original entries by a constant K has a mean of $K\,\overline{X}$.
4. Find the mean of a set of students' grades, 76, 82, 81, 84, 73, 72, 88, 78, by adding the constant −80 to each entry and after finding the mean of the resulting set of data add +80 to the mean obtained.

5.4 Measures of Variability

In collecting data it is quite possible for different sets of data to have the same mean yet differ considerably in the variability or spread of the data about the mean. A measure of the relative variability of data can be extremely important. For example, a manufacturer of $^1/_2$-in. diameter bolts may have two automatic machines producing bolts such that the mean diameter of parts produced by each machine is the desired $^1/_2$ in. However, if one machine produces bolts with a maximum variability of $\pm\,^1/_8$ in. and the other with a maximum variability of $\pm\,^1/_{100}$ in. a more uniform quality product will be produced by the machine with the smaller variability.

The simplest measure of variability of a set of data is the **range**.

DEFINITION: The **range** of a set of data is the arithmetic difference between the largest and smallest entry.

Example 1: If the heights of 5 college freshmen picked at random are 5′9″, 6′1″, 5′10″, 6′0″, 6′0″, and 5′8″, determine the range of the data.

Solution: The largest entry is 6′1″ and the smallest is 5′8″. The range is the arithmetic difference between 6′1″ and 5′8″, namely 5″.

In industrial quality-control work the range is often used to determine the acceptable tolerance span for a particular product. Samples are drawn from a production run to insure that they fall within the tolerance span. By this technique 100 percent inspection can be eliminated and the cost of inspection sharply reduced. Although useful, the range considers only the extreme values and is not sensitive to changes in the interior data. For example, the following two sets of data A and B have the same mean (50) and range (80) and yet they have a distinctly different variability.

$$\begin{array}{ll} A & 10, 12, 14, 16, 50, 84, 86, 88, 90 \\ B & 10, 44, 46, 48, 50, 52, 54, 56, 90. \end{array}$$

A more useful measure of variability would consider the relative position of each entry with respect to the mean. For example, we might find the sum of the arithmetic differences between each entry and the mean. In set A the differences between each entry and the mean would be as follows:

$$-40, -38, -36, -34, 0, 34, 36, 38, 40.$$

A quick inspection indicates that the sum of the differences is zero. In fact, in any distribution the arithmetic sum of the differences about the mean is zero. However, if we either changed the sign of the negative numbers of squared each difference before adding, we would get a nonzero sum. In practice, it is more convenient to square the difference between each entry and the arithmetic mean. The sum of these squared deviations about the mean is used to obtain a measure of variability called the **variance**, defined as follows:

DEFINITION: If $x_1, x_2, x_3, \ldots x_n$ represent the entries and $\bar{x}$ the arithmetic mean, then the ***variance**** of a set of data symbolized s^2 is given by

$$s^2 = \frac{(x_1 - \bar{x})^2 + (x_2 - \bar{x})^2 + (x_3 - \bar{x})^2 + \cdots + (x_N - \bar{x})^2}{N - 1}.$$

DEFINITION: The **standard deviation** of a set of data symbolized s, is the positive square root of the variance s^2, i.e.,

$$s = \sqrt{s^2}.$$

The variance and standard deviation are the most common and most useful measures of variability. At first glance it might appear that the standard deviation would have limited use. However, the standard deviation will be expressed in the same units as the data is given, whereas the variance will be expressed in the square of the units in which the data is given. It is highly desirable to be able to express the statistical measures of central tendency and variability in the same units as the given data.

Example 2: A baseball team scored 2, 5, 3, 1, 4, 3 runs in their last six games. Find the variance and standard deviation of the runs scored.

Solution: First we must determine the mean, $\bar{x}$.

$$\bar{x} = \frac{2 + 5 + 3 + 1 + 4 + 3}{6} = 3 \text{ runs}.$$

The variance s^2 is given by

$$s^2 = \frac{(2 - 3)^2 + (5 - 3)^2 + (3 - 3)^2 + (1 - 3)^2 + (4 - 3)^2 + (3 - 3)^2}{6 - 1}$$

$$= \frac{1 + 4 + 0 + 4 + 1 + 0}{5}$$

$$= 2 \text{ square runs}.$$

The standard deviation s is given by

$$s = \sqrt{2} \doteq 1.414 \text{ runs}.$$

Where the mean of a set of data is not a whole number, the squared deviations will be computationally awkward. If a desk calculator is not available we can simplify the computation by using one of many equivalent formulas for s^2. We will list one such formula without including the derivation pointing out, however, that it has a distinct computational advantage in that the mean need be squared only once.

*The variance, s^2, is obtained by dividing by $N - 1$, rather than N so that s^2 will more closely approximate the population variance from which the data is drawn. For a more detailed discussion see the Suggested Reading list at the end of the chapter.

ALTERNATE FORMULA FOR s^2: If $x_1, x_2, x_3, \ldots, x_N$ represent the entries of a set of data and $\overline{x}$ the mean, then the variance s^2 is given by

$$s^2 = \frac{(x_1^2 + x_2^2 + x_3^2 + \cdots + x_n^2) - N(\overline{x})^2}{N - 1}.$$

Example 3: The grade point average of 10 high school seniors picked at random from the graduating class and listed in ascending order is 1.8, 2.0, 2.2, 2.2, 2.5, 2.7, 3.0, 3.1, 3.2, 3.3. Use the alternate computation formula to find the variance and standard deviation.

***Solution*:** As a matter of convenience we list the data in tabular form.

	x_i	x_i^2
	1.8	3.24
	2.0	4.00
	2.2	4.84
	2.2	4.84
	2.5	6.25
	2.7	7.29
	3.0	9.00
	3.1	9.61
	3.2	10.24
	3.3	10.89
Total	26.0	70.20

$$\overline{x} = \frac{26.0}{10} = 2.6$$

$$s^2 = \frac{70.2 - 10(2.6)^2}{9} = \frac{70.2 - 67.6}{9}$$

$$\doteq 0.29$$

$$s = \sqrt{0.29} \doteq 0.54.$$

The variance and standard deviation provide us with a relative measure of the variability of sets of data of the same basic type. For example, we can compare the variability of the same parts produced by different machines or the performance of different groups of students on the same test or the weight of cereals of several standard brands in the same box size sold across the country.

As we have indicated, the mean and standard deviation are most often used by statisticians to characterize concisely a set of data. Knowing the mean gives us information regarding the center of the data and the value of the standard deviation give us information on how the data is spread around the mean. If the standard deviation is a small number, then the data is grouped closely about the mean and if the standard deviation is a large number, then the data is spread out about the mean.

Unfortunately, the smallness or largeness of a standard deviation as a comparative measure of variability is not a very precise mathematical concept. The relation between a mean, a standard deviation, and the proportion of data falling within a specified number of standard deviations about the mean can, however, be made precise.

If the histogram or frequency polygon for a set of data is modified in such a way that the vertical scale represents the probability of each outcome, and the base of each interval on the horizontal is one unit, then the area under the probability distribution figure formed will be 1. For example, the probability distribution representing the sum of dots appearing on the roll of a pair of dice is shown in Figure 5.4. Each rectangle has a base one unit wide and height equal to the probability of the indicated sum. The total area of the 11 rectangles is 1.

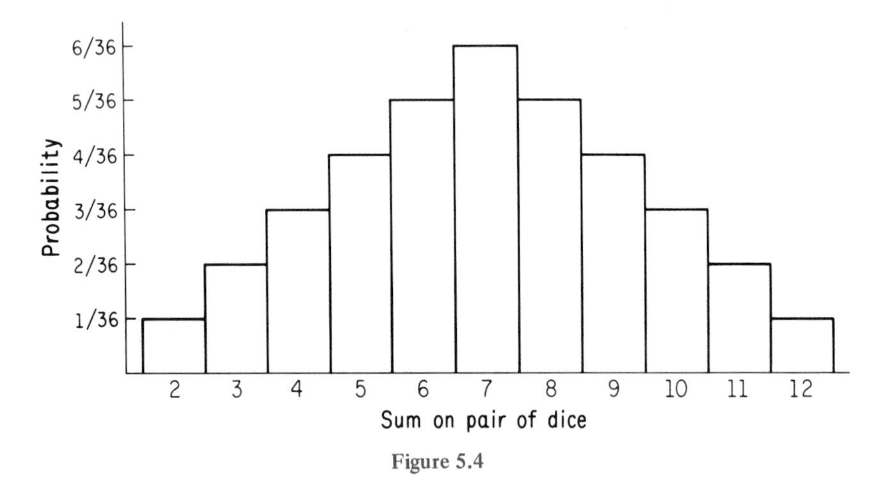

Figure 5.4

The Russian mathematician P. L. Chebyshev (1821-1894) discovered that for all probability distributions the fraction of area between any two symmetric values about the mean is a function of the standard deviation. The functional relationship between the variables is given by Chebyshev's theorem.

CHEBYSCHEV'S THEOREM: For any probability distribution, at least the amount of the data given by the fraction $\left(1 - \frac{1}{k^2}\right)$ lies within k standard deviations of the mean of the data.

The Chebyshev theorem provides us with a conservative estimate of the proportion of the data within a specified number of standard deviations from the mean. For $k = 1$ the fraction $\left(1 - \frac{1}{k^2}\right) = 0$ guaranteeing at least 0 percent of the data is within one standard deviation of the mean. For $k = 2$, the fraction $\left(1 - \frac{1}{k^2}\right) = 3/4$ guaranteeing at least 75 percent of the data is within two standard deviations of the mean.

Example 4: If a given set of data has a mean of 15 and a standard deviation of 2, at least, what proportion of the data falls between 9 and 21? At least how much between 7 and 23?

Solution: The interval 9 to 21 corresponds to 15 ± 6 or $\overline{x} \pm 3s$. Using Chebyshev's Theorem with $k = 3, \left(1 - \frac{1}{3^2}\right) = 1 - \frac{1}{9} = \frac{8}{9}$. Thus at least 8/9ths of the data fall between 9 and 21. In a similar manner 7 to 23 prepresents $\overline{x} \pm 4s$ and $\left(1 - \frac{1}{4^2}\right) =$ 15/16. At least 15/16ths of the data falls between 7 and 23.

Example 5: Using the data of Example 3 in this section, show that at least 75 percent of the data falls within two standard deviations of the mean.

Solution: The mean of the data is 2.6 and the standard deviation 0.54. A two standard deviation interval corresponds to 2.6 ± 1.08 or a span of 1.52 to 3.68. A review of the data indicates that 1.8 is the smallest entry and 3.3 the largest entry. All of the data, which is certainly at least 75 percent of the data, is within two standard deviations of the mean.

The fact that a relationship exists between the standard deviation and the proportion of data that lies about the mean is of fundamental importance to statistical inference. It is relatively easy to draw certain conclusions about a possible outcome based on the number of standard deviations that the outcome is away from the mean if we have some knowledge of the distribution form of the data.

In the next section we will fi[illegible] that much stronger statements than anticipated by Chebyshev's theorem can be [illegible] the shape of a probability distribution can be identified with a particular proba[illegible]y distribution called the normal distribution.

Problem Set 5.4

A. 1. The mileage covered on various car trips was recorded for one week. The mileages listed were

1.5	3.3	5.0	28.0	1.7	2.3	1.2
28.0	28.0	0.7	28.0	28.0	220.0	1.8

Find the (a) range, (b) variance, and (c) standard deviation.

2. Consider the numbers, 2, 6, 3, 1, 5, 7.
   (a) Find the mean and standard deviation.
   (b) Add 7 to each of the given numbers and then find the standard deviation.
   (c) Multiply each of the given numbers by 5 and find the standard deviation.
   (d) Compare the results in (b) and (c) to those found in (a). What, if any, generalizations appear to hold.

3. If a given set of data has a mean of 73 and has a standard deviation of 2, at least what proportion of the data falls between 53 and 93? How much between 63 and 83?

4. Use the data in Table 5.1 (page 170) to draw a probability distribution where the vertical scale is the probability of each grade and the horizontal scale represents each individual grade. Show that the area of the distribution formed is one.
5. A basketball team scored 109, 89, 137, 119, 93, 127, 101, and 153 in its first eight games. Find the (a) range, (b) variance, and (c) standard deviation.
6. A given set of data has a mean of 70 and we know that at least 24/25 of the data fall between 55 and 85. According to Chebyshev's theorem what is the standard deviation for the set of data?
7. A die is rolled four times. Draw a histogram where the horizontal axis represents the number of sixes that occur and the vertical axis the probability of each outcome. Show that the area of the probability distribution formed is one. What is the mean of the number of sixes that appear?
8. It can be proved that if $x_1, x_2, x_3, \ldots, x_n$ represent a set of data with a variance of s^2, then (a) a set of data formed by adding a constant k to each entry also has a variance of s^2 and (b) a set of data formed by multiplying each entry by a constant k has a variance of $k^2 s^2$. If a given set of data which has a variance of 5 is modified so that each entry has 4 added to it and is then doubled, what will be the variance of the new set of data?
9. Calculate the variance of the following sets of data by (i) the standard formula, and (ii) the alternate computation formula:
   (a) 12, 10, 8, 2, 6, 3, 8.
   (b) 7.2, 3.1, 2.4, 5.6, 4.2.
10. Compare the data given in (a) and (b) with the data given in Prob. A-9. Find the variance of the data using the principle discussed in Prob. A-8.
   (a) 6, 4, 2, −4, 0, −3, 2.
   (b) 14.4, 6.2, 4.8, 11.2, 8.4.

5.5 The Normal Distribution

The most important probability distribution used in statistical analysis is the normal distribution. The normal curve is frequently described as a continuous symmetric bell-shaped curve with the mean, median, and mode at the same central position. There is not a unique normal distribution, since the mathematical formula for the curve depends on the two variables, the mean and standard deviation. Figure 5.5 illustrates three typical normal distributions.

Each of the normal curves in Figure 5.5 have the same mean but different standard deviations. Despite the fact that the shapes are different, the total area under each curve is 1 and the area within k standard deviations of the mean in each case is the same. For example, the area under each curve within 1 standard deviation from the mean is approximately 68 percent of the total area.

The normal curves extend indefinitely in either direction about the mean but the amount of area under the curve more than 3 standard deviations away from the mean is negligible.

Since the amount of area within k standard deviations of the mean in a normal distribution is precisely determined and does not depend on an exact value for either the

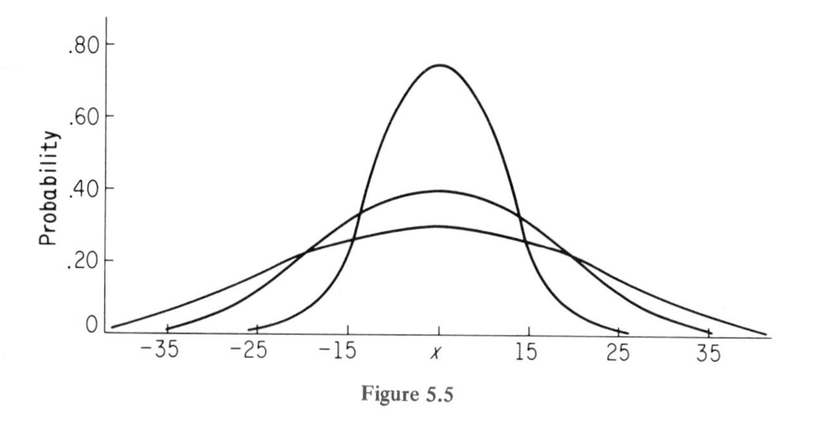

Figure 5.5

mean or the standard deviation, statisticians identify a normal curve having a mean of 0 and a standard devivation of 1 as the **standard normal distribution** or as the **z distribution.** Areas under the standard normal curve for specified standard deviations from the mean have been tabulated and provide us with a standard reference for all normal distributions (see Appendix 3).

Figure 5.6 illustrates the standard normal curve where we can readily observe that there is approximately 68 percent of the area within 1 standard deviation, 95 percent of the area within 2 standard deviations and 99.7 percent of the area within 3 standard deviations of the mean. It is perhaps important to point out that not every symmetric bell-shaped curve is a normal curve. A normal distribution has not only the general characteristics we have described but the proportion of area between any two z values is precisely determined by a mathematical formula involving x and s. If a normal distribution may be applied to a set of data, then we have a much stronger result than we would have been provided by using the Chebyshev's theorem.

The widespread usefulness of the normal distribution stems from the fact that em-

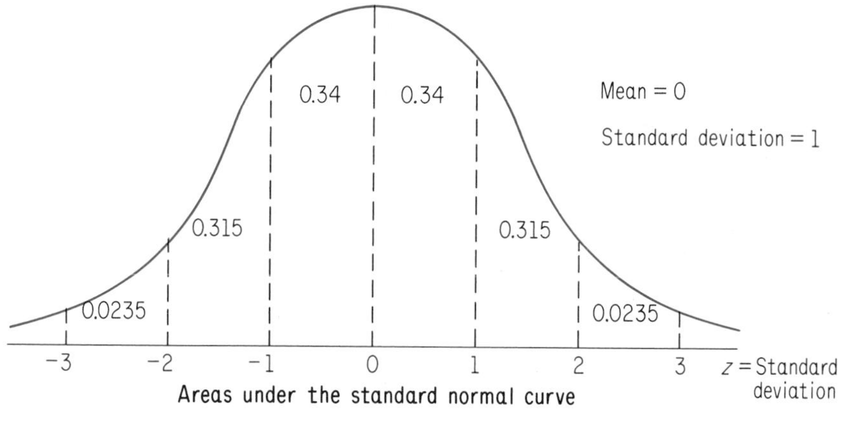

Figure 5.6

pirical measurements of a suprisingly large number of physical phenomena tend to produce distributions that conform closely to the normal curve. Distribution of heights and weights of individuals, IQ test scores, yearly income, repeated measurements of the acceleration of gravity, crop production, etc. all exhibit the characteristics of a normal distribution.

In order to use the tabulated information on the standard normal distribution (Appendix 3) it will be necessary to convert the given data into z scores which represent the deviation from the mean measured in standard deviations under the standard normal curve. For example, if a given continuous distribution has an $\overline{x}$ of 10 and an $s = 2$, then a measurement of 12 lies 1 standard deviation away from the mean and thus has a z score of +1. In general, if X represents any entry then the z score corresponding to that entry is given by

$$z = \frac{X - \overline{x}}{s}.$$

The following examples will illustrate how we may apply the normal distribution to interpret normally distributed data which is assumed to be continuous.

Example 1: The grades obtained on a standardized English test have a normal distribution with a mean of 75 and a standard deviation of 10. What proportion of the test scores fall between 70 and 90?

Solution: In order to use the tabulated areas in Appendix 3 we convert the given test grades into z scores as follows:

For $x = 70$,

$$z = \frac{70-75}{10} = -.5.$$

For $x = 90$,

$$z = \frac{90-75}{10} = 1.5.$$

The numerical z conversion may be geometrically illustrated by drawing the distribution of the data followed by the standard normal distribution.

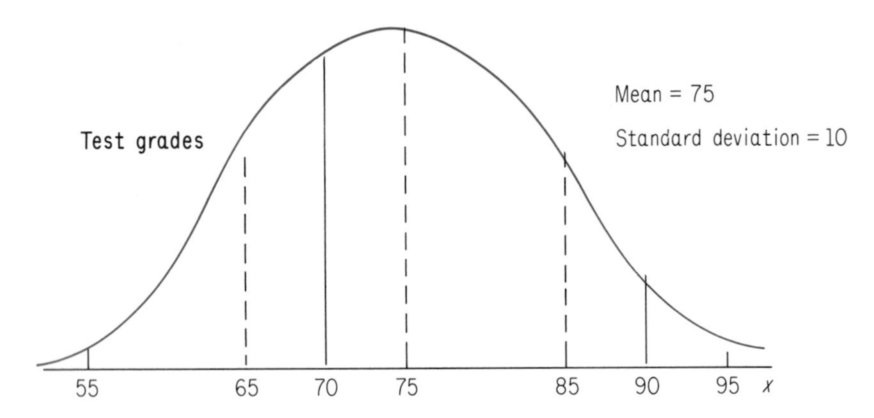

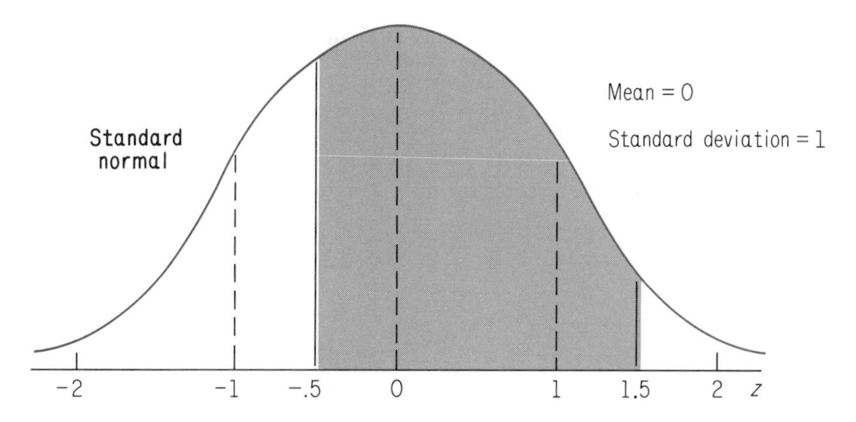

We seek the amount of shaded area under the standard normal curve between $z = -.5$ and $z = 1.5$. The tables are so constructed that at each z score we are given the area to the left or below that point. Thus, we find that at $z = 1.5$ we have .9332 of the area below that point and at $z = -.5$ we have .3085 of the area below that point. The arithmetic difference represents the area in between the two z scores. Expressing this in probability notation we have

$$\begin{aligned} \Pr(z < 1.5) - \Pr(z < -.5) &= .9332 - .3085 \\ &= .6247. \end{aligned}$$

Thus 62.47 percent of the test scores fall between 70 and 90 or alternately we can state that the probability someone has a score between 70 and 90 is .6247.

Example 2: In a given region the average family income is reported as \$8,000 per year and the standard deviation is given as \$1,200. What is the probability that a family picked at random will have an income greater than \$11,000 per year if the incomes are normally distributed?

Solution: The z score corresponding to the given information is given by

$$z = \frac{11{,}000 - 8{,}000}{1200} = 2.5.$$

We seek the amount of area above a z of 2.5. Since the tabulated z values give the area below a specified z score and the total area under the curve is 1 we can find the required result as follows:

$$\begin{aligned} \Pr(z > 2.5) &= 1 - \Pr(z < 2.5) \\ &= 1 - .9938 \\ &= .0062. \end{aligned}$$

Alternately since the curve is symmetric, the amount of area below $z = -2.5$ must be the same as the area above $z = 2.5$. The area below $z = -2.5$ may be read directly from the table as .0062.

Example 3: On a reading comprehension test 500 college freshmen average 3.2 minutes to read a short article. If the reading times are normally distributed and the standard deviation is 0.8, how many students would one expect to have reading times between 2.0 and 3.2 minutes?

Solution: In order to find the number of students having reading times between 2.0 and 3.2 minutes we must first find the area between these scores under the normal curve as shown. The z scores for the given data are as follows:

For $z = 3.2$,

$$z = \frac{3.2 - 3.2}{.8} = \frac{0}{.8} = 0.$$

For $z = 2.0$,

$$z = \frac{2.0 - 3.2}{.8} = \frac{-1.2}{.8} = -1.5.$$

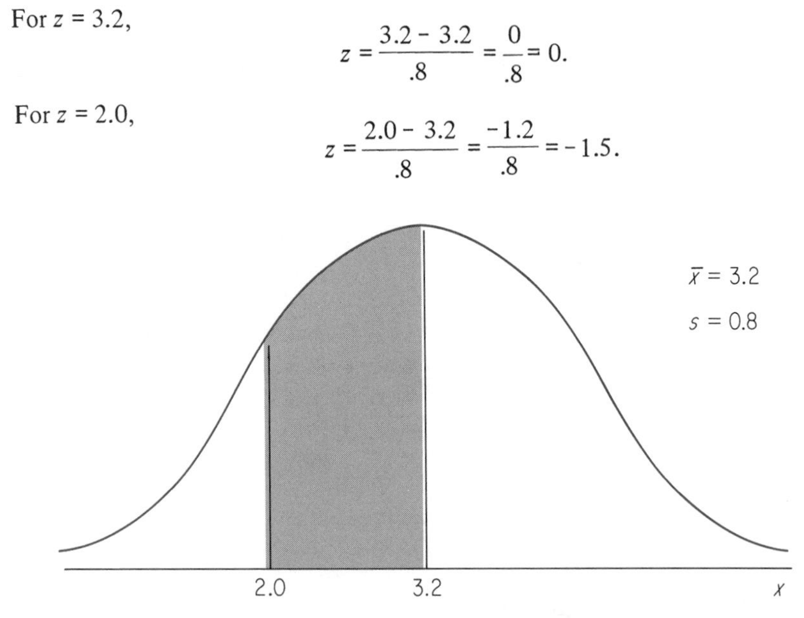

The area between the two z scores may be found by taking the arithmetic difference of the table values of the area below each z value as follows:

$$\begin{aligned}\Pr(-1.5 < z < 0) &= \Pr(z < 0) - \Pr(z < -1.5)\\ &= .5000 - .0668\\ &= .4332.\end{aligned}$$

Thus 43.32 percent of the total number taking the test should have scores in the interval 2.0 to 3.2. Rounded off to the nearest whole number 43.32 percent of the 500 freshmen is 217 freshmen.

The normal distribution is a continuous probability distribution and particularly well suited to problems involving data which is at least theoretically continuous. We have assumed in the preceding examples that the data could be measured with increasing precision so as to approach a continuous spread of data. Fortunately, the normal distribution is not limited in its use to continuous variables and may be used to obtain surprisingly close approximations to problems involving discrete variables where the general shape of the frequency distribution approaches a normal curve.

In the next section we will illustrate how the continuous normal distribution may be used to closely approximate one particularly important discrete distribution—namely, the binomial probability distribution.

Problem Set 5.5

In the following problems assume the data is continuous.

A. 1. In a normal grading distribution with mean 72 and standard deviation of 10, what percentage of the students scored 90 or higher?

2. The average weight for a certain group of students was found to be 130 pounds with a standard deviation of 10. What proportion of weights are between (a) 125 and 135 pounds and (b) 140 and 120 pounds?

3. The average IQ score for a group of 100 people was found to be 110 with a standard deviation of 20. How many people would have an IQ score between (a) 100 and 120, and (b) 130 and 90?

4. A z score of 1.5 is found to correspond to a measurement of 18 and the standard deviation for this collection of measurements is determined to be 4. Find the mean for this collection of measurements.

5. Find the area under the normal curve between $z = 0$ and $z = 1.1$. What does the area represent?

6. Find the area under the normal curve between $z = -.42$ and $z = 2.12$.

7. If $\overline{x} = 75$ percent, $s = 6$ and we specify that 10 percent of the students get A's and the grades are normally distributed, what is the lowest possible number grade that will still be an A?

8. The average weight for a group of people was found to be 125 pounds with a standard deviation of 15 pounds. Determine the probability that a person selected from the given group of people will weigh (a) more than 140 pounds, (b) less than 105 pounds, and (c) between 105 and 140 pounds.

9. Find a particular value of z, say z_0, such that
   (a) $\Pr(-z_0 < z < z_0) = .94$.
   (b) $\Pr(-z_0 < z < z_0) = .95$.
   (c) $\Pr(-z_0 < z < z_0) = .99$.

B. 1. As a part of an admissions procedure a certain college requires each student to take a placement test in mathematics, English, and social studies. The scores on each test are normally distributed with the mean and standard deviation as indicated in the table below. If Tom Smith's scores on each test are as indicated, compare his performance with other students by finding the percentage of students below his score.

	Mean	Standard Deviation	Tom Smith's Score
Mathematics . . .	53.7	7.2	48
English.	62.8	9.4	68
Social Studies. . .	73.4	10.5	75

2. A company manufactures an automobile part whose mean length is 5.0 in. As long as the length is within 4 percent of the mean the part is considered acceptable. The standard deviation for the parts produced on a given day is .010 in. If the daily production is 15,000 parts, how many parts would not be acceptable?

3. The mean life span of an electric-razor motor is normally distributed with a mean of 8 years and with a standard deviation of 1.5 years. If the manufacturer is willing to replace only 5 percent of the electric-razor motors that fail, what is the maximum guarantee period he should offer for free replacement?

5.6 The Normal Approximation to the Binomial

In Sec. 3.5 we introduced the binomial probability model and discussed its widespread applicability to experiments involving two outcomes where the probabilities of each outcome remain constant and an independent trial sequence is repeated a fixed number of times. The probability of a general term in the binomial probability model is given by

$$b(k;n,p) = \binom{n}{k,n-k} p^k q^{n-k}$$

where k represents the number of successes in n trials, p the probability of success, and q the probability of failure. The reader will recognize that even rather simple binomial experiments give rise to computational difficulties due to the necessity of raising p and q to the required power and the need to evaluate large factorial coefficients.

A convenient approximation method using the normal distribution eliminates the awkward computation of the binomial probability model and provides us with surprisingly accurate results. The larger the number of trials and the closer the probability of success is to 1/2 the more exact the approximation becomes.

To illustrate the relationship between the two probability distributions consider a binomial experiment of tossing a coin ten times where we regard the number of heads that appear as a "success." Figure 5.7 illustrates a histogram for this experiment where the

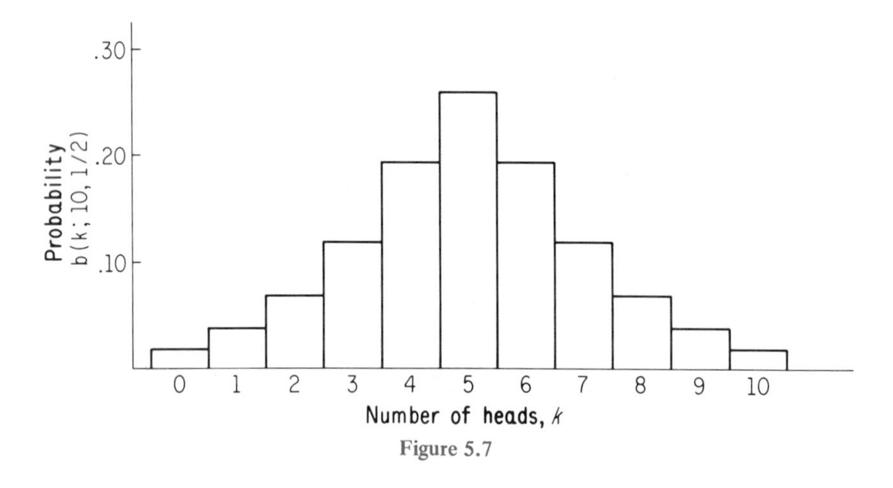

Figure 5.7

horizontal scale represents the number of heads (k) and the vertical scale the probability for each possible outcome given by $b(k;10,1/2)$. Notice that although each outcome is typical of a discrete variable that can only assume the values 0,1,2, . . . , 10, the base of each rectangle is drawn as one unit. For example, the rectangle corresponding to 3 heads has a base extending from 2.5 to 3.5 and a vertical height of .12. By making the drawing in this manner, the total area under the histrogram is 1 unit.

Now let us take the same histogram and superimpose a normal distribution on the graph as shown in Figure 5.8.

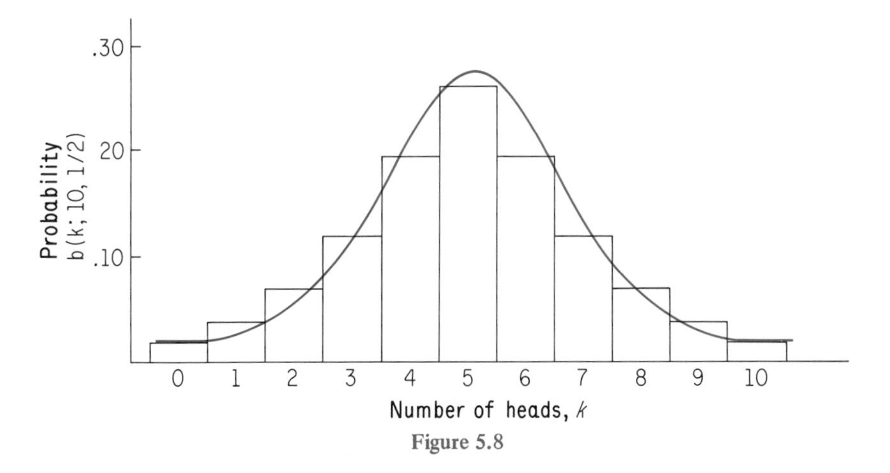

Figure 5.8

The total area under the normal distribution is also 1 and there is a close agreement between the normal curve and the histogram. In order to convert the binomial distribution to the standard normal distribution, and obtain the corresponding z score we need to know the mean and the standard deviation of the binomial distribution. The following theorem, the proof of which is omitted, gives us the necessary information.

THEOREM 5.6.1: The binomial distribution $b(k;n,p)$ where k represents the number of successes, n the number of trials, and p the probability of success, has a mean and standard deviation given by

$$\overline{x} = np \quad \text{and} \quad s = \sqrt{npq}.$$

To illustrate use of the normal approximation and display the accuracy of the approximation, let us consider the question, "What is the probability that in 10 losses of a fair coin we obtain 6 or 7 heads?" The exact binomial probability can be determined as follows:

$$b(6;10,1/2) + b(7;10,1/2) = \binom{10}{6,4}(1/2)^6(1/2)^4 + \binom{10}{7,3}(1/2)^7(1/2)^3$$

$$= \frac{10!}{6!4!} \times \frac{1}{1024} + \frac{10!}{7!3!} \times \frac{1}{1024}$$

$$= \frac{210}{1024} + \frac{120}{1024}$$

$$\doteq .3222.$$

To use the normal approximation we first determine $\overline{x}$ and s as follows:

$$\overline{x} = np = (10)(1/2) = 5$$
$$s = \sqrt{npq} = \sqrt{10(1/2)(1/2)} \doteq 1.58.$$

The outcome of six heads corresponds to the histogram interval from 5.5 to 6.5 inclusive and the outcome of seven heads to the interval 6.5 to 7.5. The overall area span for six or seven heads corresponds to the interval from 5.5 to 7.5. We find the z score for the end-points of the interval and the area between the z scores in the following manner:

For 5.5,

$$z = \frac{x - \overline{x}}{s} = \frac{5.5 - 5}{1.58} \doteq .32.$$

For 7.5,

$$z = \frac{x - \overline{x}}{s} = \frac{7.5 - 5}{1.58} \doteq 1.58$$

$$\begin{aligned}\Pr(.32 < z < 1.58) &= \Pr(z < 1.58) - \Pr(z < .32)\\ &= .9429 - .6255\\ &= .3174.\end{aligned}$$

The error between the normal approximation of .3174 and the exact binomial probability of .3222 is less than 1/2 of 1 percent.

The method described is general and very efficient since we need to make reference only to the standard normal tables.

We will complete this section by illustrating the normal approximation method as applied to another binomial experiment.

Example 1: Suppose that a geneticist finds that one-sixth of a given species of mice are born with a birth defect. What is the probability that out of 180 newborn mice of this specie that 40 or more will have birth defects?

Solution: The mean and standard deviation of the binomial distribution are given by

$$\overline{x} = np = 180\,(1/6) = 30$$
$$s = \sqrt{npq} = \sqrt{180(1/6)(5/6)} = 5.$$

To obtain the z score for 40 or more we use 39.5 in the basic formula since exactly 40 corresponds to the interval 39.5 to 40.5. Thus

$$z = \frac{x - \overline{x}}{s} = \frac{39.5 - 30}{5} = 1.90.$$

The probability of 40 or more corresponds to the area of the right of $z = 1.90$. Thus

$$\begin{aligned}\Pr(z > 1.90) &= 1 - \Pr(z < 1.90)\\ &= 1 - .9713\\ &= .0287.\end{aligned}$$

The information obtained by solving Example 1 provided us with some insight into statistical inference, if we reinterpret the question paralleling more closely scientific research experiments. Suppose the genetic researcher is attempting to isolate different causes for birth defects in mice. Under controlled experiments he may introduce a factor he believes contributes to birth defects. A sample of mice is chosen for the experiment and the treatment factor introduced. If 40 or more of the group of 180 newborn mice exhibit birth defects, what is the probability that this result could be obtained if the known rate of one-sixth of the newborn having birth defects is the only operational factor? In other words, the treatment factor has no effect. The result indicates that such an outcome should occur only 2.87 percent of the time. Thus the fact that 40 or more had birth defects is a most unlikely outcome. We have substantial reason to believe that the treatment factor does have an adverse effect on the offspring of this particular specie of mice.

We will consider statistical inference in more detail in Sec. 5.8, Testing a Statistical Hypothesis. In the next section we will extend the range of application of the standard normal distribution by introducing the concept of a sampling distribution.

Problem Set 5.6

A. 1. Let the experiment be that of rolling a die ten times. If the number of dots appearing is divisible by 3, then we regard this as a success. For this binomial experiment find the mean and standard deviation.

2. In Prob. A.1, what is the probability of three or four successes in ten rolls of the die? Calculate this probability by (a) the binomial formula, and (b) the normal distribution approximation.

3. Consider a binomial experiment with $n = 25$, $p = 0.4$. Calculate $\Pr(8 \leqslant k \leqslant 11)$ using the normal approximation to the binomial.

4. Consider a binomial experiment with $n = 10$, $p = 0.3$. Calculate $\Pr(k \leqslant 5)$ using the normal approximation to the binomial.

5. The reliability of an electirc-light bulb is the probability that a bulb, chosen at random from production, will light under the conditions for which it has been designed. A random sample of 1,000 bulbs was tested and 27 were found to be defective. Find the probability of obtaining 27 or more defectives, assuming that the reliability is .98.

6. A fair coin is tossed 1,600 times. Use the normal approximation to find the probability of obtaining (a) between 770 and 830 heads inclusive, and (b) more than 850 heads.

7. A marksman claims he can hit a target 90 percent of the time. If in his next 400 shots he hit the target 342 times, what reaction would you have to his claim based on the probability of that outcome?

8. A survey indicates that 60 percent of the homes in a given community have two television sets. What is the probability that if 200 homes are picked at random, 125 or more will have two television sets?

9. If 50 percent is considered a passing grade on a fifty-item test, determine the probability of obtaining better than a passing grade on that test if a person guesses, if there are (a) two choices for each question, (b) three choices for each question.

5.7 Sampling Distributions

The normal distribution is applicable to many continuous distributions and may be used to approximate such discrete distributions as the binomial. However, we have indicated that the distribution of the raw data must conform closely to the shape of the normal curve, otherwise significant errors may result. It would appear therefore, that a non-normal shape of the distribution of the raw data may place a serious limitation on the use of the normal distribution. To show that this limitation is not as serious as it might appear is our next goal.

We have purposely avoided, until now, making the distinction that must be made between sample data and population data. We have done this with the belief that some basic background knowledge in statistics is necessary before the distinction can be made meaningful. A group of measurements which we choose to regard as the totality of data under consideration we regard as the **population** or **universe**. Any subset of a population is regarded as a **sample** of the population. In context, the sample and population may have different meanings. For example, a local placement test used for determining advanced standing in a given subject may be given yearly to entering freshmen. All of the test results can be regarded as the population and the test results of each freshmen class regarded as a sample. However, if a nationally standardized test is used for placement purposes, the population would consist of all of the student test scores in the entire nation and this particular school's test results could be regarded as a sample.

Statistical measures of central tendency and variability are meaningful for both samples and populations although the notation and language used to describe each set of results varies. Numerical measures of sample data, such as the mean and variance, are called **statistics** and similar numerical measures of population data are called **parameters.** Arabic letter notation is used to identify such sample data measures as $\overline{x}, \tilde{x}, s^2$, and s. Greek letter notations for the respective population measures are $\mu, \tilde{\mu}, \sigma^2, \sigma$. The reader should observe that we have used sample notation throughout our previous discussion, since in practice we rarely have available the entire population data. The basic formulas for finding population parameters as opposed to sample statistics, exclusive of notation, are the same with the exception of the population variance. The population variance is given by

$$\sigma^2 = \frac{(x_1 - \mu)^2 + (x_2 - \mu)^2 + \cdots + (x_N - \mu)^2}{N}$$

where the divisor is N rather than $N - 1$.

One of the most important uses of sample statistics is to infer conclusions about the population parameters. This becomes particularly significant if it is difficult or perhaps impossible to apply the same measures to an entire population. For example, we may run some research experiments on laboratory animals and use the research results to predict how all other animals of the same type would react. It would be pointless to attempt to perform the same test on all animals of the same type and fortunately, unnecessary.

In this brief introduction to statistics we cannot examine in detail the sampling distributions of all of the common sample statistics. However, we will present information on the sampling distribution of one of the most important sample statistics, the **sample mean**.

To motivate our discussion, consider the practical difficulties we would have in attempting to determine the mean age of all the residents in the United States by actually listing the age of each citizen. The magnitude of the task would quickly prompt us to seek alternate methods for estimating the population mean.

Suppose we obtained a series of random samples from various regions of the United States and used the mean of all the sample means as our estimate of the population mean. Intuitively, we would have a degree of confidence that the result must be close to the true population mean, but how do we determine the degree of confidence and how can we be sure we are close to the population mean? In order to answer questions of this type regarding sample means, statisticians make reference to one of the most remarkable theorems in all of mathematical history, the central limit theorem.

CENTRAL LIMIT THEOREM: If a random sample of size n is drawn from any population having a mean μ and standard deviation of σ, then the sampling distribution of the sample means approaches the standard normal distribution as n increases. The mean of the sampling distribution of $\bar{x}$'s will be μ, i.e., $E(\bar{x}) = \mu$, and the standard deviation of the sampling distribution, symbolized $\sigma_{\bar{x}}$, is given by

$$\sigma_{\bar{x}} = \frac{\sigma}{\sqrt{n}}.$$

Note that no matter what shape the population distribution may assume, the sampling distribution of the means will be approximately normally distributed. In addition we can determine the mean and standard deviation of the sampling distribution of the sample means by a single formula involving μ, σ, and n. Thus, we can convert sample data statistics to z scores and obtain the probabilities of various outcomes.

Let us examine one example of a nonnormal population and show that the sampling distribution of the sample means is approximately normally distributed and the mean and standard deviation of the sample means ($\bar{x}$'s) are as predicted by the central limit theorem. Suppose a population consists of the five digits 2, 4, 6, 8, and 10. Then the mean of the population, μ, is 6 and the standard deviation of the population, σ, is $2\sqrt{2}$. The probability distribution for the population as shown in Figure 5.9, is rectangular in shape and definitely nonnormal.

Now suppose we decide to form samples of size 2 from the population by first

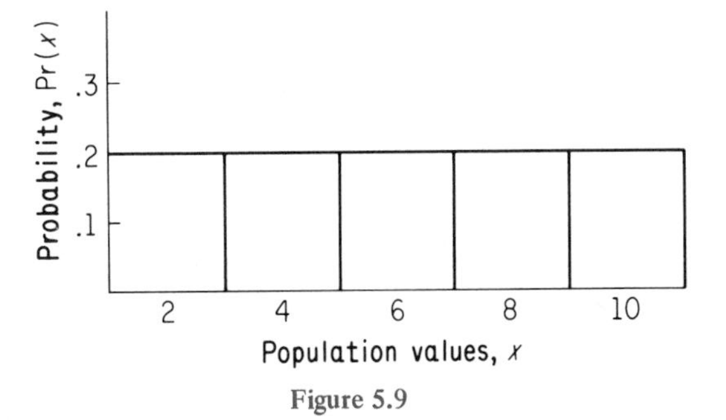

Figure 5.9

selecting any one of the five digits at random, replacing the selected digit, and then drawing another digit at random. This selection process will generate 25 possible samples of size 2. Table 5.4 lists the 25 samples and the $\overline{x}$ for each sample.

Table 5.4

Sample	$\overline{x}$	Sample	$\overline{x}$
(2, 2)	2	(6, 8)	7
(2, 4)	3	(6, 10)	8
(2, 6)	4	(8, 2)	5
(2, 8)	5	(8, 4)	6
(2, 10)	6	(8, 6)	7
(4, 2)	3	(8, 8)	8
(4, 4)	4	(8, 10)	9
(4, 6)	5	(10, 2)	6
(4, 8)	6	(10, 4)	7
(4, 10)	7	(10, 6)	8
(6, 2)	4	(10, 8)	9
(6, 4)	5	(10, 10)	10
(6, 6)	6		

Sum of 25 $\overline{x}$'s = 150.
Mean of or expected value of $\overline{x}$'s = $E(\overline{x}) = \frac{150}{25} = 6.$

Note that the average of all the sample means, $E(\overline{x}) = 6$, is the same as the population mean. Thus, the sample statistic $\overline{x}$ does reproduce the population parameter μ. Now let us organize the sample mean data into a probability distribution and observe the shape of the resulting distribution (Figure 5.10).

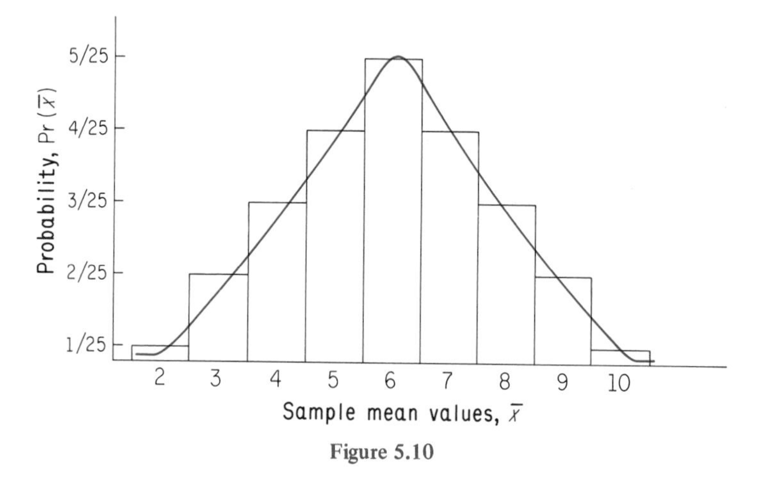

Figure 5.10

The resulting shape of the sampling distribution of the $\overline{x}$'s is even for a sample size of 2, beginning to take on the shape of the normal distribution. If the sample size were increased to size 3, we would have 125 samples to consider and the approximation to the normal approximation conforms more closely to the shape of the sampling distribution.

If each of the 25 sample means is regarded as an entry in a population distribution with a mean of 6, we can find the standard deviation of that distribution, $\sigma_{\overline{X}}$, by taking the square root of the mean of the squared deviations about the mean 6. Thus

$$\sigma_{\overline{X}}^2 = \frac{(2-6)^2 + (3-6)^2 + (4-6)^2 + \cdots + (9-6)^2 + (10-6)^2}{25}$$

$$= \frac{100}{25} = 4$$

$$\sigma_{\overline{X}} = 2.$$

This result conforms with one of the conclusions of the central limit theorem, since according to the theorem

$$\sigma_{\overline{X}} = \frac{\sigma}{\sqrt{n}}$$

which in the context of this example where $n = 2$ and $\sigma = 2\sqrt{2}$, gives

$$\sigma_{\overline{X}} = \frac{2\sqrt{2}}{\sqrt{2}} = 2.$$

In practice, we would not be likely to list all the samples or prepare a probability distribution for the $\overline{x}$'s. We would use the central limit theorem to determine $E(\overline{x})$ and $\sigma_{\overline{x}}$ then refer to the standard normal z scores for determining probability of various outcomes.

The statistician must also consider in the design of his experiment what relationship exists between the shape of the population and the sample size that will insure a minimum of error in assuming that the sampling distribution takes on the shape of the standard normal distribution. The empirical rule which the statistician uses is as follows:

> *If the population is normally distributed or is assumed to be normally distributed and σ is known, samples of any size may be drawn from the population and the $\overline{x}$ distribution may be assumed normal. If the population is nonnormal and σ is known, the sample size should be at least 20 to insure that the $\overline{x}$ distribution is normally distributed.*

The following examples will illustrate applications of the general theory.

Example 1: A nationally standardized mathematics placement test has a mean, μ = 300, and a standard deviation, σ = 60. If a random sample of the test scores of twenty five students is selected, what is the probability that the average of 25 scores will be between 282 and 318?

Solution: The mean of the sampling distribution of $\overline{x}$'s will be the same as the population mean, i.e., $E(\overline{x}) = 300$. The standard deviation of the sampling distribution of means is given by

$$\sigma_{\overline{X}} = \frac{\sigma}{\sqrt{n}} = \frac{60}{\sqrt{25}} = 12.$$

The z scores for the respective sample means are as follows:

For $\overline{x} = 282$,

$$z = \frac{\overline{x} - \mu}{\sigma_{\overline{X}}} = \frac{282 - 300}{12} = -1.5.$$

For $\overline{x} = 318$,

$$z = \frac{\overline{x} - \mu}{\sigma_{\overline{X}}} = \frac{318 - 300}{12} = 1.5.$$

The $\Pr(-1.5 < z < 1.5)$ is obtained as follows:

$$\begin{aligned} \Pr(-1.5 < z < 1.5) &= \Pr(z < 1.5) - \Pr(z < -1.5) \\ &= .9332 - .0668 \\ &= .8664. \end{aligned}$$

Thus the probability is .8664 that the mean of the 25 test scores will fall between 282 and 300.

Example 2: A company manufactures a standard light bulb that has a normally distributed life span measured in hours. The population mean is $\mu = 1200$ hours and the population standard deviation is $\sigma = 250$ hours. An inspector picks random samples of size 16 from various production runs. For a given sample what is the minimum $\overline{x}$ he can expect so that he has a .95 probability that the sample mean will exceed this minimum $\overline{x}$ value?

Solution: The mean of the $\overline{x}$ distribution will be 1200 and the $\sigma_{\overline{X}}$ is given by

$$\sigma_{\overline{X}} = \frac{\sigma}{\sqrt{n}} = \frac{250}{\sqrt{16}} = \frac{250}{4} = 62.5.$$

The z score which corresponds to a value such that above it there will be 95 percent of the area is $z = -1.64$. Using the z score form, namely,

$$z = \frac{\overline{x} - \mu}{\sigma_{\overline{X}}}$$

we can find the corresponding $\overline{x}$ as follows:

$$\begin{aligned} -1.64 &= \frac{\overline{x} - 1200}{62.5} \\ (-1.64)(62.5) &= \overline{x} - 1200 \\ -102.5 + 1200 &= \overline{x} \\ 1097.5 &= \overline{x}. \end{aligned}$$

Thus the inspector has a .95 probability that the average of the life spans of the 16 light bulbs will exceed 1097.5 hours.

Example 3: In Example 2 what increase should be made in the sample size n to increase the probability to .99 that the sample means will fall between 282 and 300.

Solution: The first step must be to determine the z values corresponding to an interval that contains .99 of the area symmetrically distributed about the mean. From the table in Appendix 3, the corresponding z values are $z = -2.58$ and $z = 2.58$, i.e.,

$$\Pr(-2.58 < z < 2.58) = .99.$$

Using one of the z values, namely 2.58 and the corresponding sample mean, 318, we can determine $\sigma_{\overline{X}}$. Thus

$$z = \frac{\overline{x} - \mu}{\sigma_{\overline{X}}}, \quad 2.58 = \frac{318 - 300}{\sigma_{\overline{X}}}$$

which when solved for $\sigma_{\overline{X}}$ will give

$$\sigma_{\overline{X}} = \frac{318 - 300}{2.58} = \frac{18}{2.58} \doteq 6.98.$$

Using the result obtained for $\sigma_{\overline{X}}$ we can then determine the required sample size n; thus

$$\sigma_{\overline{X}} = \frac{\sigma}{\sqrt{n}}, \quad 6.98 = \frac{60}{\sqrt{n}}$$

which when solved for n will give

$$\sqrt{n} = \frac{60}{6.98} = 8.59.$$

Squaring the $\sqrt{n}$, we have $n \doteq 74$.
Increasing the sample size to 74 will increase the probability to .99, the sample mean will fall between 282 and 300.

In this section we have presented a general discussion of the sampling distribution of sample means and the extended use of the standard normal distribution provided by the central limit theorem. Sampling distributions of other sample statistics, such as the sample variance, are also of considerable interest to statisticians. Investigation of other sampling distributions necessitates a review of probability distributions other than the standard normal, and is the proper subject matter of a more advanced treatment of statistics than is our intent in the text.

Before completing our discussion of statistics, however, we will use the material presented thus far to consider the testing of a hypothesis by means of a statistical analysis.

Problem Set 5.7

A. 1. A population consists of the numbers 3, 4, 6, 7. Consider all the possible samples of size two which can be drawn with replacement from this population. Find (a) the mean of the population, (b) the standard deviation of the population, (c) the mean of the sampling distribution of means, and (d) the standard deviation of the sampling distribution of means.

2. In Prob. A.1 find the sample variance for each sample and then the mean of all the sample variances. How does this value compare to the population variance. What generalization is suggested by this example?
3. (a) Draw the probability distribution for the rolling of a single die.
   (b) From the population in 3(a), namely 1, 2, 3, 4, 5, 6, form samples of size two by selecting any one of the six numbers at random. Replace this number and then select another number at random. Draw the probability distribution for sample means generated this way.
4. A basketball player makes 55 percent of his foul shots. What is the probability that in his next 100 tries he makes 65 or more? What is the probability that he makes exactly 45?
5. A producer of soft drinks places on the average 8 ounces in each bottle produced and the standard deviation of the population is known to be one ounce, i.e., $\sigma =$ 1 ounce. In a random selection of 25 bottles, what is the probability that the $\bar{x}$ of the sample will be greater than 8.5 ounces?
6. In preparing for a new labor contract a union determines that the average per hour wage in the industry is \$4.20 and that the standard deviation is \$0.48. At a particular plant a random sample of 36 workers is chosen and their average per hour rate is \$3.96. What is the probability that the average would be lower than this? Would you say that this is a rather typical plant?
7. A company produces a part which has a mean inside diameter of 0.750 in. and a standard deviation of 0.030 in. An inspector randomly picks samples of size 25 and obtains an $\bar{x}$ for each sample. The acceptable tolerance limit for the inside diameter is set as $\bar{x} \pm 2\,\sigma_{\bar{x}}$. What are the maximum and minimum acceptable measurements for the $\bar{x}$ of a sample of 25?

B. 1. A national survey shows that the heights of entering college freshmen are normally distributed with a mean of 68.4 in. and a standard deviation of 2.8 in.
   (a) What is the probability a sample of 16 students chosen at random will have an average height of between 67.7 and 69.1 inches inclusive?
   (b) What increase should be made in the sample size to increase the probability to 0.95 that the sample mean will be between 67.7 and 69.1?
2. If when a sample of size 64 is chosen the $\sigma_{\bar{x}} = 4$, what must the size of the sample be to reduce $\sigma_{\bar{x}}$ to 2?
3. A cigarette manufacturer states that his product has a mean nicotine content of 17.7 milligrams with a standard deviation of 2.8 milligrams. A random sample of four cigarettes have nicotine contents as follows: 21.6, 21.0, 20.7, 21.5. What is the probability that an $\bar{x}$ greater than this could have been obtained? What conclusion would you make regarding the manufacturer's statement?

5.8 Testing a Statistical Hypothesis

In the preceding section we applied the central limit theorem to the sampling distribution of sample means to determine the probability of various sample mean outcomes. In the process we obtained z scores corresponding to the specified $\bar{x}$'s by a basic formula that involved known population parameters μ and σ. Frequently, we may not know the exact numerical value of μ and σ and thus we must hypothesize some value or alternately our primary interest may be the testing of a hypothesis about these or other population parameters. A statistician will form a hypothesis about a population parameter and then examine random samples from the population in order to decide whether or not to accept the hypothesis. We will restrict our attention in this section to hypothesis testing relative to the population parameter μ. However, the same general approach is applicable to the hypothesis testing of other population parameters.

At the outset it is important to state that a statistical hypothesis should not be confused with a scientific hypothesis or theory. As is usually the case in statistics, we are dealing with the probabilities of events and we do not "prove" or "disprove" a statistical hypothesis. We make our decision as to the acceptance or rejection of a hypothesis on the basis of the probability of the sample outcomes. A previously accepted hypothesis may be rejected if new sample data justifies it on the basis of a probability of analysis.

We do not need to present any additional theory regarding the sampling distributions of the mean or the normal distribution. The analysis of a statistical hypothesis as compared to our work in the preceding section is primarily one of a point of view. Instead of beginning with a given value for the parameter μ and instead of determining the probability of sample mean outcomes, we use the sample mean data to decide whether the results support the hypothesis or lead us to reject the hypothesis.

The hypothesis which we form about the population parameter μ is called the **null hypothesis** and symbolized, $\mathbf{H_0}$. If the null hypothesis is rejected then we must accept some other hypothesis called the **alternate hypothesis**, symbolized $\mathbf{H_1}$. Most null hypotheses specify the exact numerical value of the parameter, e.g., $\mu = k$, where k is a constant. However the alternate hypotheses will be inexact and will vary based on the type of experiment being considered. The alternate hypothesis will, nevertheless, **take** on one of three basic forms: the mean is greater than the hypothesized constant, $\mu > k$; the mean is less than the hypothesized constant, $\mu < k$; or the mean does not equal the hypothesized constant, $\mu \neq k$. The "greater than" and "less than" alternative hypothesis are called one-sided or **one-tailed tests** and the "not equal" alternative is called a **two-tailed test**.

In the design of an experiment to test a statistical hypothesis consideration must also be given to two other matters. First, what sample statistic and associated probability distribution will be used to test the hypothesized population parameter. In tests concerning μ we naturally will use the sampling distribution of $\overline{x}$ and the standard normal probability distribution. The second important matter to be considered is the specifying of a reject region. This will involve the selection of critical value(s) beyond which we will reject the hypothesis.

When statisticians reject a hypothesis they wish to have a high degree of confidence in their action. The sample data outcome leading to rejection of the hypothesis must be very unlikely. Two **levels of significance** are commonly used by statisticians. If an outcome occurs with a probability of .05 or less, when H_0 is assumed true, then the outcome is said to be **statistically significant**. If an outcome occurs with a probability of .01 or less, when H_0 is assumed true, then the outcome is said to be **highly significant**.

In practice the statistician decides on a level of significance for a hypothesis test based on the risk involved if an error is made in accepting or rejecting the null hypothesis. Since not all of the population data is examined, judgments made on the basis of the sample data can be in error in two distinct ways. The null hypothesis may be true and we reject it based on an outcome that occurs very infrequently, or the null hypothesis may be false and we accept it as true based on an outcome that does not fall in our reject region. The level of significance is often chosen to minimize the effect of one of the two types of errors. For example, if an experiment concerning possible cures for cancer is undertaken, the statistician may wish to insure that certain results are highly significant (.01 level) before any claims of success are made. In both the examples which we will present and in the problems at the end of this section, the level of significance is arbitrarily chosen.

The preliminary remarks regarding a hypothesis test can be summarized into a six-step procedure for testing a hypothesis of the form $\mu = k$.

1. State a null hypothesis, H_0.
2. Determine an appropriate alternate hypothesis, H_1.
3. Select a level of significance for the test (.05 or .01 level).
4. Based on the sampling distribution of $\bar{x}$ and the standard normal distribution, identify the critical region.
5. Compute an observed value of the z score based on the sample data.
6. Accept or reject the null hypothesis.

The following examples illustrate how we may apply the six-step procedure.

Example 1: A manufacturer of ball point pens has automatic equipment that produces pens of reasonably uniform diameter. An inspector hypothesizes that the mean diameter of all the pens produced is 0.250 in. and the standard deviation is .10. A sample of 64 pens is chosen at random and the $\bar{x}$ of the sample is .268. Does the sample result lead us to accept the hypothesized value?

Solution:

1. $H_0: \mu = .250$.
2. $H_1: \mu \neq .250$. The two-tailed alternative is preferred. Before the data is collected, we have no reason to believe that the diameter is in one direction only.
3. Select a level of significance of .05.
4. The sampling distribution of $\bar{x}$'s and the standard normal distribution is used. For two-tailed tests the .05 level of significance is split and .025 assigned to each tail to determine the critical z values. From the standard normal tables $\Pr(z > 1.96) = .025$ and $\Pr(z < -1.96) = .025$, thus $\Pr(-1.96 < z < 1.96) = .95$. The critical z values are $z = 1.96$ and $z = -1.96$.

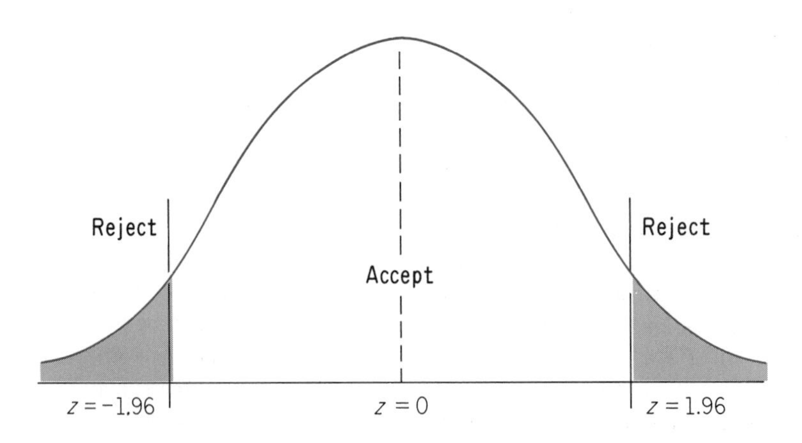

5. Using the sample data the value for $\sigma_{\overline{x}}$ is given by

$$\sigma_{\overline{x}} = \frac{\sigma}{\sqrt{n}} = \frac{.10}{\sqrt{64}} = .0125.$$

Hence the observed value of the z score is

$$z_{obs} = \frac{x - \mu}{\sigma_{\overline{x}}} = \frac{.268 - .250}{.0125} = \frac{.018}{.0125} = 1.44.$$

6. Since the observed value of z falls in the acceptance region, we accept the null hypothesis, $H_0: \mu = 0.250$.

Example 2: Over a period of several years students at a given college have received an average grade of 76 in their first college-level English course. The grades have been normally distributed with a standard deviation of 8. An educator claims that the average grade of each student can be significantly improved by the use of audio-visual materials. To test the claim a random sample of 25 students is assigned to a class using the audiovisual materials. At the end of the course, the class average is 81. Do the results justify the claim that audiovisual materials will improve student performance?

Solution:

1. $H_0: \mu = 76$.
2. $H_1: \mu > 76$. A one-tailed test is indicated since the claim is made that the average will be significantly improved.
3. Select a level of significance of .01.
4. The sampling distribution of $\overline{x}$'s will be used and the standard normal distribution. We assign .01 to the upper tail to determine the critical z value. From the standard normal tables $\Pr(z > 2.33) = .01$. Thus the critical z value is $z = 2.33$.

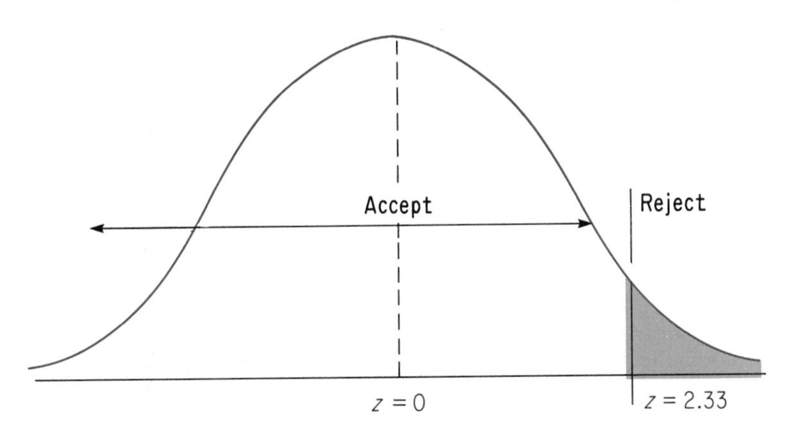

5. Using the sample data the value for $\sigma_{\overline{x}}$ is given by

$$\sigma_{\overline{x}} = \frac{\sigma}{\sqrt{n}} = \frac{8}{\sqrt{25}} = 1.6.$$

Hence, the observed value of the z score is

$$z_{obs} = \frac{x - \mu}{\sigma_{\overline{x}}} = \frac{81 - 76}{1.6} = \frac{5}{1.6} = 3.125.$$

6. Since the observed value of z falls in the reject region we reject H_0 which means we accept the alternative hypothesis. The use of audiovisual materials does significantly improve the students' performance.

The type of hypothesis testing presented in this section has been restricted to tests regarding the parameter μ. A more general theory exists that extends the basic concepts discussed here to testing of other parameters which in turn requires a consideration of probability distributions other than the standard normal.

Generally, the person who collects data is interested in a great deal more than simply reporting the mean or standard deviation of a set of sample data. He may wish to compare these results with other results and be able to decide whether or not the differences are statistically significant.

Statistical theory of hypothesis testing provides anyone who collects sample data from population's with a technique for making decisions. Decisions made on the basis of statistical analysis are, of course, subject to error. However, the type of error and the probability of error can be precisely determined and we can state conclusions and make decisions at known levels of probability. Thus we have a control mechanism applicable in most human endeavors that can eliminate haphazard decision making.

Problem Set 5.8

In the hypothesis test questions which follow, use the six-step procedure described in the text.

A. 1. A national magazine indicates that cars in the United States are driven an average of 8,000 miles per year with a standard deviation of 1,600 miles. To test this hypothesis a random sample of 100 automobile owners is surveyed. Use a level of significance of .05. What decision would be made if the $\overline{x}$ of the sample is 7,500 miles?

2. A stock broker claims that he can improve the earnings of his clients by more than the current average of $2.50 per share. The population standard deviation is $0.35. If a random sample of 49 of his clients have an average earnings per share of $2.70, would you accept his claim at an .01 level of significance?

3. A chemistry instructor states that a class run without a laboratory will not have a significantly different level of performance than a class with a laboratory. Past records indicate students with a laboratory have had an average grade of 74.0 with a standard deviation of 8. A class of 25 students, randomly selected, is assigned to the same chemistry course without a laboratory. Their final class average is 71.6. At an .05 level of significance, determine whether the results obtained are statistically significant.

4. In order to insure that a fruit-juice manufacturer is placing at least 15 ounces in

the can as stated on the label, a government inspector makes periodic checks. The standard deviation of the assumed normal population is 2.4 ounces. A random sample of 16 cans have a mean of 14.25 ounces. At an .01 level of significance is there reason to believe the manufacturer is not meeting the standard?

5. Internal Revenue records show that the average deduction claimed for property taxes is $450 per year with a standard deviation of $120. Taxpayers in a given area claim their property taxes are significantly higher than the national average. As evidence they cite a survey which showed that a random sample of 36 taxpayers in their area had an average property tax of $500 per year. At an .05 level of significance would you accept their claim?
6. A new fertilizer is claimed to give a significantly higher crop production. A potato farmer has been averaging 20 bushels per acre with a standard deviation of three bushels. After applying the fertilizer the farmer obtains a yield of 21.5 bushels per acre for his 50-acre farm. Do the results justify the claim for the fertilizer? Use a .01 level of significance.
7. College students at a given college read an average of 48 books per year with a standard deviation of 12 books. A random sample of 25 science majors read an average of 45 books per year and a random sample of 25 humanities majors read an average of 52 books per year. At the .05 level of significance are either of the sample groups different from the college-wide average?
8. As part of their hiring procedure a certain company gives all prospective typist employees a speed-typing test. Records indicate the average time taken for the test is 4.5 minutes with a standard deviation of 42 seconds. A business shcool sends 9 of their graduates to the company for possible employment and they take an average time of 4.0 minutes. Are these graduates significantly different (.05) than others who have taken the same test?
9. A golf-ball testing machine is calibrated to read in yards the distance a ball is driven. Over a long period of time Brand X golf balls average a distance of 250 yards with a standard deviation of 20 yards. Brand X changes the material of the golf-ball core and claims a longer driving distance will result. A random sample of 25 is chosen and the average obtained is 260 yards. At the .01 level of significance is the driving distance significantly improved?
10. A clothing industry survey shows that the average cost to manufacture men's suits is $45 with a standard deviation of $7.20. A new production procedure is adopted at a certain company aimed at cutting production costs. A random sample of 36 men's suits is selected from the new production and their average cost is $43.20. At an .05 level of significance, does the result indicate the cost of production has been significantly reduced?

Review Exercises

5.1 Define the following terms:

(a) Descriptive statistics.
(b) Inferential statistics.
(c) Measures of central tendency.
(d) Measures of variability.

(e) Discrete data.
(f) Continuous data.
(g) Frequency polygon.
(h) Parameters.
(i) Central limit theorem.
(j) Critical region.
(k) Level of significance.

5.2 For a given thirty-day period, the maximum temperature readings rounded to nearest degree were as follows:

55, 59, 62, 66, 69, 50, 56, 60, 65, 70, 58, 67, 63, 54, 59,
64, 60, 60, 53, 55, 62, 66, 59, 52, 53, 61, 63, 64, 56, 57.

Using an interval of 3 beginning with 50–52 inclusive, develop a frequency distribution, histogram, and frequency polygon.

5.3 Using the data in Prob. 5.2 find the mean of the total number of entries by using the midpoint of each interval. Find the median and mode.

5.4 If the mean of a set of data is 12 what will be the mean of a new set of data formed by multiplying each entry in the original set of data by 4 and then adding 5 to the product obtained.

5.5 A population consists of the entries 1, 1, 2, 3, 3, 5. Find the (a) variance, (b) standard deviation, (c) mean and standard deviation of the sampling distribution of the means of the samples of size 4 drawn with replacement from the population.

5.6 A sample of persons drawn from a large population have the following ages: 5, 11, 5, 87, 11, 10, 11, 20. Find the (a) range, (b) median, (c) mode, (d) mean, (e) sample variance.

5.7 A sample of 5, 10, 12, 51 is drawn from a population. Find the sample variance and use this information to find the variances of the samples.
(a) 7, 12, 14, 17.
(b) 25, 50, 60, 75.

5.8 With reference to the standard normal distribution, find (a) $\Pr(z > 1.57)$, (b) $\Pr(z < 2.81)$, (c) $\Pr(-1.5 < z < 1.5)$, (d) a z_0 such that $\Pr(-z_0 < z < z_0) = .88$.

5.9 A national survey indicates that the average amount in bank savings accounts is \$220. and the standard deviation is \$70. What percentage of the population have savings accounts in excess of \$360?

5.10 A coin is weighted so that the probability of a head is 1/4, i.e., $\Pr(H) = 1/4$. Using the normal approximation find the probability that in the next 27 tosses between 5 and 8 heads inclusive occur?

5.11 An opinion poll taken claims that 80 percent of the population favor the President's legislative program. If this claim is correct, what is the probability that in a random sample of 50 (a) at least 30 favor the President's program, and (b) less than 30 favor the President's program.

5.12 A company is considering for purchase a new machine which makes bolts. They will buy the machine unless the fraction of defects exceeds 10 percent. If $H_0: p = .10$ and $H_1: p > .10$, and if for a sample of size 20 the critical or reject region corresponds to 5 or more defective parts, what is the probability H_0 will be rejected even if H_0 is true?

5.13 If all possible samples of size 36 are drawn from a large population having a mean of 112 and standard deviation of 24, what is the probability that a given sample mean will fall between 112 and 120.

5.14 A certain type of rat shows a normally distributed mean weight gain of 65 grams during the first three months of life. A sample of 16 rats is fed a different diet from birth until they are three months old. Using a .05 level of significance, set up a hypothesis test using the information that the average weight gain of the sample was 61 grams.

5.15 The daily wages in a particular industry average $26.40 per day with a standard deviation of $5.60. A company is suspected of underpaying their employees. A random sample of 49 workers is chosen and their average daily pay is $25.20. At an .01 level of significance, can this company be accused of paying inferior wages?

Suggested Reading

Adams, J. K., *Basic Statistical Concepts.* McGraw-Hill, New York, 1955.

Crouch, Ralph, *Finite Mathematics with Statistics for Business.* McGraw-Hill, New York, 1968.

Freund, John E., *Modern Elementary Statistics.* Prentice-Hall, Englewood Cliffs, N.J., 1970.

Hoel, P. G., *Introduction to Mathematical Statistics.* Wiley, New York, 1954.

Neyman, J., *First Course in Probability and Statistics.* Holt, Rinehart and Winston, New York, 1950.

Spiegel, Murray R., *Theory and Problems of Statistics.* Schaum Outline Series, McGraw-Hill, New York, 1961.

Someone who had begun to read geometry with Euclid, when he had learned the first proposition, asked Euclid, "But what shall I get by learning these things?" Whereupon Euclid called his slave and said, "Give him three-pence since he must make gain out of what he learns."

—STOBAEUS

CHAPTER **6**

Systems of Linear Equations, Vectors, and Matrices

6.1 Preliminaries

A common human experience is to observe the relationship that exists between variables. For example, distance traveled is observed to vary both as the speed and the time an object is in motion. Expenses are related to the amount of income whether it be a single individual or a large corporation. In almost every human activity man seems compelled to uncover fundamental relationships that exist among the objects he observes.

In order to make the relationships that may exist between variables explicit, we frequently attempt to make a mathematical model that will accurately reflect the real life situation. Many mathematical models have the same basic structure although different symbolic notations may be utilized. For example, the algebra of sets and symbolic logic introduced earlier were models of the abstract mathematical system, Boolean algebra. If the significant features which are commonly shared are recognized and an abstraction process is initiated that explicitly identifies the primitive terms, formulates definitions, states what assumptions are being used, and theorems are logically deduced, then we have a mathematical system. In this chapter and the next we will be considering such mathematical systems as linear equations, vectors, matrices, and linear programming. Each of these systems will have its own special characteristics, and each will be defined for our purposes over the real number system. It is appropriate, therefore, that before proceeding we state what are known as the field properties which are applicable to **R**, the set of real numbers.

Field Axioms

1. Every pair of numbers a and b in R have a unique sum $a + b$ which is also in R and a unique product ab which is also in R. — **Closure laws**
2. For any a and b in R — **Commutative laws**

$$a + b = b + a \quad \text{and} \quad ab = ba$$

3. For any a, b and c in R — **Associative laws**
   $(a + b) + c = a + (b + c)$
   $(ab)c = a(bc)$.
4. There is a unique real number in R symbolized "0" and called "zero" such that for each real number a in R — **Additive identity**
   $a + 0 = a$.
5. There is a unique real number in R symbolized, "*1*", $1 \neq 0$, such that for each real number a in R — **Multiplicative identity**
   $a \cdot 1 = a$.
6. For every a in R there is a unique number in R, symbolized $-a$, such that — **Additive inverse**
   $a + (-a) = 0$.
7. For every nonzero a in R there is a unique number, symbolized a^{-1} $\left(\text{or } \frac{1}{a}\right)$, such that — **Multiplicative inverse**
   $a \cdot a^{-1} = 1$.
8. For any a, b, and c in R — **Distributive law of multiplication over addition**
   $a(b + c) = ab + ac$.

The following axioms specify the meaning we will attach to the equality relation. Let a, b, and c denote real numbers.

Equality Axioms

1. $a = a$. — **Reflexive property**
2. If $a = b$ then $b = a$. — **Symmetric property**
3. If $a = b$ and $b = c$, then $a = c$. — **Transitive property**
4. If $a = b$ and $a + c = d$, then $b + c = d$. Also if $a = b$ and $ac = d$, then $bc = d$. — **Substitution property**
5. If $a = b$ then $a + c = b + c$. — **Addition property**
6. If $a = b$ then $ac = bc$. — **Multiplication property**

Many of the topics which we will discuss will involve algebraic expressions using the field properties. Considerable insight can be gained, however, if these algebraic expressions are related to geometric diagrams. A convenient geometric diagram is a one- or a two-dimensional coordinate system.

A one-dimensional coordinate system of the real numbers may be related to a number line where we assume that every point on the line can be paired with exactly one real number and conversely, every real number can be paired with exactly one point on the line. The line shown in Figure 6.1 designates one point as the real number "*0*" and a second point as the real number "*1*." The number "*1*" lies to the right of "*0*" with the distance between these points being regarded as the unit distance.

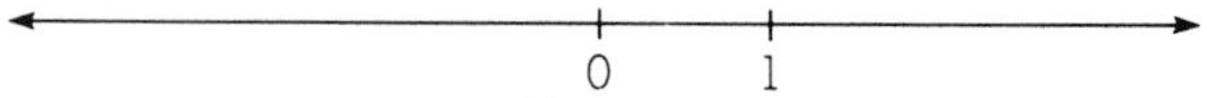

0 1

Figure 6.1

We can now mark off unit distances both to the right of zero which we will regard as "*positive*" and to the left of zero which we will regard as "*negative*." In this manner, as illustrated in Figure 6.2, we can establish a one-to-one correspondence between the integers and the end points of the unit line segments placed on the line.

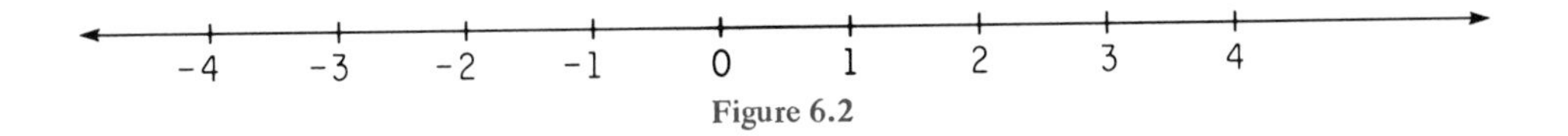

Figure 6.2

Rational numbers such as 1/2, 2/5, 15/7, etc. can also be associated with points on the real number line by subdividing the appropriate unit distance. For example, the rational number 15/7 corresponds to a point 1/7 of the distance between 2 and 3 on the number line.

Irrational numbers can also be paired with points on the real number line. One method for doing this involves a construction technique. For example, the irrational number $\sqrt{2}$ can be identified with a point on the number line by constructing a right triangle with the "legs" each equal to the unit distance. The hypotenuse will then have a length of $\sqrt{2}$ (Pythagorean Theorem). If the right triangle is placed on the number line as shown in Figure 6.3 with the length of the hypotenuse representing the radius of a circle

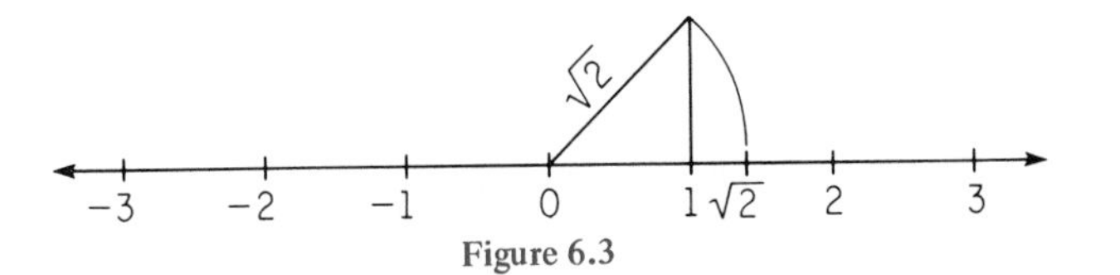

Figure 6.3

with the center at "*0*", the circle will intercept the number line at a point which we designate as $\sqrt{2}$. The real number that corresponds to a unique point on the line is called the coordinate of the point.

The one-dimensional coordinate system of a line can be extended to a two-dimensional rectangular coordinate system of the plane by defining a one-to-one correspondence between ordered pairs of real numbers and points in a plane.

The rectangular coordinate system or what is also called the Cartesian coordinate system (after the mathematician René Descartes) is formed by placing two real number lines, usually having the same scale, perpendicular to each other in such a manner that their point of intersection is the zero of each line. This point of intersection is called the **origin**. The "positive" numbers lie to the right of the origin on the horizontal line and above the origin on the vertical line. As illustrated in Figure 6.4, the horizontal line is identified as the x-axis and the vertical line is called the y-axis.

If P is an arbitrary point in the plane, we will associate with this point a unique ordered pair of real numbers designated (x,y). The first component of the ordered pair is the horizontal **directed distance** from the y-axis to the point P (see Figure 6.4) and the second component is the vertical directed distance from the x-axis to the point P. By directed distance we mean that we will use signed numbers to indicate that from a fixed

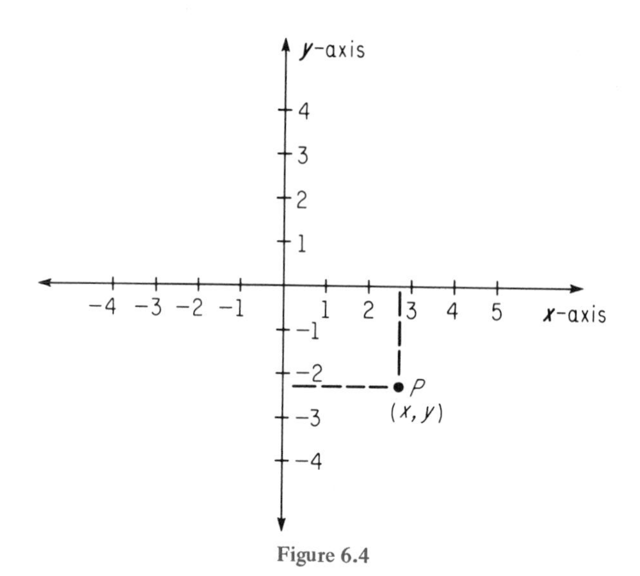

Figure 6.4

point horizontal distance to the "right" and vertical distance "upward" in the plane will be regarded as "positive," whereas horizontal distance to the "left" and vertical distance "downward" will be regarded as "negative."

Example 1: Let P_1 be the point (-3,5), P_2 be the point (-4,-3), P_3 be the point (2,-1), P_4 be the point (1,3), P_5 be the point (0,6), and P_6 be the point (0,0). Locate these points in a two-dimensional coordinate system.

Solution: The respective points corresponding to each ordered pair are shown in Figure 6.5. For example, P_1 (-3,5) is located 3 units to the left of the y-axis and 5 units above the x-axis.

One of the most important uses of a two-dimensional coordinate system is to develop a "**graph**" of an algebraic equation involving two variables. The graph will consist of the geometric representation of the set of all ordered pairs of real numbers in the plane which satisfy a particular equation. Of prime consideration to our work in this text is the graph of **linear equations** in two variables of the form

$$\mathbf{Ax + By = C}$$

where **A**, **B**, and **C** are real numbers, **A** and **B** not both zero.

It can be shown that the geometric form of every linear equation in two variables is a straight line in the plane and conversely, every straight line in the plane has an algebraic form which is a linear equation in two variables.

In order to sketch the graph of a linear equation of the $Ax + By = C$ form, it is sufficient to find two distinct ordered pairs that satisfy the equation and draw a line through the corresponding points in the plane. Two points that are particularly convenient and easy to find are known as the x- and y-intercepts of a line. The x-intercept is the point of

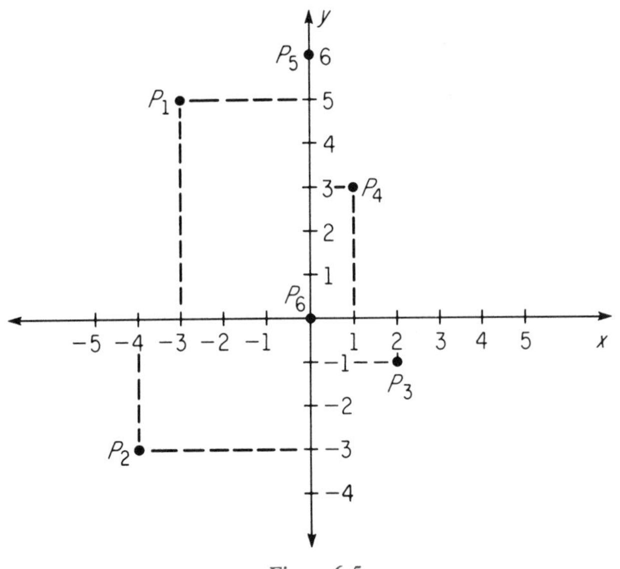

Figure 6.5

intersection of a line with the x-axis and may be found by letting $y = 0$ in the given equation; i.e., the x-intercept will be of the form $(x_1,0)$ where x_1 is a point on the x-axis. Similarly, the y-intercept will correspond to the point where the line intersects the y-axis. The y-intercept may be found by letting $x = 0$ in the given equation and will be of the form $(0,y_1)$, where y_1 is a point on the y-axis. Of course, if a line is parallel to either of the axes only one of the intercepts will exist.

Using the general form of a linear equation; namely, $Ax + By = C$, we can express the x-intercept or y-intercept in terms of A, B, and C as follows:

Let $y = 0$, then

$$Ax + B(0) = C \qquad Ax = C \qquad \text{and} \qquad x = C/A.$$

Thus, the x-intercept is $(C/A,0)$.

Let $x = 0$, then

$$A(0) + By = C \qquad By = C \qquad \text{and} \qquad y = C/B.$$

Thus, the y-intercept is $(0,C/B)$.

Example 2: Sketch the graph of $2x - 3y = 5$ by using the x- and y-intercepts.

Solution: The x-intercept is $(C/A,0)$ or $(5/2,0)$ and the y-intercept is $(0,C/B)$ or $(0,-5/3)$. Drawing a line through the two intercepts we obtain the graph of $2x - 3y = 5$. See Figure 6.6.

If a line is parallel to the y-axis, then the directed distance from the y-axis will be a constant, say k_1, for all points on the line. Since the x-coordinate of any point on the line is always k_1, we can express the equation of the line as $x = k_1$. Although each point

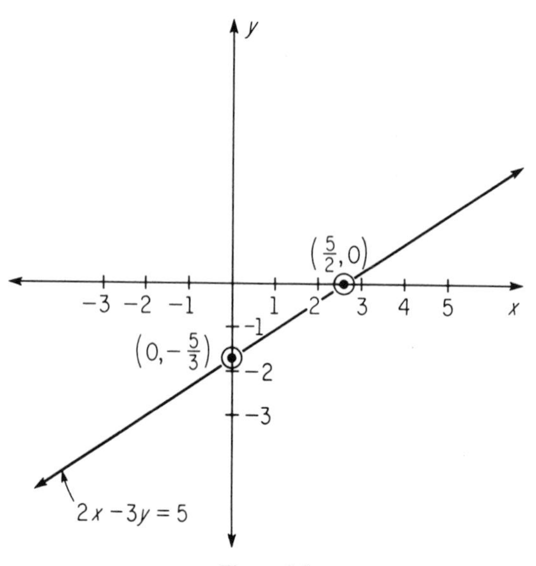

Figure 6.6

on the line has a constant x-coordinate, the y-coordinates of points on the line will vary and, in fact, may be any real number. To relate this to the general form of a linear equation in two variables, i.e., $Ax + By = C$, we observe that in order for the general equation to be satisfied with a fixed x-value and any real number value for y, the coefficient of y, namely B, must be zero. Thus, the equation $Ax + By = C$ becomes $Ax = C$ or $x = C/A$ (the x-intercept) and in this discussion k_1 equals C/A.

Similarly, a line parallel to the x-axis will have a directed distance from the x-axis which will be a constant, say k_2, for all points on the line. The intersection of this line with the y-axis will have an intercept of $(0,k_2)$ and the equation of the line will be $y = k_2$.

Example 3: Sketch the graphs of $y = 4$ and $x = -3$ on a two dimensional coordinate system.

Solution: In Example 3 the two lines drawn in the plane intersect at the point $(-3,4)$. See Figure 6.7. The two linear equations may be regarded as a system of equations and the point in the plane common to both lines as an ordered pair which is the solution to the system of equations. In the next section we will discuss the various geometric possibilities for two lines drawn in the plane as well as the corresponding algebraic techniques for finding the solution of a system of equations.

We have shown that when a linear equation in two unknowns is given, we are able to sketch the graph of the equation, which is a straight line, in a two-dimensional coordinate plane. Conversely, situations arise where we are given a straight line graph and we wish to determine the linear equation of the line. From plane geometry we know that two points determine a unique line. Also, since every nonhorizontal line in the xy-plane will cross the x-axis, it forms an angle, say θ, with the positive x-axis as shown in Figure

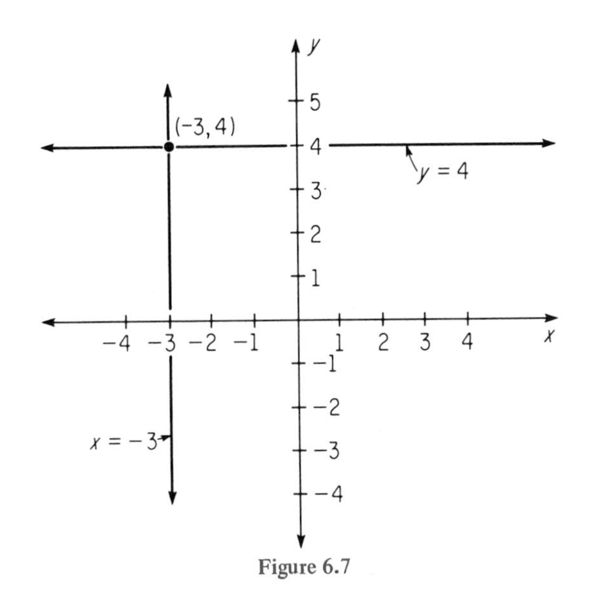

Figure 6.7

6.8. We should further note that the range of θ (in degrees) is from 0 to 180 degrees, i.e., $0^\circ \leqslant \theta < 180^\circ$. The angle θ is called the **angle of inclination** of the given line. We may now give the definition of slope of a nonvertical line.

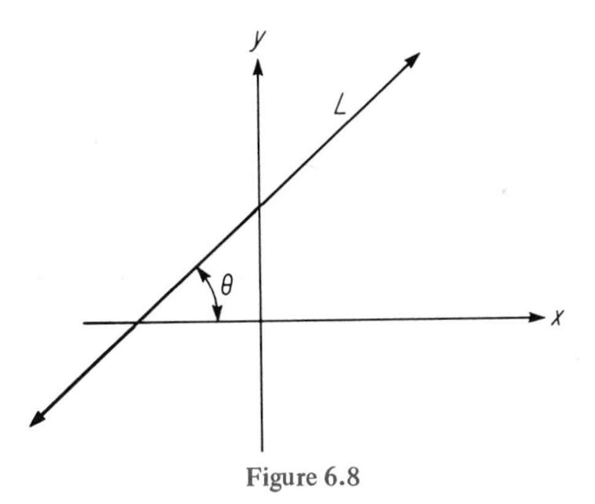

Figure 6.8

DEFINITION: Let L be a nonvertical line whose angle of inclination is θ. **The slope of line L**, denoted m_L, is defined by the following trigonometric equation:

$$m_L = \tan \theta.$$

Note: For $\theta = 90^\circ$, $\tan \theta$ is not defined.

However, to find the value of the slope of a nonvertical line which passes through two given points, we use the following equivalent definition:

DEFINITION: Let L be a nonvertical line which passes through the points P_1,P_2 whose coordinates are respectively (x_1,y_1) and (x_2,y_2). Then the **slope of the line** L, symbolized m_L, is given by

$$m_L = \frac{y_2 - y_1}{x_2 - x_1} \quad \text{or} \quad m_L = \frac{y_1 - y_2}{x_1 - x_2}.$$

In the above definition the points P_1 and P_2 are arbitrarily chosen distinct points on the line. Thus, the slope of a specific line is a unique real number. Also, if two distinct lines have the same slope, they are parallel.

Let L be a line determined by the points P_1,P_2 whose coordinates are (x_1,y_1) and (x_2,y_2), respectively. Furthermore, let P be an arbitrary point, distinct from P_1 and P_2, whose coordinates are (x,y). Then, since the slope of a line is independent of the points chosen, we may choose the coordinates of say, P and P_1 (or P and P_2) to also determine the slope of L. Thus, the slope of L is also given by the following equation:

$$m_L = \frac{y - y_1}{x - x_1}.$$

Since $x \neq x_1$, the equation

$$m_L = \frac{y - y_1}{x - x_1}$$

may be written in an equivalent form as $y - y_1 = m_L(x - x_1)$ which we obtain by multiplying both sides of the equation by $(x - x_1)$.

The equation, $y - y_1 = m_L(x - x_1)$, is called the **point-slope** form of the linear equation that represents the line L. Note that P_1 and P_2 are given fixed points and are used to determine m_L, the slope of the line.

Example 4: Let P_1 and P_2 be two points whose coordinates are $(1,-1)$ and $(4,1)$, respectively. Find the equation of the line determined by P_1 and P_2.

Solution: Let L be the line determined by points P_1 and P_2 as shown in the Figure 6.9. The slope of L is given by

$$\begin{aligned} m_L &= \frac{y_2 - y_1}{x_2 - x_1} \\ &= \frac{1 - (-1)}{4 - 1} \\ &= \frac{1 + 1}{3} \\ &= 2/3. \end{aligned}$$

Now let P be an arbitrary point on the line L. Using the point-slope form with the point P_1, namely, $y - y_1 = m_L(x - x_1)$, we obtain $y - (-1) = 2/3\ (x - 1)$. If we wish to express the result in the $Ax + By = C$ form, we can perform the following algebraic manipulations.

$$y + 1 = 2/3\ (x - 1)$$
$$3(y + 1) = 2(x - 1)$$
$$3y + 3 = 2x - 2$$
$$3y - 2x = -5$$

thus, $3y - 2x = -5$ is the linear equation of the line determined by P_1 and P_2 (Figure 6.9).

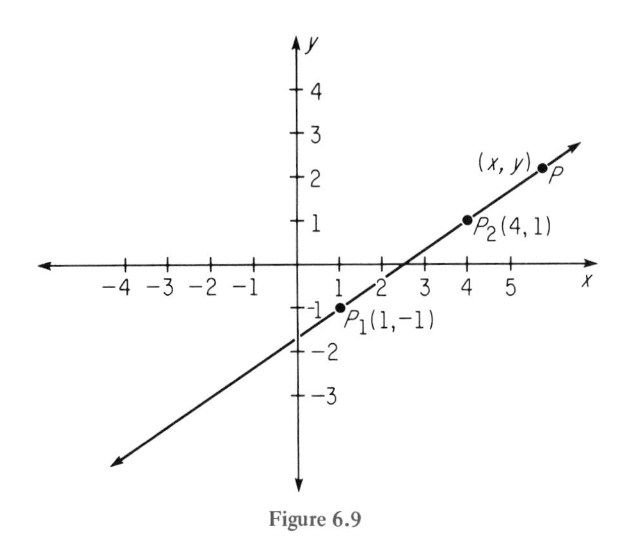

Figure 6.9

Two final observations need to be made. Although convention dictates that "x" and "y" are used to designate the respective horizontal and vertical axes, no alphabetic restriction is necessarily implied by the coordinate system development given in this section. In many applied problems it is more convenient to use letters which represent the variables being considered. For example, the equation denoting the velocity (V) per unit time (t) of an object thrown downward from the top of a building with an initial velocity V_o is given by the expression $V = 32t + V_o$. In sketching a graph it would be convenient to let "V" correspond to the vertical axis and "t" correspond to the horizontal axis. Note that V_o is an arbitrary constant and corresponds on the graph to the "V" intercept.

Finally, coordinate systems could be extended beyond a two-dimensional rectangular form to a three-dimensional rectangular form. In this text, we will discuss linear equations and systems of linear equations in more than two variables from an algebraic point of view. However, we will limit our geometric interpretations to the two-dimensional Cartesian plane.

Problem Set 6.1

A. 1. Discuss how you would determine the points on the real line that correspond to the real numbers 3 and $-\sqrt{5}$.

2. Draw a two-dimensional Cartesian coordinate system that uses the same unit distance for each axis. Locate the points P_1, P_2, P_3, P_4, and P_5 whose coordinates are $(-2,3)$, $\left(-1, -\frac{7}{2}\right)$, $\left(2\sqrt{2}, \frac{1}{2}\right)$, $(-5,-1)$, and $\left(\frac{3}{2}, -\frac{5}{3}\right)$, respectively.
3. Determine the x- and y-intercepts of each line. Sketch the line determined by each of the following equations.
(a) $2x - 3y = 6$. (b) $x + y = 10$. (c) $x - y = 1$.
(d) $x - y = 10$. (e) $4x + 5y = 7$.
4. Find the slope, and the x- and y-intercepts of the line determined by the equation $3y - 2x = 6$.
5. Write in $Ax + By = C$ form the equations of the lines determined by the following pairs of points:
(a) $(3,0), (0,1)$. (b) $(2,3), (1,1)$. (c) $(6,6), (-1,-1)$.
(d) $(-5,5), (3,-3)$. (e) $(1,-3), (-2,-1)$.
6. For the equation, $3y = 5$, sketch the line determined by the given equation. Also find the slope, and if they exist, the x- and y-intercepts.
7. Sketch in the same coordinate system, the lines, determined by the equations $2x - y = 5$ and $x + y = 1$. If these lines intersect, what are the coordinates of this point? Does this point satisfy both given equations? Is the point of intersection the only point that will satisfy both equations?
8. Sketch in the same coordinate plane, the lines, determined by the equations $2x - y = 3$ and $4x - 2y = 12$. Do these lines intersect?
9. Give a reason for each statement in the proof of the following theorem. For every real number a, $a \cdot 0 = 0$.

PROOF:
(a) $0 + 0 = 0$.
(b) $a \cdot (0 + 0) = a \cdot 0$.
(c) $a \cdot 0 + a \cdot 0 = a \cdot 0$.
(d) $(a \cdot 0 + a \cdot 0) + (-(a \cdot 0)) = a \cdot 0 + (-(a \cdot 0))$.
(e) $(a \cdot 0) + ((a \cdot 0) + (-(a \cdot 0))) = a \cdot 0 + (-(a \cdot 0))$.
(f) $a \cdot 0 + 0 = 0$.
(g) $a \cdot 0 = 0$.
10. Give a reason for each statement in the proof of the following theorem. If a is a real number, then $-(-a) = a$.
(a) $a + (-a) = 0$.
(b) $[a + (-a)] + (-(-a)) = 0 + (-(-a))$.
(c) $a + ((-a) + (-(-a))) = 0 + (-(-a))$.
(d) $a + 0 = 0 + (-(-a))$.
(e) $a = -(-a)$.
(f) $-(-a) = a$.
11. In a given physics experiment, pressure was measured with respect to temperature. The following table was obtained:

Temperature (T) in °C	5	10	15	20	25	30
Pressure (P) in pounds per square inch	10	20	30	40	50	60

From the table, plot the ordered pairs (T,P) in a TP coordinate system, where T is the horizontal axis and P is the vertical axis. Assume that these points give a linear relationship between pressure and temperature. Draw this line in the coordinate system and then write the equation that determines this line.

6.2 Linear Systems of Equations in Two Unknowns

In the previous section the reader was introduced to the basic concepts of a one- and a two-dimensional coordinate system as well as the associated graphs of linear equations. We are now prepared to discuss the geometric significance of a system of linear equations drawn in a two-dimensional coordinate system together with the algebraic techniques that aid us in determining solutions to such systems if they exist.

Let us begin by recalling that the graph of a linear equation in two unknowns will be a straight line when drawn in the Cartesian plane. If two linear equations in two unknowns are drawn on the same set of axes, three distinct geometric situations may arise.

1. **The two lines may intersect at a unique point.**
2. **The two lines may coincide.**
3. **The two lines may be parallel to each other and have no point in common.**

To each of the three geometric possibilities an algebraic interpretation can be given. Particularly important to us is the case in which two lines intersect at a unique point since the algebraic counterpart will represent a unique solution of a system of equations. Before proceeding further with our analysis, however, let us make explicit what is meant by a system of equations and solutions of the system.

DEFINITION: Let $a_1x + b_1y = c_1$, and $a_2x + b_2y = c_2$ be two linear equations in the variables x and y where the a_i's, b_i's, and c_i's are constants. When these equations are treated simultaneously they are called a system of two linear equations in two unknowns, and the system is denoted as

$$\begin{cases} a_1x + b_1y = c_1 \\ a_2x + b_2y = c_2. \end{cases}$$

DEFINITION: A **solution** to a system of two linear equations in two unknowns is an ordered pair (x_1,y_1) of real numbers which satisfies both equations. The set of all solutions to a given system is known as the **solution set** of the system.

We will illustrate the concepts involved in the preceding definitions by consideration of the following example.

Example 1: Determine geometrically the solution set of the system of equations

$$\begin{cases} 3x + 2y = -6 \\ x + 2y = \ \ 2. \end{cases}$$

Solution: Draw the graph of each linear equation in the xy-plane (Figure 6.10).
The point of intersection of the two lines in the plane is (−4,3). Since this ordered pair satisfies each of the given linear equations, (−4,3) is a solution of the system of equation. This solution is unique since two distinct lines which are not parallel can intersect in one and only one point.

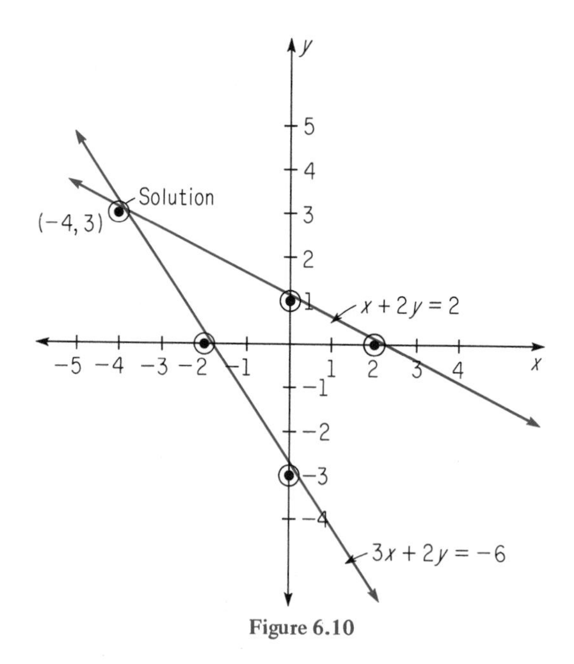

Figure 6.10

In the previous example we were able to find the solution geometrically without difficulty since each component of the ordered pair was an integer. However, solutions to linear systems are not restricted to ordered pairs having integers for each component. For example, if the solution of some system of linear equations is the ordered pair (2,0.717117) then the geometric method will only afford us a close approximation to the actual solution. To obtain precise solutions to systems of equations geometric methods must be abandoned and algebraic techniques employed.

Let us reconsider the system of equations in **Example 1**,

$$\begin{cases} 3x + 2y = -6 \\ x + 2y = \ \ 2 \end{cases}$$

and illustrate how we may use algebraic methods to obtain a solution.

Let (x_1,y_1) represent a solution of the system. We will attempt to find the numerical value of x_1 and y_1. If (x_1,y_1) is a solution, then this ordered pair will satisfy both equations. Thus,

(1) $3x_1 + 2y_1 = -6$

and

(2) $x_1 + 2y_1 = 2.$

Solving Eq. (2) for x_1 we have

(3) $x_1 = 2 - 2y_1.$

Now let us substitute the value of x_1 back into Eq. (1) which will give us a single equation in one unknown (y_1) that can be solved as follows:

$$\begin{aligned} 3(2 - 2y_1) + 2y_1 &= -6 \\ 6 - 6y_1 + 2y_1 &= -6 \\ -4y_1 &= -12 \\ y_1 &= 3. \end{aligned}$$

To obtain the value of x_1 we can substitute y_1 back into Eqs. (1), (2), or (3). Using Eq. (3) we have

$$\begin{aligned} x_1 &= 2 - 2(3) \\ &= 2 - 6 \\ &= -4. \end{aligned}$$

Thus, the ordered pair solution to the system of equations is (−4,3).

The algebraic method which we have just described is known generally as the **substitution method**. There are many methods for solving linear systems. Most of the alternate methods depend in some way on the transformation of a given system into an **equivalent system** which has the same solution set as the original system. The transforming of one system into another, each with identically the same solutions, is accomplished by the repeated application of one or more of the following **elementary operations**.

E_1 **Each term in an equation may be multiplied by the same nonzero real number.**

E_2 **Any equation in the system may be interchanged with any other equation in the system.**

E_3 **Any equation multiplied by a real number may then be added to any other equation in the system.**

Example 2: Solve the system of equations

$$\begin{cases} 3x - 5y = 7 \\ 2x - 4y = 6 \end{cases}$$

by the use of the elementary operations.

Solution: The system

$$\begin{cases} 3x - 5y = 7 \\ 2x - 4y = 6 \end{cases}$$

is equivalent to

$$\begin{cases} 3x - 5y = 7 \\ x - 2y = 3 \end{cases}$$

by use of E_1 where the second equation is multiplied by the real number 1/2. The system

$$\begin{cases} 3x - 5y = 7 \\ x - 2y = 3 \end{cases}$$

is equivalent to

$$\begin{cases} y = -2 \\ x - 2y = 3 \end{cases}$$

by the use of E_3 where the second equation is multiplied by -3 and added to the first equation. We now have a solution for y, namely $y = -2$. Continuing, the system

$$\begin{cases} y = -2 \\ x - 2y = 3 \end{cases}$$

is equivalent to

$$\begin{cases} y = -2 \\ x = -1 \end{cases}$$

by the use of E_3 where the first equation is multiplied by 2 and added to the second equation. The solution for the system is $(-1,-2)$.

The strategy used in the preceding example was to find a multiplier such that when a given equation is multiplied by a real number and added to a second equation then one of the variables is eliminated. By repeated application of this principle an equivalent system can be obtained in which if there is an ordered pair solution (x_1,y_1) then the equivalent system form will be

$$\begin{cases} x = x_1 \\ y = y_1. \end{cases}$$

In the examples considered thus far we have found a single ordered pair solution for a system of two linear equations. When there exists a unique solution to a system of equations, the system is said to be **consistent** and the equations are **independent**. From a geometric point of view the equations would represent straight lines that intersect at a single point.

If one line in a plane should coincide with another line then they have an infinite number of points in common.

The linear equations representing the two coincident lines are said to be **dependent** equations.

If two lines in a plane should be parallel to each other then they would have no points in common. The system of equations representing the parallel lines is said to be **inconsistent**.

Before giving a precise algebraic meaning to the terms inconsistent, and dependent, we will illustrate both the algebraic and the geometric approach to solving a system of equations.

Example 3: Graph and attempt to solve algebraically the system of equations

$$\begin{cases} 4x + 3y = 6 \\ 8x + 6y = 20. \end{cases}$$

Solution: The graph of each linear equation is shown in Figure 6.11. The drawing suggests that the lines are parallel and thus have no point in common.

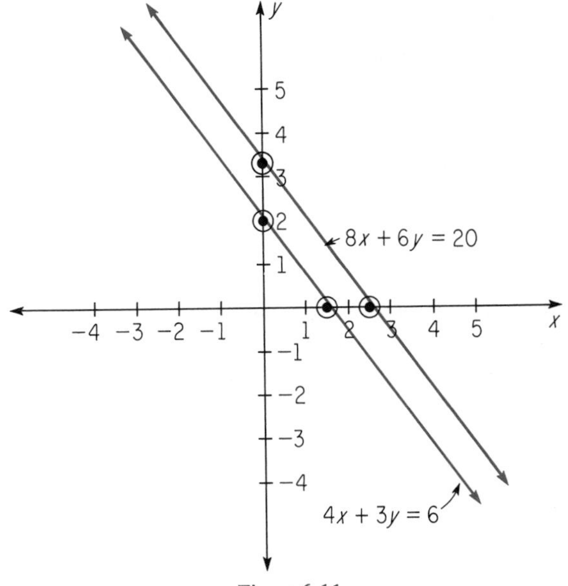

Figure 6.11

However, let us assume that a solution does exist and attempt to apply the elementary operations.

$$\begin{cases} 4x + 3y = 6 \\ 8x + 6y = 20 \end{cases}$$

is equivalent to

$$\begin{cases} 4x + 3y = 6 \\ \qquad 0 = 8 \end{cases}$$

by E_3 where the first equation is multiplied by −2 and added to the second equation. The statement $0 = 8$ is false. Thus, our assumption that a solution exists has led us to a contradiction. Our assumption must be false. Therefore, there is no solution to this system of equations. This system of linear equations is inconsistent.

In general, if

$$\begin{cases} a_1x + b_1y = c_1 \\ a_2x + b_2y = c_2 \end{cases}$$

is a system of equations, where

$$a_1 = ka_2 \qquad \text{and} \qquad b_1 = kb_2$$

but

$$c_1 \neq kc_2$$

for some nonzero real number k, then the system of equations will be **inconsistent.** In other words, if a multiplier can be found that will make the coefficients of the respective variables in each equations the same, but the constant terms are not equal, then the system must be inconsistent. Knowing the system is inconsistent insures that the lines representing the linear equations will be parallel to each other when drawn in the xy-plane.

Example 4: Graph and attempt to solve algebraically the system of equations

$$\begin{cases} x + 3y = 2 \\ 3x + 9y = 6. \end{cases}$$

Solution: The graph of each linear equation is the same straight line in the plane. An infinite number of points lie on both lines and satisfy the system. Let us assume that a solution exists to this system and apply the elementary operations. The sys-

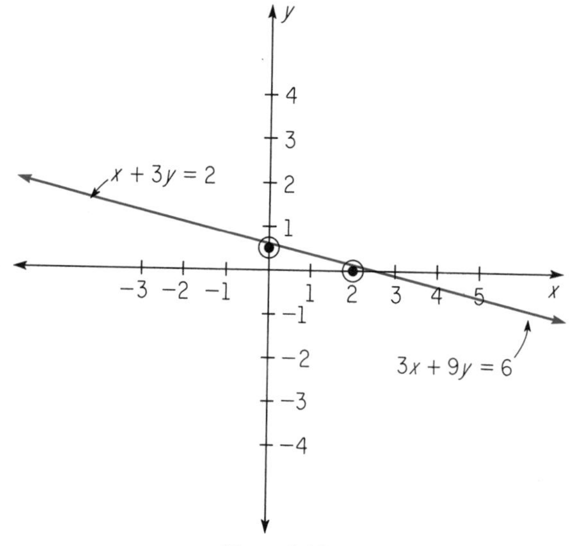

Figure 6.12

tem of equations

$$\begin{cases} x + 3y = 2 \\ 3x + 9y = 6 \end{cases}$$

is equivalent to

$$\begin{cases} x + 3y = 2 \\ 0 = 0 \end{cases}$$

by multiplying the first equation by -3 and adding it to the second equation.

The equation $0 = 0$ is an identity, true for all real numbers. There exists an infinite number of ordered pair solutions which will satisfy both equations and the equations are dependent.

In general, if

$$\begin{cases} a_1x + b_1y = c_1 \\ a_2x + b_2y = c_2 \end{cases}$$

is a system of equations where

$$a_1 = ka_2, b_1 = kb_2, \quad \text{and} \quad c_1 = kc_2$$

for some nonzero real number k then the equations are said to be **dependent**. In other words, if one equation may be transformed term by term into the other equation by multiplication of a single nonzero constant, then the equations are dependent and equivalent to each other. The lines representing these dependent equations will coincide when drawn in the xy-plane.

In order to insure that a system of equations

$$\begin{cases} a_1x + b_1y = c_1 \\ a_2x + b_2y = c_2 \end{cases}$$

does have a unique solution (i.e., is a consistent system) it is sufficient to show that if

$$a_1 = ka_2$$

then

$$b_1 \neq kb_2.$$

In other words, it is impossible to make the respective coefficients of the variables the same by using a single nonzero multiplier. Geometrically, this will mean that the lines are not parallel and do not coincide, so they must intersect in a single point.

Problem Set 6.2

A. 1. Solve the following systems of equations graphically; and determine if the equations are dependent or independent, and if the system of equations is consistent or inconsistent.

(a) $\begin{cases} 3x - y = 5 \\ x + y = 3. \end{cases}$ (b) $\begin{cases} 3x + 3y = 6 \\ x + \frac{3}{2}y = 3. \end{cases}$ (c) $\begin{cases} x + 3y = 2 \\ x - y = 5. \end{cases}$

(d) $\begin{cases} 2y + x = -2 \\ -4y - 2x = -4. \end{cases}$ (e) $\begin{cases} x = -y \\ x - y = 0. \end{cases}$ (f) $\begin{cases} \frac{x}{3} + \frac{y}{5} = 1 \\ \frac{x}{2} - \frac{y}{3} = -1. \end{cases}$

2. Solve each system of equations by the substitution method.

(a) $\begin{cases} 2x - y = 9 \\ x - 5y = 0. \end{cases}$ (b) $\begin{cases} 2x + 3y = -6 \\ 3x - 2y = 12. \end{cases}$ (c) $\begin{cases} 4x - y = 1 \\ 3x - 2y = 4. \end{cases}$

(d) $\begin{cases} \frac{5x}{3} + \frac{y}{4} = \frac{1}{3} \\ \frac{-y}{2} + \frac{2x}{3} = \frac{2}{5}. \end{cases}$ (e) $\begin{cases} x + y = \frac{1}{3} \\ x - y = \frac{2}{3}. \end{cases}$

(f) $\begin{cases} 4x - 3y + 3 = 3x - y + 6 \\ 5x - y - 7 = 2x - 4y + 3. \end{cases}$

3. Solve each of the following system of equations using the elementary operations.

(a) $\begin{cases} 2y - 5x = -7 \\ -3y + 7x = 5. \end{cases}$ (b) $\begin{cases} y - 5x = 3 \\ -10x + 2y = 6. \end{cases}$ (c) $\begin{cases} 2x - 3y = 1 \\ x + 5y = 9. \end{cases}$

(d) $\begin{cases} 6x - 3y = 10 \\ -2x + y = -2. \end{cases}$ (e) $\begin{cases} 3 = 2x + 5y \\ -1 = 3x - 2y. \end{cases}$ (f) $\begin{cases} x = -y \\ x - y = 0. \end{cases}$

4. A direct current experiment results in the following equations for currents (in amperes) i_1 and i_2 in an electrical network: $3i_1 + 2i_2 = 6$, $i_1 - 2i_2 = 2$. Find i_1 and i_2.

5. A piece of rope is 25 feet. Where must it be cut for one piece to be 7 feet longer than the other piece? (*Hint*: Form two equations in two variables, where the variables represent the lengths, and then solve these equations simultaneously.)

6.3 Vectors Over the Reals

In the preceding sections in this chapter we have presented many of the basic concepts of coordinate geometry and we have displayed the close association that exists between an algebraic equation and the graph of that equation. Much of the advance that has taken place in mathematics and science can be directly attributed to the fusion of the algebraic and geometric techniques into analytic geometry and the availability of this model for the exposition and simplification of increasingly complex problems.

In this section we will look at some of the basic concepts of vectors which in a somewhat analagous manner may be studied from both a geometric as well as an analytic point of view. In the sciences, particularly in physics, certain quantities such as force and velocity are said to have both a **magnitude** and a **direction**. Geometrically, these quantities may be represented by an arrow with the length of the arrow representing the magnitude and the position of the arrow in a coordinate system representing the direction. Physicists call the arrows **vectors** and those quantities which may be represented by vectors are referred to as vector quantities.

From an analytic point of view the vector may be considered as part of a mathematical system defined over the real numbers. The abstract system formed is known as a vector space and may be studied independently of any geometric interpretation.

Our primary purpose will be in the development of the properties of a vector space and the extension of these concepts into a study of matrices. Where appropriate, we will indicate the geometric significance of certain vector space concepts.

Let us begin with a fundamental analytic definition of what is meant by the term **vector**.

DEFINITION: An **n-dimensional vector** is an ordered n-tuple of real numbers symbolized, $(a_1,a_2,a_3, \ldots, a_n)$. The numbers $a_1, a_2, \ldots, a_n$ are called **components** of the vector.

The number of components for a given vector determines the dimension of the vector. Thus $(2,-3)$ is a two-dimensional vector and $(-1,0,1)$ represents a three-dimensional vector.

DEFINITION: Let α and β be n-dimensional vectors. **α equals β**, $\alpha = \beta$, if and only if the corresponding components of each vector are equal.

Example 1: Let $\alpha = (a_1,a_2,a_3,a_4)$ and $\beta = (b_1,b_2,b_3,b_4)$. $\alpha = \beta$ if and only if $a_1 = b_1$, $a_2 = b_2$, $a_3 = b_3$, and $a_4 = b_4$.

It is important to observe that $(0,0,0) = (0,0,0)$ and $(1,0,1,0) = (1,0,1,0)$ but $(1,0,0) \neq (0,1,0)$.

DEFINITION: Let $\alpha = (a_1,a_2,a_3, \ldots, a_n)$ be a vector. **The length of vector** α, symbolized $|\alpha|$, is given by

$$|\alpha| = \sqrt{a_1^2 + a_2^2 + a_3^2 + \cdots + a_n^2}.$$

Note, that since $|\alpha|$ is given as the positive square root the length of a vector is always a nonnegative quantity.

Example 2: Let $\alpha = (4,3)$. Find the length of α.

Solution:

$$\begin{aligned} |\alpha| &= \sqrt{4^2 + 3^2} \\ &= \sqrt{16 + 9} \\ &= \sqrt{25} \\ &= 5. \end{aligned}$$

DEFINITION: Let $\alpha = (a_1,a_2,a_3,\ldots,a_n)$ and $\beta = (b_1,b_2,b_3,\ldots,b_n)$ be vectors. **The vector sum** $\alpha + \beta$ is given by $\alpha + \beta = (a_1 + b_1, a_2 + b_2, a_3 + b_3,\ldots,a_n + b_n)$.

In other words, the vector sum is found by adding the corresponding components. Observe, however, that although we can always add any two real numbers, we cannot add two vectors unless they are of the same dimension. This requirement is often described by indicating that the vectors must be **conformable** for addition.

Example 3: Let $\alpha = (1,-2,1/4,\pi)$ and $\beta = (-1,3,1/3,\sqrt{2})$. Find the vector sum $\alpha + \beta$.

Solution:

$$\begin{aligned}\alpha + \beta &= (1 - 1, -2 + 3, 1/4 + 1/3, \pi + \sqrt{2})\\ &= (0,1,7/12,\pi + \sqrt{2}).\end{aligned}$$

Before proceeding further with the development of a mathematical system in which the elements are vectors, let us consider a geometric interpretation of vectors of two-dimensions, i.e., an ordered pair of real numbers. We will regard the two-dimensional vector as a directed line segment with its initial point at the origin of a two-dimensional coordinate system and the ordered pair representing the vector as the terminal point of the vector. For example, if $\alpha = (-1,3)$ then the vector is drawn as shown in Figure 6.13

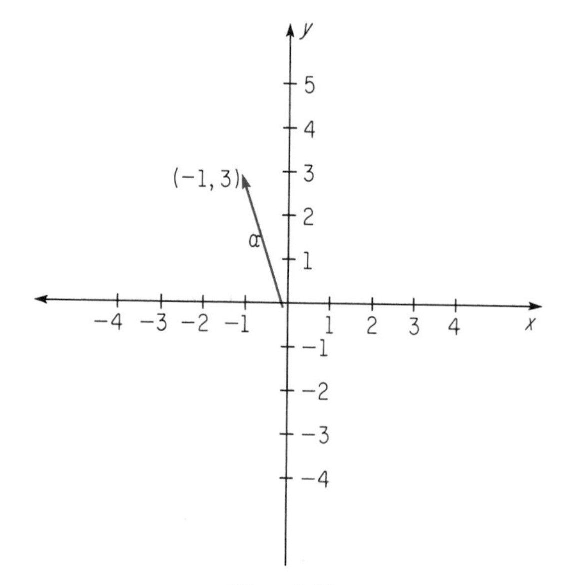

Figure 6.13

with the arrow at the terminal point indicating the direction of the vector.

It is convenient to regard two vectors as being equal in a geometric sense if they have the same length and the same direction. Thus, a vector is unchanged if it is moved parallel to itself and is regarded in this context as a **free** vector.

In general, if we are given a pair of vectors in a two-dimensional coordinate system, we can apply the following definition to determine whether or not the vectors have the same length and the same direction.

DEFINITION: If vector α has its initial point at (a,b) and its terminal point at (c,d) and a vector β has its initial point at (e,f) and its terminal point at (g,h), then $\alpha = \beta$ i.e., α has the same length and direction as β, if and only if

$$c - a = g - e \quad \text{and} \quad d - b = h - f.$$

For example, the vector α having an initial point at (3,4) and a terminal point at (7,1) is equal to the vector β in the standard position having its initial point at the origin (0,0) and its terminal point at (4,−3), since

$$7 - 3 = 4 - 0 \quad \text{and} \quad 1 - 4 = -3 - 0.$$

In other words, by moving the vector α parallel to its given position it can be made to coincide with the vector β. Sliding or moving vectors in the plane parallel to their given position is a very useful technique for displaying the geometric concept of a vector sum.

Example 4: Let $\alpha = (2,5)$ and $\beta = (6,4)$. Find the vector sum and graph α, β and the sum $\alpha + \beta$.

Solution:

$$\alpha + \beta = (2 + 6, 5 + 4)$$
$$\alpha + \beta = (8,9).$$

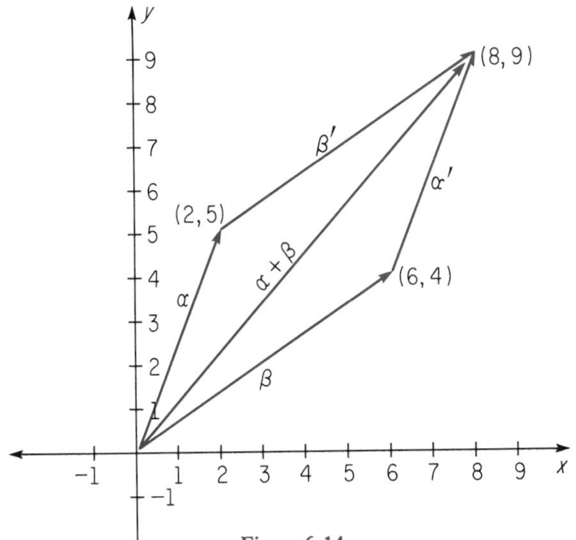

Figure 6.14

In the preceding example, the vector sum is a new vector which is the diagonal of the parallelogram formed by sliding α parallel to its initial position so that α' has its initial point at the terminal point of β and β' is positioned so that its initial point is at the terminal point of α. This technique is frequently employed by engineers and physicists where the two vectors may denote forces acting on an object and the vector sum is called the **resultant** of the forces. The resultant force represents both the magnitude and the direction of a single vector that is equivalent to the combined effect of the two given vectors acting on a given object.

As we have indicated, a vector is an ordered n-tuple of real numbers. A mathematical system based on vectors defined over the reals and designated as a **vector space** will have many properties which are similar to the field properties applicable to the real number system. However, certain vector properties, particularly those related to "multiplication" are somewhat different.

The following properties will precisely define what is meant by a **vector space**.

V1. Closure of addition of vectors The sum of two vectors of the same dimension n, will be a unique vector of dimension n.

V2. Commutativity of addition of vectors If α and β are n-dimensional vectors, then $\alpha + \beta = \beta + \alpha$.

V3. Associativity of addition of vectors If α, β, γ, are n-dimensional vectors, then $(\alpha + \beta) + \gamma = \alpha + (\beta + \gamma)$.

V4. Existence of an additive identity for vectors There exists an additive identity, denoted o_n, called the zero vector, such that for any n-dimensional vector α, $\alpha + o_n = o_n + \alpha = \alpha$. **Note:** The o_n indicates that o_n is a n-dimensional vector such that each component is the real number zero.

V5. Existence of additive inverses For any n-dimensional vector α, there exists a unique vector denoted $-\alpha$, such that,

$$\alpha + (-\alpha) = (-\alpha) + \alpha = o_n.$$

If each component of a n-dimensional vector is multiplied by the same real number we have what is called **scalar multiplication** and the real number multiplier is called a **scalar quantity**.

DEFINITION: If c is a scalar and $\alpha = (a_1, a_2, a_3, \ldots, a_n)$ is a vector, then

$$c\alpha = (ca_1, ca_2, ca_3, \ldots, ca_n).$$

V6. Closure of scalar multiplication If c is a scalar and α is a n-dimensional vector, $c\alpha$ is a unique n-dimensional vector.

V7. Distributive Properties

1. If c is a scalar and α, β are vectors, then

$$c(\alpha + \beta) = c\alpha + c\beta.$$

2. If c and d are scalars and α is a vector, then

$$(c + d)\alpha = c\alpha + d\alpha.$$

V8. Scalar multiplication associativity If c, d are scalars and α is a vector, then

$$(cd)\alpha = c(d\alpha).$$

V9. Identity laws

1. For the scalar zero, 0, and the vector α

$$0 \cdot \alpha = o_n.$$

2. For the scalar one, 1, and the vector α

$$1 \cdot \alpha = \alpha.$$

Each of the vector space properties can be related to a geometric interpretation where the vectors are considered directed line segments. For example, the associativity property of vector addition is illustrated in Figure 6.15. Vector β is drawn with its initial

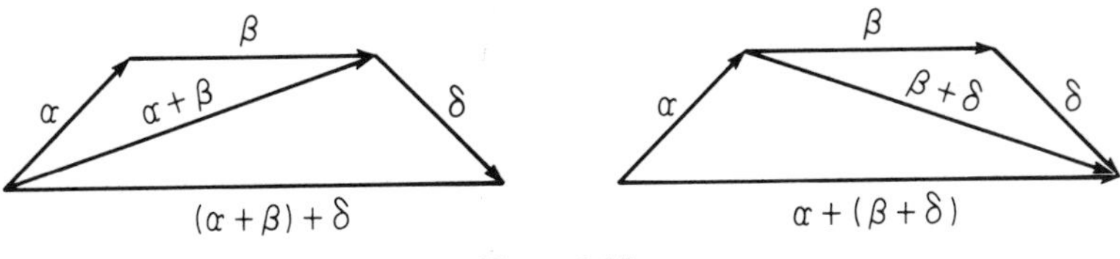

Figure 6.15

point of α and vector δ is drawn with its initial point at the terminal point of β.

Multiplication of a vector by a scalar in a geometric sense affords us a method of changing the magnitude or direction of a given vector. For example, if the two-dimensional vector $\alpha = (3,4)$, which has a magnitude, $|\alpha| = \sqrt{3^2 + 4^2} = 5$, is multiplied by "$-2$" we obtain a new vector $\beta = -2\alpha = (-6,-8)$. The vector β has twice the magnitude of α since $|\beta| = \sqrt{(-6)^2 + (-8)^2} = 10$ and is in the opposite direction from the vector α. In a two-dimensional coordinate system the multiplication of a vector by a scalar, other than the real number 1, will either "stretch" or "shrink" the length of the original vector. If the scalar multiplier is positive, the new vector is in the same direction as the given vector; if the scalar multiplier is negative, the new vector is in the opposite direction.

Thus far, we have symbolized a vector as an ordered n-tuple of real numbers written in a horizontal row. However, it is frequently convenient to write a vector in a vertical column. The terminology expressed in the following definition will be used to distinguish between the two cases.

DEFINITION: A vector α will be called a **row vector** if $\alpha = (a_1,a_2,a_3, \ldots, a_n)$ where the ordered n-tuple is written horizontally. A vector α will be called a **column vector** if

$$\alpha = \begin{pmatrix} a_1 \\ a_2 \\ a_3 \\ \cdot \\ \cdot \\ \cdot \\ a_n \end{pmatrix}$$

where the components are written in a vertical column.

We have previously discussed multiplication of a vector by a scalar. There are other types of multiplication which can also be defined on vectors. One of these, known as the **dot product** of vectors, is defined as follows.

DEFINITION: Let $\alpha = (a_1, a_2, a_3, \ldots, a_n)$ and

$$\beta = \begin{pmatrix} b_1 \\ b_2 \\ b_3 \\ \cdot \\ \cdot \\ \cdot \\ b_n \end{pmatrix}$$

then the **dot product** of α with β is given by

$$\alpha \cdot \beta = (a_1, a_2, a_3, \ldots, a_n) \cdot \begin{pmatrix} b_1 \\ b_2 \\ b_3 \\ \cdot \\ \cdot \\ \cdot \\ b_n \end{pmatrix}$$

$$\alpha \cdot \beta = a_1 b_1 + a_2 b_2 + a_3 b_3 + \cdots + a_n b_n.$$

As a matter of convenience, we have shown the dot product in the form of a row vector multiplied by a column vector. However, in a more general sense, it is not necessary to restrict the definition in this manner. The dot product of two vectors having the same number of components may be considered as the algebraic sum of the products of the corresponding components.

Observe that the dot product is a real number or scalar and that α and β must be of the same dimension.

Example 5: Let $\alpha = (2,-1,0,5)$ and

$$\beta = \begin{pmatrix} 3 \\ -1/2 \\ 2 \\ -1 \end{pmatrix}.$$

Find the dot product $\alpha \cdot \beta$.

Solution:

$$\begin{aligned} \alpha \cdot \beta = (2,-1,0,5) \cdot \begin{pmatrix} 3 \\ -1/2 \\ 2 \\ -1 \end{pmatrix} \\ &= (2)(3) + (-1)(-1/2) + (0)(2) + (5)(-1) \\ &= 6 + 1/2 + 0 - 5 \\ &= 3/2. \end{aligned}$$

One extremely important application of the dot product for vectors of dimension 2 is to determine whether or not two geometric vectors are perpendicular, based on the fact that the numerical value of the dot product is either zero or non-zero. The method employed is expressed in the following theorem.

THEOREM 1: If $\alpha = (a_1,a_2)$ and $\beta = (b_1,b_2)$ are nonzero vectors, then α and β are perpendicular if and only if $\alpha \cdot \beta = 0$, i.e., $a_1b_1 + a_2b_2 = 0$.

PROOF: Applying the Pythagorean Theorem to two vectors as shown in the drawing, we have

$$|\alpha|^2 + |\beta|^2 = |\alpha + \beta|^2$$

if and only if the vectors are perpendicular. Since

$$|\alpha|^2 = a_1^2 + a_2^2$$
$$|\beta|^2 = b_1^2 + b_2^2$$

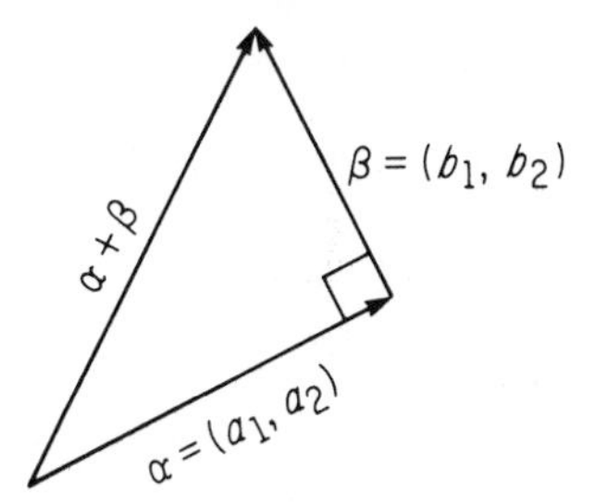

and

$$|\alpha + \beta|^2 = (a_1 + b_1)^2 + (a_2 + b_2)^2$$

then

$$\begin{aligned} a_1^2 + a_2^2 + b_1^2 + b_2^2 &= a_1^2 + 2a_1b_1 + b_1^2 + a_2^2 + 2a_2b_2 + b_2^2 \\ 2a_1b_1 + 2a_2b_2 &= 0 \\ a_1b_1 + a_2b_2 &= 0. \end{aligned}$$

Example 6: Show that the vectors $\alpha = (2,6)$ and $\beta = (-3,1)$ are perpendicular.

Solution:

$$\begin{aligned} \alpha \cdot \beta &= (2,6) \cdot \begin{pmatrix} -3 \\ 1 \end{pmatrix} \\ &= -6 + 6 = 0. \end{aligned}$$

Since the dot product is zero the vectors are perpendicular.

Problem Set 6.3

A. 1. Let $\alpha = (-3,5)$, $\beta = (3,2)$ be vectors. Find

(a) $\alpha + \beta$. (b) $-\beta$. (c) $(-1)\beta$.
(d) $\alpha + (-2)\beta$. (e) $|\alpha|$. (f) $|\beta|$.
(g) $|\alpha + \beta|$.

2. Let $a = (-1,4,-3,-9)$ and

$$b = \begin{pmatrix} 2 \\ -4 \\ -3 \\ -1 \end{pmatrix}$$

be a four-dimensional row and column vectors, respectively. Find
(a) $a \cdot b$. (b) $|a|$. (c) $|b|$.

3. Let $c = (-3,2)$, $d = (-1,5)$ be two-dimensional vectors, respectively. Represent these vectors in a coordinate plane and then find the sum of the vectors $c + d$, by the parallelogram method. Also, find $-d$ and the vector sum $c + (-d)$ by the parallelogram method.

4. Let $a = (-3,4)$ and $b = (4,3)$ be two-dimensional vectors. Find the dot product $a \cdot b$ and the dot product $b \cdot a$. What can be said about the dot product of these two vectors? Is this true for two arbitrary vectors of the same dimension? Prove your answer.

5. Let $\alpha = (a_1, a_2, \ldots, a_n)$ be a vector. Show that $|\alpha| = |-\alpha|$.

6. Let α be an n-dimensional vector. Show that $-\alpha = (-1)\alpha$, where -1 is a scalar.

7. Prove that the addition of n-dimensional vectors is associative.

8. A sailboat is heading downstream on a river whose current is 2-1/2 miles per hour. A 5 mile per hour wind is blowing across the river acting at a right angle to the heading of the boat. What is the resultant velocity of the sailboat if these are the only two forces one regards as acting on the sailboat?

9. Let $a = (4,-7)$ be a two-dimensional vector. What is the relationship between the components of a vector $b = (b_1, b_2)$ so that the dot product of a and b equals zero?

B. 1. Verify by use of a geometric drawing that for two-dimensional vectors α, β, and scalar c, that $c(\alpha + \beta) = c\alpha + c\beta$.

2. If $\alpha = (-2,3)$ and $\beta = (1,-2)$ find the components a, b of vector $\gamma = (a,b)$, such that
(a) $2\alpha + \gamma = \beta$ (b) $\alpha \cdot \gamma = 4$ and $\beta \cdot \gamma = -3$

3. Let α be a vector in the plane with an initial point at the origin and terminal point at $(2,3)$; β has an initial point at $(2,3)$ and terminal point at $(4,-1)$ and γ is a vector with an initial point at $(4,-1)$ and terminal point at the origin. What is the resultant of the vector sum $(\alpha + \beta) + \gamma$?

4. If the line segments joining the points $A(0,0)$, $B(-3,4)$, $C\left(\frac{21}{5}, \frac{68}{5}\right)$, and $D\left(\frac{36}{5}, \frac{48}{5}\right)$ are drawn in the plane show that the figure $ABCD$ is a rectangle. (*Hint*: First show the figure is a parallelogram and then that the dot product of the appropriate vectors is zero.)

5. Two forces act at right angles to each other. The resultant of these forces was found to be 15 pounds. One of the forces was 12 pounds. Find the other force.

6.4 Matrices

Thus far, in this chapter, we have discussed systems of linear equations and vectors. We will now turn our attention to the study of matrices. We will discover that many of the properties of a vector space carry over to matrices and that matrix theory will provide us with a simplified technique for finding solutions (if they exist) of systems of linear equations. Although a relatively new branch of mathematics, matrix theory has an extremely wide range of application. For example, nuclear physicists use certain basic matrix forms to succinctly describe the sub-atomic electron shells and modern business corporations use matrices for such tasks as inventory control. We will present matrices from the mathematical system point of view without restricting our discussion to a few specific applications. Let us begin with a definition of a matrix that will suit our purposes in this text.

DEFINITION: A **matrix** is any collection of real numbers arranged in a rectangular array.

For symbolic convenience we display the real numbers within brackets as illustrated in the following examples.

(a) $\begin{bmatrix} 2 & 3 \\ -1 & 0 \end{bmatrix}$ (b) $\begin{bmatrix} 5 & 7 & -3 \end{bmatrix}$ (c) $\begin{bmatrix} 2 & 3 & -10 \\ 0 & 1 & 2 \end{bmatrix}$ (d) $\begin{bmatrix} 3 \\ -1 \\ 5 \end{bmatrix}$.

The preceding examples illustrate that a matrix may consist of any number of rows or columns provided the display is rectangular. Notice particularly that (b) and (d) bear a striking resemblance to row and column vectors. Example (a) which has the same number of rows and columns is designated as a **square matrix.** In order to describe the shape and size of a matrix as well as identifying particular elements in a matrix, we introduce the following definition.

DEFINITION: A matrix of **dimension** ***mxn***, (read "m by n")

$$\begin{bmatrix} a_{11} & a_{12} & \cdots & a_{1n} \\ a_{21} & a_{22} & \cdots & a_{2n} \\ \cdot & \cdot & \cdots & \cdot \\ a_{m1} & a_{m2} & \cdots & a_{mn} \end{bmatrix}_{m \times n}$$

consists of m rows (each row having n elements) and n columns (each column having m elements).

REMARKS:

1. m and n are natural numbers not necessarily distinct.
2. An $m \times n$ matrix consists of mn elements (each element is a real number).
3. The subscripts used for each element identify the location of each entry. The first subscript indicates the row in which the element appears and the second subscript specifies the column in which the element appears.

Example 1: Specify the dimension of the matrix

$$\begin{bmatrix} 1 & -2 & 1/2 & \pi \\ 0 & 5 & -1 & 3/4 \\ 1/3 & 2 & 0 & 0 \end{bmatrix}$$

and identify the elements a_{13}, a_{33}, a_{34}, and a_{24}.

Solution: The dimension of the matrix is 3×4 with $a_{13} = 1/2$, $a_{33} = 0$, $a_{34} = 0$, and $a_{24} = 3/4$.

When we are considering a matrix as part of a system it will be convenient to have a more condensed notation to describe the essential characteristics. One method is to denote a matrix by a capital Arabic letter A, B, C, etc., with the $m \times n$ dimension specified as a subscript. For example, $A_{5 \times 3}$ will represent a matrix having 5 rows and 3 columns. An alternate method is to let the symbol a_{ij} represent an arbitrary element of the matrix. The notation $[a_{ij}]_{m \times n}$ will be used to denote a matrix where i will vary over the natural numbers *1* to m inclusive and j will vary over the natural numbers from *1* to n inclusive.

We are now ready to explicitly define some of the fundamental properties of a mathematical system with matrices as elements.

DEFINITION: Let A, B be two matrices of dimension $m \times n$. **Matrix A will equal matrix B, $A = B$,** if and only if the corresponding entries in each matrix are equal. Symbolically:

$$[a_{ij}]_{m \times n} = [b_{ij}]_{m \times n}$$

if and only if $a_{ij} = b_{ij}$ for all i and j.

Observe that matrix equality can be discussed only if the given matrices are of the same dimension.

DEFINITION: If every entry of a matrix A is the real number zero, the matrix is called the **zero matrix** and is denoted as either $0_{m \times n}$ or $[0_{ij}]_{m \times n}$.

DEFINITION: If a matrix has dimension $1 \times n$, $n > 1$, then the matrix is called a **row matrix.** Also, if a matrix has dimension $m \times 1, m > 1$ then the matrix is called a **column matrix.**

Operations such as addition and multiplication are defined on matrices and they parallel closely the operations defined on vectors in the preceding section.

DEFINITION: Let $A = [a_{ij}]_{m \times n}$, $B = [b_{ij}]_{m \times n}$, then the **matrix sum of A with B, $A + B$,** is given by

$$A + B = [a_{ij} + b_{ij}]_{m \times n}.$$

In other words to add two matrices, the matrices must first be of the same dimension, i.e., **conformable for addition**, and the elements in the sum matrix are the sum of the corresponding elements in the matrices A and B.

Example 2: Let

$$A = \begin{bmatrix} 2 & 0 & 3 \\ 1 & 4 & -1 \end{bmatrix}_{2 \times 3} \qquad B = \begin{bmatrix} -1 & 0 & -3 \\ -1 & -4 & 2 \end{bmatrix}_{2 \times 3}.$$

Find the matrix sum $A + B$.

Solution:

$$A + B = \begin{bmatrix} 2 + (-1) & 0 + 0 & 3 + (-3) \\ 1 + (-1) & 4 + (-4) & -1 + (\ 2) \end{bmatrix}_{2 \times 3}$$

$$A + B = \begin{bmatrix} 1 & 0 & 0 \\ 0 & 0 & 1 \end{bmatrix}_{2 \times 3}.$$

If A, B, and C are $m \times n$ matrices and thus conformable for addition, the following matrix addition properties will hold.

Closure of Matrix Addition The matrix sum $A + B$ will be a unique matrix of dimension $m \times n$.

Commutativity of Matrix Addition

$$A + B = B + A.$$

Associativity of Matrix Addition

$$(A + B) + C = A + (B + C).$$

Existence of Additive Identity Matrix There exists a matrix $[0_{ij}]_{m\times n}$, called the zero matrix, such that for any matrix $A_{m\times n}$,

$$A_{m\times n} + 0_{m\times n} = A_{m\times n}.$$

Existence of Additive Inverses For any matrix $A_{m\times n}$ there exists a unique $m \times n$ matrix $-A_{m\times n}$, such that

$$A_{m\times n} + (-A_{m\times n}) = 0_{m\times n}.$$

Note: Each element in the additive inverse matrix is a real number which is the additive inverse of the real number appearing in the corresponding position in matrix A.

The operation of multiplication as defined on matrices may be either of the type where each entry in a matrix is multiplied by the same real number (scalar) or alternately we can define the product of one matrix by another matrix. There is considerable difference between the two types of multiplication and a separate discussion of each type is justified.

DEFINITION: Let $A = [a_{ij}]_{m\times n}$ and c be a scalar then the product of the scalar c and the matrix A, cA, is given by

$$c[a_{ij}]_{m\times n} = [ca_{ij}]_{m\times n}.$$

If A and B are $m \times n$ matrices and c and d are scalars, then the following properties of multiplication of a matrix by a scalar hold.

Commutativity of Scalar Multiplication

$$cA = Ac.$$

Associativity of Scalar Multiplication

$$c(dA) = (cd)A.$$

Distributivity of Scalar Multiplication

1. $c(A + B) = cA + cB$.
2. $(c + d)A = cA + dA$.

Example 3: Let

$$A = \begin{bmatrix} 2 & -3 & 4 \\ -1 & 0 & 1 \end{bmatrix}_{2\times 3} \quad \text{and} \quad B = \begin{bmatrix} -1 & 3 & 5 \\ 0 & 2 & -4 \end{bmatrix}_{2\times 3}.$$

Find (a) the additive inverse of A, (b) $-2B$, (c) $5(A+B)$.

Solution:

(a) $$-A = \begin{bmatrix} -2 & 3 & -4 \\ 1 & 0 & -1 \end{bmatrix}_{2\times 3}.$$

(b) $$-2B = \begin{bmatrix} 2 & -6 & -10 \\ 0 & -4 & 8 \end{bmatrix}_{2\times 3}.$$

(c) $$5(A+B) = 5\left(\begin{bmatrix} 2 & -3 & 4 \\ -1 & 0 & 1 \end{bmatrix} + \begin{bmatrix} -1 & 3 & 5 \\ 0 & 2 & -4 \end{bmatrix}_{2\times 3}\right)$$

$$= 5\left(\begin{bmatrix} 1 & 0 & 9 \\ -1 & 2 & -3 \end{bmatrix}_{2\times 3}\right)$$

$$= \begin{bmatrix} 5 & 0 & 45 \\ -5 & 10 & -15 \end{bmatrix}_{2\times 3}.$$

For some insight into multiplication of a matrix by a matrix let us recall the dot product of vectors. For example, if $\alpha = (2,-1,3)$ and

$$\beta = \begin{pmatrix} 1 \\ 4 \\ -2 \end{pmatrix}$$

then $\alpha \cdot \beta = (2)(1) + (-1)(4) + (3)(-2) = -8$. In an analogous manner, we can define multiplication of a row matrix times a column matrix as follows:

DEFINITION: Let A be a $1 \times n$ matrix, $A = [a_{11}\ a_{12} \cdots a_{1n}]_{1\times n}$ and B be a $n \times 1$ matrix,

$$B = \begin{bmatrix} b_{11} \\ b_{21} \\ \vdots \\ b_{n1} \end{bmatrix}_{n\times 1}.$$

The ***matrix product*** AB is given by

$$AB = [a_{11}\ a_{12}\ \cdots a_{1n}]_{1\times n} \cdot \begin{bmatrix} b_{11} \\ b_{21} \\ \vdots \\ b_{n1} \end{bmatrix}_{n\times 1}$$

$$= [a_{11}\, b_{11} + a_{12}\, b_{21} + \cdots + a_{1n}\, b_{n1}]_{1\times 1}.$$

Observe that the preceding definition specifies that there must be the same number of elements in the row matrix A as there are in the column matrix B in order to form the product AB. Also, the product of a matrix of dimension $1 \times n$ by a matrix of dimension $n \times 1$ is a matrix of dimension 1×1; in other words, a matrix with a single element which is the sum of the indicated products.

Before introducing a definition of matrix multiplication applicable to matrices conformable for multiplication, let us consider an example which involves other than row and column matrices. Let

$$A = \begin{bmatrix} 1 & 0 & 2 \\ 3 & -1 & 1 \end{bmatrix}_{2\times 3} \quad \text{and} \quad B = \begin{bmatrix} 1 & -1 & 2 & 0 \\ 0 & 1 & 3 & 2 \\ -2 & 3 & 1 & -3 \end{bmatrix}_{3\times 4}.$$

To find the matrix product AB we will consider each row of A as a row vector and each column of B as a column vector. The entry in the first row and the first column of the product will be the dot product of

$$(1 \quad 0 \quad 2) \cdot \begin{pmatrix} 1 \\ 0 \\ -2 \end{pmatrix} = (1)(1) + (0)(0) + (2)(-2)$$

which is the real number "-3." The entry in the product AB which appears in the first row, second column is found by determining the dot product of the first row of A by the second column of B, i.e.,

$$(1 \quad 0 \quad 2) \cdot \begin{pmatrix} -1 \\ 1 \\ 3 \end{pmatrix} = (1)(-1) + (0)(1) + (2)(3) = 5.$$

If the same process is continued, we find that the matrix product AB gives us a 2×4 matrix with each entry as shown below.

$$AB = \begin{bmatrix} 1 & 0 & 2 \\ 3 & -1 & 1 \end{bmatrix}_{2\times 3} \cdot \begin{bmatrix} 1 & -1 & 2 & 0 \\ 0 & 1 & 3 & 2 \\ -2 & 3 & 1 & -3 \end{bmatrix}_{3\times 4}$$

$$= \begin{bmatrix} -3 & 5 & 4 & -6 \\ 1 & -1 & 4 & -5 \end{bmatrix}_{2\times 4}.$$

A generalization of the preceding example suggests the following definition for the multiplication of one matrix by another matrix.

DEFINITION: Let A be a $m \times n$ matrix and B a $n \times p$ matrix. The **product of A by B, $AB = C$,** in that order, is a $m \times p$ matrix where each entry c_{ij} of C is obtained by finding the sum of the products of the corresponding entries of the ith row of A by the jth column of B.

Example 4: If A is a $m \times n$ matrix, B a $n \times p$ matrix, develop a general expression for the c_{23} entry in the product $AB = C$.

Solution: Let

$$A = \begin{bmatrix} a_{11} & a_{12} & a_{13} & \cdots & a_{1n} \\ a_{21} & a_{22} & a_{23} & \cdots & a_{2n} \\ \cdot & \cdot & \cdot & \cdots & \cdot \\ a_{m1} & a_{m2} & a_{m3} & \cdots & a_{mn} \end{bmatrix}_{m \times n}$$

$$B = \begin{bmatrix} b_{11} & b_{12} & b_{13} & \cdots & b_{1p} \\ b_{21} & b_{22} & b_{23} & \cdots & b_{2p} \\ \cdot & \cdot & \cdot & \cdots & \cdot \\ b_{n1} & b_{n2} & b_{n3} & \cdots & b_{np} \end{bmatrix}_{n \times p}$$

$$AB = C = \begin{bmatrix} c_{11} & c_{12} & c_{13} & \cdots & c_{1p} \\ c_{21} & c_{22} & c_{23} & \cdots & c_{2p} \\ \cdot & \cdot & \cdot & \cdots & \cdot \\ c_{m1} & c_{m2} & c_{m3} & \cdots & c_{mp} \end{bmatrix}_{m \times p}.$$

The entry c_{23} is obtained by multiplying the elements in the second row of A by the corresponding elements in the third column of B and then finding the sum of the products. Thus, $c_{23} = a_{21} b_{13} + a_{22} b_{23} + a_{23} b_{33} + \cdots + a_{2n} b_{n3}$.

It should be emphasized that in order for the matrices A and B to be conformable for multiplication (in the order AB) it is necessary that the number of columns of A equal the number of rows of B. The dimension of the product matrix, AB, will have the same number of rows as A and the same number of columns as B. This means that the matrix product of many rectangular matrices is not defined. For example, if A is a 2×3 matrix and B a 1×4 matrix, neither the product AB nor the product BA is defined. Also, even if the matrix product AB does exist, the matrix product BA may not exist. We may not, in general, arbitrarily change the order of matrix multiplication which means the operation is not commutative. In listing the properties of matrix multiplication that do hold we must clearly state the conditions under which the operations are defined.

Let A, B, and C be matrices and d a scalar. Then the following matrix multiplication properties hold.

M_1 **Closure of Matrix Multiplication** If $A = [a_{ij}]_{m \times n}$, $B = [b_{ij}]_{n \times p}$ then the matrix product AB is a unique matrix $C = [c_{ij}]_{m \times p}$.

M_2 **Associativity of Matrix Multiplication** If $A = [a_{ij}]_{m \times n}$, $B = [b_{ij}]_{n \times p}$, and $C = [c_{ij}]_{p \times q}$, then $(AB)C = A(BC)$.

M_3 **Distributivity of Matrix Multiplication over Addition** If $A = [a_{ij}]_{m \times n}$, $B = [b_{ij}]_{n \times p}$, and $C = [c_{ij}]_{n \times p}$, then $A(B + C) = AB + AC$.

M_4 **Scalar and Matrix Multiplication Associativity** If $A = [a_{ij}]_{m \times n}$, $B = [b_{ij}]_{n \times p}$, and d is a scalar, then

1. $d(AB) = (dA)B$.
2. $(Ad)B = A(dB)$.

M_5 **Existence of a Multiplicative Identity** For every square matrix $A = [a_{ij}]_{m \times m}$ there exists an identity matrix

$$I = \begin{bmatrix} 1 & 0 & \dots & 0 \\ 0 & 1 & \dots & 0 \\ \cdot & \cdot & \dots & \cdot \\ 0 & 0 & \dots & 1 \end{bmatrix}_{m \times m}$$

such that, $AI = IA = A$.

The reader will recognize that we have omitted any reference to a multiplicative inverse in the listing of the properties. Our reasoning for doing so is that all matrices do not possess a multiplicative inverse and the conditions under which an inverse will exist lie beyond the scope of our objectives in this text.

At this point let us move from the abstract mathematical system of matrices to a concrete example of how these concepts may be applied.

Example 5: A clothing store sells one brand of sweaters in two styles; turtleneck and plain, and three colors; red, green, and blue. Currently the store has in stock 30 red, 20 green, and 10 blue in each of the two styles. Normal pricing for the sweaters is \$8 each for a turtleneck and \$6 each for the plain style. The owner is contemplating a special sale where the turtleneck price would be reduced to \$7 each and the plain type to \$4 each. Assuming all sweaters would be sold, what would be the difference between the total amount received at the normal pricing and the special sale pricing?

Solution: Although many methods might be used to solve this problem, we will arrange the data in a matrix form. One matrix will be used to represent the sweater inventory differentiated by color and style and the other matrix will represent the possible pricing levels. The matrix product will give us entries corresponding to gross income.

Sweater Matrix

	Turtleneck	**Plain**
Red	30	30
Green	20	20
Blue	10	10

$$\begin{bmatrix} 30 & 30 \\ 20 & 20 \\ 10 & 10 \end{bmatrix}_{3 \times 2}$$

Price Matrix

	Normal	**Sale**
Turtleneck	\$8	\$7
Plain	6	4

$$\begin{bmatrix} \$8 & \$7 \\ 6 & 4 \end{bmatrix}_{2 \times 2}$$

The matrix product will represent gross income.

$$\begin{bmatrix} 30 & 30 \\ 20 & 20 \\ 10 & 10 \end{bmatrix}_{3 \times 2} \cdot \begin{bmatrix} \$8 & \$7 \\ 6 & 4 \end{bmatrix}_{2 \times 2} = \begin{bmatrix} \$240 + 180 & \$210 + 120 \\ 160 + 120 & 140 + 80 \\ 80 + 60 & 70 + 40 \end{bmatrix}_{3 \times 2} .$$

Thus, the product matrix becomes

$$\begin{bmatrix} \$420 & \$330 \\ 280 & 220 \\ 140 & 110 \end{bmatrix}_{3 \times 2} .$$

Where the first column is the income at regular prices and the second column the income at the sale prices. The sum of the first column is \$840 and the sum of the second column is \$660. Thus, the difference in total income is \$180.

At the beginning of this section we indicated that matrix theory could be useful in solving systems of linear equations. The matrix procedure which we will use will be closely related to the elementary transformations which are performed on linear equations in Sec. 6.2. However, in applying these concepts to a matrix where the entries are real numbers, we will interpret the acceptable operations as **elementary row operations.**

DEFINITION: The **elementary row operations** which may be performed on a matrix A are

1. interchange any two rows of A
2. multiply any row of A by a nonzero scalar
3. multiply any row of A by a nonzero scalar and add term-by-term to any other row of A.

DEFINITION: If a matrix B can be obtained from a matrix A by a finite sequence of elementary row operations performed on A, then A and B are said to be **row equivalent,** which we denote as $A \sim B$.

By repeated applications of the elementary row operations, we can transform any given matrix into a variety of matrices which are all row equivalent to the given matrix. For reasons which will become clear in the next section it is very desirable that we attempt to transform a matrix representing systems of equations into a special type of matrix called an **echelon matrix.**

DEFINITION: An $m \times n$ matrix is called an **echelon matrix** if

1. each of the first k rows has some nonzero entries and the entries in the remaining m-k rows are all zeros $(1 \leqslant k \leqslant m)$
2. the first nonzero entry in each of the first k rows is the real number 1

3. in any one of the first k rows the number of zeros preceding the first non-zero entry is less than the number of zeros in the next row.

An example in which a systematic procedure is used will help to clarify the definition.

Example 6: Let

$$A = \begin{bmatrix} 2 & 3 & 1 & 0 & 1 \\ 1 & 6 & 3 & -2 & 4 \\ 3 & 2 & -1 & 4 & 0 \end{bmatrix}_{3\times 5}.$$

Transform A into a row equivalent echelon matrix.

Solution: The initial step will be to introduce a "1" as the first entry in the first row. Then if possible we will obtain a "1" in the second row, second column; and finally a "1" in the third row, third column. Alternate sequences are possible but the following is a typical pattern.

$$A = \begin{bmatrix} 2 & 3 & 1 & 0 & -1 \\ 1 & 6 & 3 & -2 & 4 \\ 3 & 2 & -1 & 4 & 0 \end{bmatrix}_{3\times 5} \underset{\text{Interchange } R1 \text{ and } R1}{\sim} \begin{bmatrix} 1 & 6 & 3 & -2 & 4 \\ 2 & 3 & 1 & 0 & -1 \\ 3 & 2 & -1 & 4 & 0 \end{bmatrix}_{3\times 5}$$

$$\underset{-2R1+R2}{\sim} \begin{bmatrix} 1 & 6 & 3 & -2 & 4 \\ 0 & -9 & -5 & 4 & -9 \\ 3 & 2 & -1 & 4 & 0 \end{bmatrix}_{3\times 5} \underset{-1/9\,R2}{\sim} \begin{bmatrix} 1 & 6 & 3 & -2 & 4 \\ 0 & 1 & 5/9 & -4/9 & 1 \\ 3 & 2 & -1 & 4 & 0 \end{bmatrix}_{3\times 5}$$

$$\underset{-3R1+R3}{\sim} \begin{bmatrix} 1 & 6 & 3 & -2 & 4 \\ 0 & 1 & 5/9 & -4/9 & 1 \\ 0 & -16 & -10 & 10 & -12 \end{bmatrix}_{3\times 5} \underset{16R2+R3}{\sim} \begin{bmatrix} 1 & 6 & 3 & -2 & 4 \\ 0 & 1 & 5/9 & -4/9 & 1 \\ 0 & 0 & -10/9 & 26/9 & 4 \end{bmatrix}_{3\times 5}$$

$$\underset{-9/10R3}{\sim} \begin{bmatrix} 1 & 6 & 3 & -2 & 4 \\ 0 & 1 & 5/9 & -4/9 & 1 \\ 0 & 0 & 1 & -13/5 & -18/5 \end{bmatrix}_{3\times 5}$$

This matrix will satisfy the definition of an echelon matrix, and represents the solution to the given problem. However, it is often convenient to continue the process further and for this example make the first three rows and the first three columns correspond to a 3×3 identity matrix. Continuing with elementary row operations we obtain this useful row equivalent form.

$$\begin{bmatrix} 1 & 6 & 3 & -2 & 4 \\ 0 & 1 & 5/9 & -4/9 & 1 \\ 0 & 0 & 1 & -13/5 & -18/5 \end{bmatrix}_{3\times 5} \underset{\sim}{-6R2+R1} \begin{bmatrix} 1 & 0 & -1/3 & 2/3 & -2 \\ 0 & 1 & 5/9 & -4/9 & 1 \\ 0 & 0 & 1 & -13/5 & -18/5 \end{bmatrix}_{3\times 5}$$

$$\underset{\sim}{1/3R3+R1} \begin{bmatrix} 1 & 0 & 0 & -1/5 & -16/5 \\ 0 & 1 & 5/9 & -4/9 & 1 \\ 0 & 0 & 1 & -13/5 & -18/5 \end{bmatrix}_{3\times 5} \underset{\sim}{-5/9R3+R2} \begin{bmatrix} 1 & 0 & 0 & -1/5 & -16/5 \\ 0 & 1 & 0 & 1 & 3 \\ 0 & 0 & 1 & -13/5 & -18/5 \end{bmatrix}_{3\times 5}.$$

The rationale for this procedure may seem at this point somewhat obscure. However, if we regard each of the real numbers in the equivalent matrices as either representing coefficients of the variables in a system of linear equations or constants which appear in such equations, then the last matrix obtained is considerably simpler. More significantly, the last matrix form could be used to directly produce the solutions (if they exist) of a system of linear equations.

In the next section we will demonstrate that although the process of reducing matrices to an echelon form may be cumbersome at times, the advantages of having a standardized method for finding solutions will outweigh any slight computational difficulties.

Problem Set 6.4

A. 1. Let

$$A = \begin{bmatrix} 2 & -3 & -1 & 0 \\ 1 & 0 & 2 & -1 \\ -1 & 2 & 3 & -4 \end{bmatrix}_{3\times 4}.$$

Determine the dimension of matrix A and also identify a_{32}, a_{13}, a_{24}, a_{33}, and a_{21}.

2. Find the matrix B such that $A + B = C$ when

$$A = \begin{bmatrix} 4 & -1 & 0 & 2 & 3 \\ 3 & 1 & -1 & -2 & 5 \\ 2 & 0 & 0 & 1 & -2 \\ -3 & 1 & -2 & 6 & 4 \end{bmatrix}_{4\times 5} \quad \text{and} \quad C = \begin{bmatrix} 3 & 0 & 5 & 4 & 4 \\ 1 & 2 & 3 & 0 & 2 \\ 1 & 2 & 0 & 4 & -2 \\ 0 & 6 & 3 & 5 & 9 \end{bmatrix}_{4\times 5}.$$

3. Let

$$B = \begin{bmatrix} 2 & -1 & 0 & 3 & 5 \\ -1 & 1 & 0 & 2 & -5 \\ 3 & 0 & 0 & 1 & 4 \\ 4 & 1 & -2 & -3 & 3 \end{bmatrix}_{4\times 5}.$$

Find (a) the additive inverse of B, (b) $(-1)B$, (c) $B + B$, (d) $(-2)B$.

4. Prove that matrix addition is commutative.
5. Let

$$A = \begin{bmatrix} 3 \\ 7 \\ -5 \end{bmatrix}_{3 \times 1}$$

$B = [1 \quad -1/7 \quad 0.2]_{1 \times 3}$. Find (a) AB, (b) BA. (c) Does $AB = BA$?

6. Let

$$A = \begin{bmatrix} 6 & 5 & -1 & 3 \\ 1 & -5 & -4 & 0 \\ 0 & -2 & 3 & -7 \end{bmatrix}_{3 \times 4} \qquad B = \begin{bmatrix} 2 & 3 \\ -6 & -5 \\ 0 & 1 \\ -1 & -7 \end{bmatrix}_{4 \times 2}.$$

Find (a) AB, (b) $A((2)B)$, (c) $(A(2))B$.

7. Let

$$A = \begin{bmatrix} 1 & -3 & 2 \\ 0 & 5 & -4 \\ -1 & 4 & 7 \end{bmatrix}_{3 \times 3}.$$

Place a 3×3 identity matrix,

$$I = \begin{bmatrix} 1 & 0 & 0 \\ 0 & 1 & 0 \\ 0 & 0 & 1 \end{bmatrix}_{3 \times 3}$$

next to matrix A to form a matrix

$$\begin{bmatrix} 1 & -3 & 2 & 1 & 0 & 0 \\ 0 & 5 & -4 & 0 & 1 & 0 \\ -1 & 4 & 7 & 0 & 0 & 1 \end{bmatrix}_{3 \times 6}.$$

Now perform elementary row operations on this 3×6 matrix to obtain an equivalent matrix in echelon form where the first three columns are of the identity matrix form. Regard the last three columns as a new 3×3 matrix, say B. Find AB and BA. What can be said about the matrices A, B?

8. Let

$$A = \begin{bmatrix} 1 & -3 & 2 \\ 0 & 5 & -4 \\ -1 & 4 & 7 \end{bmatrix}_{3 \times 3} \quad \text{(as in the previous problem)}$$

$$X = \begin{bmatrix} x_1 \\ x_2 \\ x_3 \end{bmatrix}_{3 \times 1} \quad \text{and} \quad C = \begin{bmatrix} 2 \\ -1 \\ 3 \end{bmatrix}_{3 \times 1}.$$

The matrix equation $AX = C$ represents a system of three equations in three unknowns. Since the multiplicative inverse of A, A^{-1} was found in Prob. 7, we may multiply on the left both sides of the matrix equation $AX = C$ [i.e., $A^{-1}(AX) = A^{-1}C$] to obtain $X = A^{-1}C$. From this we can find the solution (if it exists) to the system of equations

$$\begin{cases} x_1 - 3x_2 + 2x_3 = 2 \\ \quad\;\; 5x_2 - 4x_3 = -1 \\ -x_1 + 4x_2 + 7x_3 = 3. \end{cases}$$

Find the solution to this system of equations.

9. Using the process described in Probs. 7 and 8 of this problem set, find the solution set (if it exists) to the system of equations.

$$\begin{cases} 2x_1 + 3x_2 = 7 \\ 4x_1 - \;\; x_2 = 3. \end{cases}$$

10. Transform each of the following matrices into a row equivalent echelon matrix.

(a) $\begin{bmatrix} 2 & 3 & -1 & 5 \\ 1 & 5 & 3 & -2 \\ 3 & -1 & -4 & 0 \end{bmatrix}_{3\times 4}$. (b) $\begin{bmatrix} 4 & 1 & -5 \\ -3 & 0 & 1 \\ 5 & -2 & 3 \\ -7 & -3 & -1 \end{bmatrix}_{4\times 3}$.

(c) $\begin{bmatrix} 2 & -5 & -1 & 0 \\ 5 & 10 & 0 & -15 \end{bmatrix}_{2\times 4}$. (d) $\begin{bmatrix} 3 & 0 & 2 \\ 1 & 5 & 7 \\ 4 & -1 & 1 \end{bmatrix}_{3\times 3}$.

(e) $\begin{bmatrix} -4 & 3 & 0 & 8 \\ -1 & 3 & 2 & -4 \\ 0 & 4 & -1 & 7 \\ 2 & 0 & 3 & 5 \end{bmatrix}_{4\times 4}$. (f) $\begin{bmatrix} 5 & -1 \\ 1 & 1 \\ 2 & 3 \\ 3 & 2 \\ 4 & 4 \end{bmatrix}_{5\times 2}$.

11. A fruit produce company delivers crates of fruit to the Good Food and Consumers Choice stores. The accompanying table indicates the number of crates of fruit of each grade as well as the price per crate for produce sent to each store. Find the following:

Table 6.1

	Oranges		Grapefruit		Apples		
Grade	**No. of Crates**	**Price per Crate ($)**	**No. of Crates**	**Price per Crate ($)**	**No. of Crates**	**Price per Crate ($)**	**Store**
A	20	2.00	15	3.00	10	3.75	GF
B	20	1.25	15	1.75	10	2.75	GF
A	10	2.25	20	3.50	5	4.25	CC
B	15	1.50	30	2.00	10	3.00	CC

(a) $A = [a_{ij}]$, where $i = 1, 2$ and 1 represents grade A, 2 represents grade B, and $j = 1, 2, 3$ where 1 represents oranges, 2 represents grapefruit and 3 represents apples, a_{ij} represents the number of crates of a certain kind of fruit the produce company delivered to the Good Food Store.

(b) $B = [b_{ij}]$ where b_{ij} represents the number of crates of certain kind of fruit that Consumers Choice Store received from the produce company. [Let i and j have the same designations as that in part (a) of this problem.]

(c) Find $A + B$. What does the sum of the matrices A, B represent?

(d) Find cost matrices $C = [c_{jk}]$, $D = [d_{jk}]$ so that AC and BD will be defined and meaningful. What do AC and BD represent?

(e) Find $AC + BD$. What does $AC + BD$ represent?

6.5 Systems of Linear Equations

In Sec. 6.2 we discussed the use of elementary operations to obtain solutions to a system of linear equations in two unknowns. In the preceding section we discussed elementary row operations which may be performed on a matrix to obtain an echelon matrix. The close relationship that exists between these two techniques is not accidental. By using elementary row operations on a matrix of real numbers representing the coefficients and constant terms in a system of linear equations we can obtain a general and simplified method for finding solutions (if they exist) of systems of equations which involve any number of unknowns. Before we undertake a systematic development of the procedure, however, it will be convenient to introduce some abbreviated notation and some new terms which will precisely describe the various matrix configurations which will be considered.

DEFINITION: A system of m linear equations in n unknowns where the a_{ij}'s and the c_i's are real numbers and the x_i's are the variables will be denoted

$$\begin{cases} a_{11}\,x_1 + a_{12}\,x_2 + \cdots + a_{1n}\,x_n = c_1 \\ a_{21}\,x_1 + a_{21}\,x_2 + \cdots + a_{2n}\,x_n = c_2 \\ \cdot \qquad \cdot \qquad \cdot\cdot\cdot\cdot \qquad = \cdot \\ a_{m1}\,x_1 + a_{m2}\,x_2 + \cdots + a_{mn}\,x_n = c_n \end{cases}$$

m and n are positive integers where m may be greater than, less than, or equal to n.

DEFINITION: An ordered n-tuple of real numbers which when replaced for unknowns $(x_1, x_2, \ldots, x_n)$ in a system of linear equations, satisfies each equation in the system is called a **solution**. The set of all solutions to a system of linear equations is called the **solution set.**

DEFINITION: If a system of m linear equations in n unknowns has one or more solutions, then the system is said to be **consistent**, otherwise it is said to be **inconsistent**.

DEFINITION: Given the system of equations

$$\begin{cases} a_{11}x_1 + a_{12}x_2 + \cdots + a_{1n}x_n = c_1 \\ a_{21}x_1 + a_{21}x_2 + \cdots + a_{2n}x_n = c_2 \\ \cdot \quad \cdot \quad \cdot \quad \cdot\cdot\cdot\cdot\cdot \quad \cdot \quad = \cdot \\ a_{m1}x_1 + a_{m2}x_2 + \cdots + a_{mn}x_n = c_n. \end{cases}$$

The matrix $A = [a_{ij}]_{m \times n}$, $1 \leqslant i \leqslant m$ and $1 \leqslant j \leqslant n$ is called the **coefficient matrix** of a system of equations. The matrix formed by adjoining on the right the column

$$\begin{bmatrix} c_1 \\ c_2 \\ \vdots \\ c_n \end{bmatrix}$$

is called the **augmented matrix** and is denoted $[A \vdots C]$.

At this point let us consider a specific example which will illustrate how we may employ matrix methods to obtain a solution of a system of linear equations.

Example 1: Let

$$\begin{cases} 2x_1 - x_2 + 2x_3 = 3 \\ x_1 - x_2 + 2x_3 = 0 \\ -3x_1 + 3x_2 - 5x_3 = -2 \end{cases}$$

be a given system of equations. Find a solution, if one exists, by the use of matrices.

Solution: The coefficient matrix A of the system is

$$\begin{bmatrix} 2 & -1 & 2 \\ 1 & -1 & 2 \\ -3 & 3 & -5 \end{bmatrix}$$

and the constant column matrix C is

$$\begin{bmatrix} 3 \\ 0 \\ -2 \end{bmatrix}.$$

Observe that the matrix form of the given system is

$$\begin{bmatrix} 2 & -1 & 2 \\ 1 & -1 & 2 \\ -3 & 3 & -5 \end{bmatrix}_{3 \times 3} \cdot \begin{bmatrix} x_1 \\ x_2 \\ x_3 \end{bmatrix}_{3 \times 1} = \begin{bmatrix} 3 \\ 0 \\ -2 \end{bmatrix}_{3 \times 1}$$

i.e., $AX = C$ where X represents the column matrix of unknowns.

Now if we could transform by the use of elementary row operations the coefficient matrix A into a row reduced echelon identity matrix I, while at the same time applying the same row operations to C, we would have the resulting form.

$$\begin{bmatrix} 1 & 0 & 0 \\ 0 & 1 & 0 \\ 0 & 0 & 1 \end{bmatrix}_{3\times 3} \cdot \begin{bmatrix} x_1 \\ x_2 \\ x_3 \end{bmatrix}_{3\times 1} = \begin{bmatrix} c_1 \\ c_2 \\ c_3 \end{bmatrix}.$$

Forming the matrix product on the left we would obtain

$$\begin{aligned} 1(x_1) + 0(x_2) + 0(x_3) &= c_1 \\ 0(x_1) + 1(x_2) + 0(x_3) &= c_2 \\ 0(x_1) + 0(x_2) + 1(x_3) &= c_3 \end{aligned}$$

which would give us by inspection the solutions $x_1 = c_1, x_2 = c_2$, and $x_3 = c_3$.

By performing the elementary row operations on the augmented matrix, we simplify the procedure and obtain a solution (if it exists) in the last column.

A step-by-step transformation of the augmented matrix to the echelon identity form may be obtained as follows:

$$\left[\begin{array}{rrr:r} 2 & -1 & 2 & 3 \\ 1 & -1 & 2 & 0 \\ -3 & 3 & -5 & -2 \end{array}\right] \underset{R1 \text{ and } R2}{\overset{\text{interchange}}{\sim}} \left[\begin{array}{rrr:r} 1 & -1 & 2 & 0 \\ 2 & -1 & 2 & 3 \\ -3 & 3 & -5 & -2 \end{array}\right]$$

$$\overset{\substack{-2R1+R2 \\ 3R1+R3}}{\sim} \left[\begin{array}{rrr:r} 1 & -1 & 2 & 0 \\ 0 & 1 & -2 & 3 \\ 0 & 0 & 1 & -2 \end{array}\right] \overset{1R2+R1}{\sim} \left[\begin{array}{rrr:r} 1 & 0 & 0 & 3 \\ 0 & 1 & -2 & 3 \\ 0 & 0 & 1 & -2 \end{array}\right]$$

$$\overset{2R3+R2}{\sim} \left[\begin{array}{rrr:r} 1 & 0 & 0 & 3 \\ 0 & 1 & 0 & -1 \\ 0 & 0 & 1 & -2 \end{array}\right].$$

Note: The vertical dotted line is used to separate the coefficient matrix from the constant matrix. The solution set is the column matrix

$$\begin{bmatrix} 3 \\ -1 \\ -2 \end{bmatrix}.$$

Thus, $x_1 = 3$, $x_2 = -1$ and $x_3 = -2$ which may be checked by direct substitution into the original system of equations.

The essence of the procedure illustrated in Example 1 is that the elementary row operations performed on the augmented matrix represent elementary transformations on a system of equations. These transformations produce equivalent systems having the same solution set. In the reduced echelon form, the system is so simplified that the solutions may be easily obtained. Although there are many other methods for solving systems of linear equations, the step-by-step systematized procedure we have described is easy to apply and this iterative approach is well suited for processing by modern computers.

Since the system of equations in Example 1 was shown to have a solution, the system is consistent. A question remains, however. Is the solution unique? Or more generally, under what conditions can we conclude that a solution is indeed a unique solution to a system of equations. The following theorem, which we state without proof, will give us the necessary criteria for the determining uniqueness of solutions.

THEOREM 1: A solution to a system of m linear equations in n unknowns will be unique if

1. $m = n$, i.e., there are exactly as many equations as there are unknowns, and
2. the reduced echelon-identity coefficient matrix has as many "ones" on the main diagonal as there are unknowns.

The solution to the problem given in Example 1 was obtained from a system of three equations in three unknowns and the reduced coefficient matrix had 3 "ones" down the main diagonal. Thus, the solution is unique.

Although most of our interest in this text is involved in solving systems of equations which do have unique solutions, it is also important to know when either more than one solution exists or when no solution exists. Fortunately, no new techniques are necessary. By employing the elementary row operations on the augmented matrix to obtain a row reduced echelon matrix, we will also be able to determine if the system is inconsistent or consistent with more than one solution. An example illustrating each of these possibilities will help to clarify the basic concepts.

Example 2: Let

$$\begin{cases} x_1 + x_2 + x_3 + 3x_4 = 5 \\ -2x_1 + 3x_3 + x_3 + 2x_4 = -2 \\ x_1 + 3x_2 + x_3 + 5x_4 = 4 \\ 2x_1 + 2x_2 + x_3 + 5x_4 = 8 \end{cases}$$

be a given system of linear equations. Find a solution, if it exists, by the use of matrices.

Solution: Let us begin by transforming the augmented matrix using elementary row operations into an echelon form matrix.

$$\left[\begin{array}{cccc:c} 1 & 1 & 1 & 3 & 5 \\ -2 & 3 & 1 & 2 & -2 \\ 1 & 3 & 1 & 5 & 4 \\ 2 & 2 & 1 & 5 & 8 \end{array}\right] \underbrace{\begin{array}{c} 2R1 + R2 \\ -1R1 + R3 \\ -2R1 + R4 \end{array}} \left[\begin{array}{cccc:c} 1 & 1 & 1 & 3 & 5 \\ 0 & 5 & 3 & 8 & 8 \\ 0 & 2 & 0 & 2 & -1 \\ 0 & 0 & -1 & -1 & -2 \end{array}\right]$$

$$\frac{1}{5}R2 \sim \begin{bmatrix} 1 & 1 & 1 & 3 & \vdots & 5 \\ 0 & 1 & 3/5 & 8/5 & \vdots & 8/5 \\ 0 & 2 & 0 & 2 & \vdots & -1 \\ 0 & 0 & -1 & -1 & \vdots & -2 \end{bmatrix} \quad \underset{\sim}{-2R2+R3} \begin{bmatrix} 1 & 1 & 1 & 3 & \vdots & 5 \\ 0 & 1 & 3/5 & 8/5 & \vdots & 8/5 \\ 0 & 0 & -6/5 & -6/5 & \vdots & -21/5 \\ 0 & 0 & -1 & -1 & \vdots & -2 \end{bmatrix}$$

$$-\frac{5}{6}R3 \sim \begin{bmatrix} 1 & 1 & 1 & 3 & \vdots & 5 \\ 0 & 1 & 3/5 & 8/5 & \vdots & 8/5 \\ 0 & 0 & 1 & 1 & \vdots & 7/2 \\ 0 & 0 & -1 & -1 & \vdots & -2 \end{bmatrix} \quad \underset{\sim}{R3+R4} \begin{bmatrix} 1 & 1 & 1 & 3 & \vdots & 5 \\ 0 & 1 & 3/5 & 8/5 & \vdots & 8/5 \\ 0 & 0 & 1 & 1 & \vdots & 7/2 \\ 0 & 0 & 0 & 0 & \vdots & 3/2 \end{bmatrix}$$

At this point, although we could continue to further reduce the matrix observe that we will not have all "ones" down the main diagonal of the coefficient matrix. In fact, the last matrix represents a system of equations of the form

$$\begin{cases} x_1 + x_2 + x_3 + 3x_4 = 5 \\ x_2 + 3/5x_3 + 8/5x_4 = 8/5 \\ x_3 + x_4 = 7/2 \\ 0 = 3/2. \end{cases}$$

The last equation $0 = 3/2$ is a false statement, or a contradiction. When a contradiction is obtained, then the system of equations is inconsistent. Since the reduced system is equivalent to the original system, the given system of equations is inconsistent and there are no solutions.

The preceding example may be generalized. If any of the rows in the reduced echelon coefficient matrix contains only zeros and the entry in the same row in the constant column matrix is non-zero, then the system of equations is inconsistent.

Example 3: Let

$$\begin{cases} 5x_1 + 2x_2 + 2x_3 + 7x_4 = 14 \\ 3x_1 + x_2 + x_3 + 4x_4 = 8 \\ x_1 - x_2 + x_3 - 2x_4 = -2 \end{cases}$$

be a given system of linear equations. Find a solution, if it exists, by the use of matrices.

Solution: Let us begin by row reducing the augmented matrix.

$$\begin{bmatrix} 5 & 2 & 2 & 7 & \vdots & 14 \\ 3 & 1 & 1 & 4 & \vdots & 8 \\ 1 & -1 & 1 & -2 & \vdots & -2 \end{bmatrix} \quad \underset{\sim}{\text{Interchange } R1 \text{ and } R3} \begin{bmatrix} 1 & -1 & 1 & -2 & \vdots & -2 \\ 3 & 1 & 1 & 4 & \vdots & 8 \\ 5 & 2 & 2 & 7 & \vdots & 14 \end{bmatrix}$$

$$\underset{\sim}{\begin{matrix} -3R1+R2 \\ -5R1+R3 \end{matrix}} \begin{bmatrix} 1 & -1 & 1 & -2 & \vdots & -2 \\ 0 & 4 & -2 & 10 & \vdots & 14 \\ 0 & 7 & -3 & 17 & \vdots & 24 \end{bmatrix} \quad \frac{1}{4}R2 \sim \begin{bmatrix} 1 & -1 & 1 & -2 & \vdots & -2 \\ 0 & 1 & -1/2 & 5/2 & \vdots & 7/2 \\ 0 & 7 & -3 & 17 & \vdots & 24 \end{bmatrix}$$

$$\underset{\sim}{-7R2+R3}\begin{bmatrix} 1 & -1 & 1 & -2 & \vdots & -2 \\ 0 & 1 & -1/2 & 5/2 & \vdots & 7/2 \\ 0 & 0 & 1/2 & -1/2 & \vdots & -1/2 \end{bmatrix} \quad \underset{\sim}{2R3}\begin{bmatrix} 1 & -1 & 1 & -2 & \vdots & -2 \\ 0 & 1 & -1/2 & 5/2 & \vdots & 7/2 \\ 0 & 0 & 1 & -1 & \vdots & -1 \end{bmatrix}$$

$$\begin{matrix} -1R3+R1 \\ \frac{1}{2}R3+R2 \\ \sim \end{matrix}\begin{bmatrix} 1 & -1 & 0 & -1 & \vdots & -1 \\ 0 & 1 & 0 & 2 & \vdots & 3 \\ 0 & 0 & 1 & -1 & \vdots & -1 \end{bmatrix} \quad \underset{\sim}{R2+R1}\begin{bmatrix} 1 & 0 & 0 & 1 & \vdots & 2 \\ 0 & 1 & 0 & 2 & \vdots & 3 \\ 0 & 0 & 1 & -1 & \vdots & -1 \end{bmatrix}.$$

The final row reduced echelon matrix represents a system of equations of the form

$$\begin{cases} x_1 \qquad\quad + \ x_4 = 2 \\ \quad x_2 \qquad + 2x_4 = 3 \\ \qquad\quad x_3 - \ x_4 = -1. \end{cases}$$

We may now write x_1, x_2 and x_3 in terms of x_4.

$$\begin{cases} x_1 = 2 - x_4 \\ x_2 = 3 - 2x_4 \\ x_3 = -1 + x_4. \end{cases}$$

For every value which we assign to x_4 we will obtain corresponding values for x_1, x_2, and x_3. Since there is an infinite number of values which x_4 may assume there are an infinite number of solutions to the given system of equations.

Again, we may form some general conclusions based on the preceding example. If the given equations are *linearly independent**, and the number of unknowns n exceeds the number of equations m, then if there is a solution, there will be an infinite number of solutions to the system. By the use of echelon form matrices, the system can be reduced to the form where some of the variables can be expressed in terms of the other remaining variables. Arbitrary substitution of real numbers for the remaining variables will generate an infinite number of solutions.

There are other specific logical possibilities which we might consider. For example, there are inconsistent systems which have the same number of variables as equations and there are both consistent and inconsistent systems which have more equations than variables. However, the examples we have chosen illustrate all the solution possibilities that can occur. We may have exactly one unique solution, no solutions, or an infinite number of solutions.

The matrix method which we have used is applicable to any system of linear equations. A significant advantage of the general matrix method is that in the reduced echelon form we can always determine quickly and conveniently whether or not solutions exist.

In this chapter we began with a discussion of one- and two-dimensional coordinate systems with which we associated linear equations. The general concepts involving linear equations in two unknowns were generalized and in the process led us to consider both vectors and matrices. Finally, in this last section we used matrix techniques to obtain

*For further information on linear independence, refer to the Suggested Reading list at the end of this chapter for appropriate references.

solutions of systems of linear equations having any number of unknowns. The emphasis throughout the chapter has been focused on linear equations. In the next chapter we will turn our attention to systems of linear inequalities and one of the modern mathematical developments based on linear inequalities; namely, linear programming.

Problem Set 6.5

A. 1. Solve the following systems of equations by the use of matrices.

(a) $\begin{cases} 2x_1 + 3x_2 = 6 \\ 5x_1 + 4x_2 = 20. \end{cases}$

(b) $\begin{cases} x_1 + 2x_2 + 3x_3 = 4 \\ -x_1 + 3x_2 - 3x_3 = 1 \\ 2x_1 + 4x_2 - 6x_3 = 2. \end{cases}$

(c) $\begin{cases} x_1 + x_2 - x_3 + x_4 = 1 \\ -x_1 - x_2 + x_3 + x_4 = 3 \\ 2x_1 + x_2 - x_3 - x_4 = 0 \\ 3x_1 + 2x_2 - 2x_3 + 2x_4 = 4. \end{cases}$

(d) $\begin{cases} x_1 - 2x_2 - 2x_3 - 2x_4 = 4 \\ x_1 - 3x_2 - 3x_3 - 3x_4 = 6 \\ 2x_1 - 5x_2 - 3x_3 - 3x_4 = 8 \\ x_1 - 2x_2 - 2x_3 - 4x_4 = 10. \end{cases}$

2. Determine solutions, if they exist, for the following systems of equations by using matrices.

(a) $\begin{cases} x_1 + x_2 = 3 \\ \dfrac{x_1}{2} + \dfrac{x_2}{2} = -1. \end{cases}$

(b) $\begin{cases} x_1 + x_2 = 3 \\ -\dfrac{x_1}{2} - \dfrac{x_2}{2} = -\dfrac{3}{2}. \end{cases}$

(c) $\begin{cases} -x_1 + 2x_2 + 4x_3 = -3 \\ 3x_1 - 3x_2 - 6x_3 = 6 \\ 4x_1 - 8x_2 - 12x_3 = 12. \end{cases}$

(d) $\begin{cases} x_1 + x_2 - x_3 = 1 \\ x_1 + x_2 - x_3 = 1 \\ x_1 - x_2 - x_3 = 5. \end{cases}$

(e) $\begin{cases} 2x_1 + x_2 - x_3 = 2 \\ x_1 - x_2 + x_3 = -3 \\ -4x_1 - 2x_2 + 2x_3 = 4. \end{cases}$

3. For the given system of equations, sketch them in an x_1x_2 Cartesian plane where x_1 is the horizontal axis. Determine if the given systems have a solution or not.

(a) $\begin{cases} x_1 + 2x_2 = 3 \\ 2x_1 - x_2 = 1 \\ 3x_1 - 4x_2 = -1. \end{cases}$

(b) $\begin{cases} x_1 + 2x_2 = 3 \\ 2x_1 - x_2 = 1 \\ 3x_1 - 4x_2 = -12. \end{cases}$

(c) $\begin{cases} x_1 + 2x_2 = 3 \\ \dfrac{x_1}{2} + x_2 = \dfrac{3}{2} \\ \dfrac{2}{3}x_1 + \dfrac{4}{3}x_2 = 2. \end{cases}$

4. For each given system of equations, determine by matrix methods whether a solution exists. If it is unique, find its value.

(a) $\begin{cases} 2x_1 - x_2 = -1 \\ x_1 + 3x_2 = -11 \\ -3x_1 + 2x_2 = -12 \\ 3x_1 - 2x_2 = 0. \end{cases}$

(b) $\begin{cases} 2x_1 - x_2 + 3x_3 = 8 \\ -3x_1 - 5x_2 + 2x_3 = 1 \\ 4x_1 + 6x_2 - 3x_3 = -1. \end{cases}$

(c) $\begin{cases} -3x_1 - 5x_2 + 2x_3 = 1 \\ 4x_1 + 6x_2 - 3x_3 = -1. \end{cases}$

(d) $\begin{cases} 2x_1 - 3x_2 - x_3 = 2 \\ -6x_1 + 9x_2 + 3x_3 = 6. \end{cases}$

(e) $$\begin{cases} x_1 - 2x_2 + x_3 - x_4 - x_5 = -2 \\ 2x_1 + 3x_2 - x_3 + 2x_4 - 3x_5 = 7 \\ 3x_1 - 5x_2 + 2x_3 - 3x_4 + x_5 = -5 \\ -x_1 + 2x_2 + x_3 + x_4 - 3x_5 = 0 \\ -2x_1 - 4x_2 + 2x_3 - x_4 + 2x_5 = -6. \end{cases}$$

(f) $$\begin{cases} -2x_1 + x_2 - 3x_3 + x_4 = -3 \\ 3x_1 - x_2 - 2x_3 - 2x_4 = -1 \\ 4x_1 - 2x_2 + 3x_4 = 0 \\ -x_1 - x_2 + 2x_3 - 5x_4 = -1. \end{cases}$$

5. To find the currents (in amperes) in a complex electrical network, it is necessary to solve the following equations.

$$\begin{aligned} 2i_1 + 3i_2 - i_3 &= 9 \\ 3i_1 + i_2 + i_3 &= 13 \\ i_1 - i_2 + 2i_3 &= 7. \end{aligned}$$

Find i_1, i_2, and i_3.

Review Exercises

6.1 State definitions for the following:
(a) slope of a line.
(b) equivalent equations.
(c) consistent system of equations.
(d) dependent equations.
(e) vector.
(f) matrix.
(g) dot product of matrices.
(h) conformable matrices for multiplication.
(i) multiplication of one matrix by another matrix.

6.2 Sketch the equation $y = 3x - 6$ in the xy-plane. Find the y-intercept and also the slope. Compare the y-intercept and slope that you found with the numbers -6 and 3 in the given equation.

6.3 Determine solution sets to the given system of equations. Are the equations in each system dependent or independent? Is each system consistent or inconsistent?

(a) $\begin{cases} 2x_1 + x_2 = 0 \\ 3x_1 + 5x_2 = 7. \end{cases}$ (b) $\begin{cases} 3x_1 - 6x_2 = 1 \\ x_1 - 2x_2 = 1. \end{cases}$ (c) $\begin{cases} \dfrac{x_1}{3} + \dfrac{2x_2}{3} = \dfrac{4}{3} \\ \dfrac{x_1}{12} + \dfrac{x_2}{6} = \dfrac{1}{3}. \end{cases}$

6.4 Let $a = (-1,2,-3,5,0)$, $b = (1,-2,3,-5,0)$, $c = (3,0,1,-4,3)$ be vectors. Find (a) $a + b$, (b) a vector p such that $a + p = c$, (c) $-3(a + b)$, (d) $\dfrac{1}{3}c$.

6.5 Let

$$A = \begin{bmatrix} 1 & -2 & 3 & -1 \\ 2 & -2 & 0 & 1 \\ 1 & 0 & -3 & 4 \end{bmatrix}_{3 \times 4} \qquad B = \begin{bmatrix} -1 & 2 \\ 3 & 1 \\ -2 & 0 \\ 0 & -1 \end{bmatrix}_{4 \times 2} \qquad C = \begin{bmatrix} 2 & 1 \\ 1 & -2 \\ 4 & 3 \\ -1 & 5 \end{bmatrix}_{4 \times 2}.$$

Find (a) $2A$, (b) AB, (c) $B + C$, (d) AC, (e) $A(B + C)$, (f) $AB + AC$.

6.6 Determine solution sets to the following system of equations.

(a) $\begin{cases} 2x_1 - 3x_2 + x_3 = 4 \\ 3x_1 - 2x_2 + 3x_3 = -2 \\ 4x_1 + 3x_2 - 2x_3 = 3. \end{cases}$

(b) $\begin{cases} 2x_1 - x_2 + x_3 = 2 \\ 3x_1 + 2x_2 + 4x_3 = 9 \\ 5x_1 + x_2 + 5x_3 = 11. \end{cases}$

(c) $\begin{cases} x_1 + 2x_2 - x_3 = 4 \\ x_1 + 4x_2 - 3x_3 = 6. \end{cases}$

(d) $\begin{cases} 2x_1 + 3x_2 = 1 \\ 5x_1 + 4x_2 = -1 \\ x_1 + x_2 = 0. \end{cases}$

6.7 Give reasons for each step in the proof of the following theorem.

THEOREM: If a, b, c are real numbers, such that $a + b = c + b$ then $a = c$.

PROOF:

1. a, b, c are real numbers.
2. $-b$ is a real number.
3. $(a + b) + (-b) = (c + b) + (-b)$.
4. $a + (b + (-b)) = c + (b + (-b))$.
5. $a + 0 = c + 0$.
6. $a = c$.

6.8 Solve the system

$$\begin{cases} 2x_1 + 3x_2 = 5 \\ 3x_1 - 4x_2 = 7 \end{cases}$$

by the substitution method.

6.9 Let $a = (3, -1, 4, -5, 6)$. Find a vector b such that $a \cdot b = 0$.

6.10 Prove or disprove that the product of two $n \times n$ matrices is commutative.

6.11 A helicopter heads due east and has an air speed of 50 miles per hour. Simultaneously, a wind of 30 miles per hour blows from the north. What is the resultant speed of the helicopter?

6.12 One person can read twice as fast as another person. In one hour they read a total of 360 pages. How many pages did each person read in the given hour?

Suggested Reading

Bush, Grace A., and Young, John E., *Foundations of Mathematics with Applications to the Social and Management Sciences*. McGraw-Hill, New York, 1968.

Campbell, Hugh G., *An Introduction to Matrices, Vectors and Linear Programming*. Appleton-Century-Crofts, New York, 1965.

Hohn, Franz, *Elementary Matrix Algebra*. Macmillan, New York, 1952.

Kemeny, John G., Snell, J. Laurie, and Thompson, Gerald L., *Introduction to Finite Mathematics*. Prentice-Hall, Englewood-Cliffs, N.J., 1966.

Marcus, Marvin, and Minc, Henryk, *Introduction to Linear Algebra*. Macmillan, New York, 1965.

McCoy, Neal H., *Introduction to Modern Algebra*. Allyn and Bacon, Mass., 1960.

How can it be that mathematics, being after all a product of human thought independent of experience, is so admirably adopted to the objects of reality?

–ALBERT EINSTEIN

CHAPTER 7

Linear Programming

7.1 The Inequality Relation on the Real Numbers

At the beginning of the preceding chapter we introduced the field axioms applicable to the real numbers and the properties of the equality relation defined on the reals. Systematically, one- and two-dimensional coordinate systems with their associated linear equations were then discussed. By the end of that chapter we had presented in considerable detail systems of linear equations and specialized methods for finding solutions.

In this chapter our discussion will follow a parallel development but will be based instead on inequalities. The inequality relations of greater than ($>$) and less than ($<$) are of fundamental importance to any serious study of mathematics. In fact some mathematicians have gone so far as to state that most of advanced mathematics and in particular calculus is essentially an exposition of inequalities.

Before we introduce any explicit definitions of the inequality relations, we must first state some assumptions which apply to the ordering of real numbers and identify one primitive term. For our purposes we will regard **positiveness** as undefined or primitive and the following to be our **order axioms** on the reals.

Order Axioms

01 Some real numbers are positive.

02 For any real number a, one and only one of the following statements is true.

$$a \text{ is zero} \quad (a = 0);$$

$$a \text{ is positive}; \quad \text{or} \quad -a \text{ is positive}.$$

03 The sum of two positive numbers is positive.

04 The product of two positive numbers is positive.

Furthermore we define a real number a to be a **negative number** if and only if $-a$ is positive.

We can now define the basic inequality relations.

DEFINITION: Let a and b be any real numbers. ***a* is less than *b***, $a < b$, if and only if $b - a$ is positive. Also, ***a* is greater than *b***, $a > b$, if and only if $a - b$ is positive.

DEFINITION: Let a and b be real numbers, $a < b$ if and only if $b > a$.

The symbolic expression $a \leqslant b$ read "a is less than or equal to b," will be used to represent the statement $a < b$ or $a = b$. Similarly, $a \geqslant b$ will be used to represent $a > b$ or $a = b$.

Any real number a which is greater than zero will be considered positive and it will be convenient to symbolize the fact that a is positive by use of the form $a > 0$. Similarly, if a is a negative number we use the symbol $a < 0$.

Many important theorems on inequalities over the reals may be proved by examining the logical consequences of the definitions and axioms given thus far. We will introduce three of these theorems which are extremely important. The first theorem will be proved and we leave to the reader the proof of the other two theorems.

THEOREM 1: Transitive laws

1. If $a < b$ and $b < c$, then $a < c$.
2. If $a > b$ and $b > c$, then $a > c$.

PROOF: 1. By hypothesis $a < b$ and $b < c$. Since $a < b$, $b - a$ is positive, i.e., $b - a > 0$ and since $b < c, c - b$ is positive, i.e., $c - b > 0$.

By the additive order axiom on positive numbers

$$(c - b) + (b - a) > 0.$$

The associative and additive inverse properties of the real numbers give us $(c - b) + (b - a) = c - a$. Thus,

$$c - a > 0$$

which implies $c > a$. Finally $a < c$ by the definitions of greater than and less than.

PROOF: 2. By hypothesis $a > b$ and $b > c$. Since $a > b$, $a - b > 0$ and since $b > c, b - c > 0$.

$(a - b) + (b - c) > 0$	Order axiom
$a - c > 0$	Real number properties
$a > c$	Definition of greater than.

THEOREM 2: Addition laws

1. If $a < b$ and c is any real number, then $a + c < b + c$.
2. If $a > b$ and c is any real number then $a + c > b + c$.

THEOREM 3: Multiplication laws

1. If $a < b$ and $c > 0$ then $ac < bc$.
2. If $a > b$ and $c > 0$ then $ac > bc$.
3. If $a < b$ and $c < 0$ then $ac > bc$.
4. If $a > b$ and $c < 0$ then $ac < bc$.

Theorem 2 states that adding any real number to both sides of an inequality preserves the direction of the given inequality. **Theorem 3** indicates that multiplication of both sides of an inequality by a positive number will preserve the direction of the inequality while multiplication by a negative real number will reverse the direction of the given inequality.

Frequently, we wish to restrict our attention to an interval of real numbers and a number that is between two given real numbers. We employ the following definition in such cases.

DEFINITION: For real numbers a, b, c the symbolic expression $a < b < c$ will mean that $a < b$ and $b < c$.

Each of the various inequality properties may be associated with a geometric diagram which will assist us in illustrating basic concepts as well as helping us to obtain solutions of algebraic inequalities. In a one-dimensional coordinate system, or number line, we use an expression of the form $x < a$, where x is a variable to represent all of the real numbers less than a specified real number a. In other words, $x < a$ corresponds to the solution set of $\{x \mid x \in R \text{ and } x < a\}$. The diagram which we associate with $x < a$ is of the form

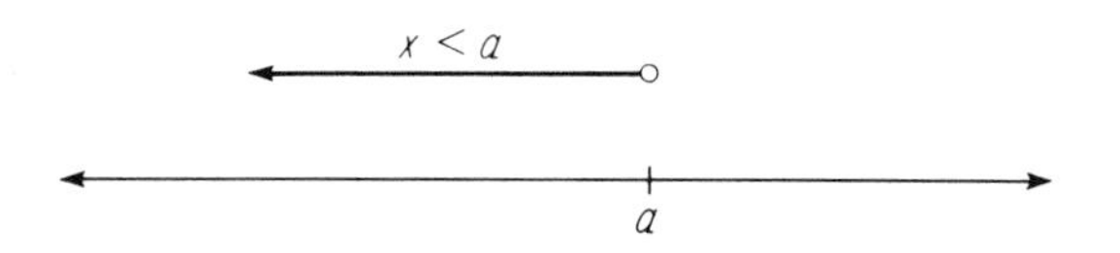

The specified real number a is placed on the number line and the ray or half-line $x < a$ is drawn above the real number line with the open dot at a indicating that a is not included in the set $\{x | x < a\}$. The arrows are used to indicate that the real number line and the ray continue indefinitely in the direction of the arrow.

If the solution set to an inequality is of the form $x \leqslant a$ we indicate that a is an element of the set by using a solid dot rather than open dot at the initial point of the ray. See the accompanying illustrations of this diagram as well as other typical forms.

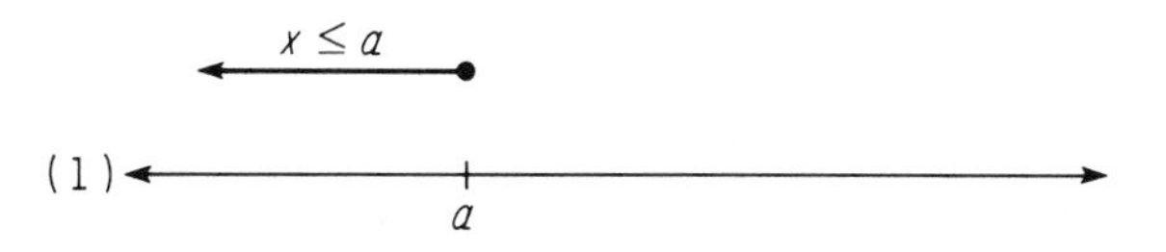

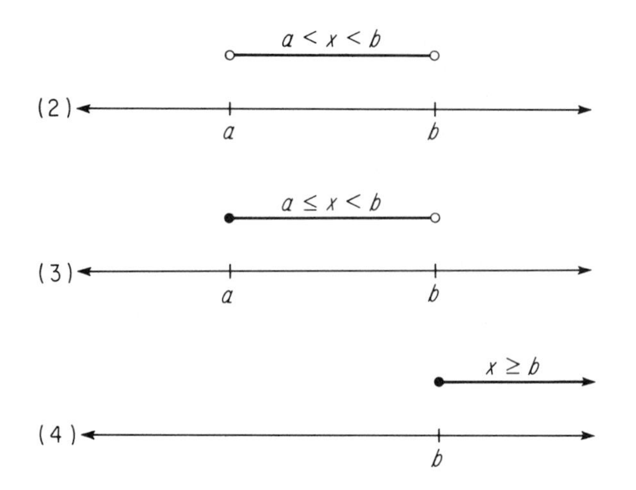

Diagram 2 shows that the interval $a < x < b$ does not include either of the end-points a or b and is called an **open interval**. Diagram 3 representing $a \leqslant x < b$ includes a but does not include b and is said to be **closed** at a and **open** at b. If both end-points are included in a real interval, the interval is said to be **closed**.

Let us now consider several examples which will illustrate how we may apply the order axioms, definitions, and theorems to find the solution of linear inequalities involving a single variable.

Example 1: Solve the inequality $5x - 8 < 2x + 1$ and justify each step by reference to an axiom, definition, or theorem. Diagram the solution set on the real-number line.

Solution:

Statement	Reason
$5x - 8 < 2x + 1$	Given
$(5x - 8) + 8 < (2x + 1) + 8$	If $a < b$, and $c \in R$ then $a + c < b + c$ (Theorem 2)
$5x + (-8+8) < 2x + (1 + 8)$	Associative property
$5x + 0 < 2x + 9$	Additive inverse, $1 + 8 = 9$
$5x < 2x + 9$	Additive identity
$5x + (-2x) < (2x + 9) + (-2x)$	If $a < b, c \in R$ then $a + c < b + c$ (Theorem 2)
$5x + (-2)x < (-2x + 2x) + 9$	Associative and commutative property
$3x < 0 + 9$	Distributive, additive inverse
$3x < 9$	Additive identity
$x < 3$	If $a < b$, and $c > 0$ then $ac < bc$ (Theorem 3)

By application of the basic properties we have shown that the solution set of $5x - 8 < 2x + 1$ is equivalent to the solution set of $x < 3$, i.e., $\{x | x \in R, x < 3\}$. The diagram for the $\{x | x < 3\}$ is in the accompanying illustration.

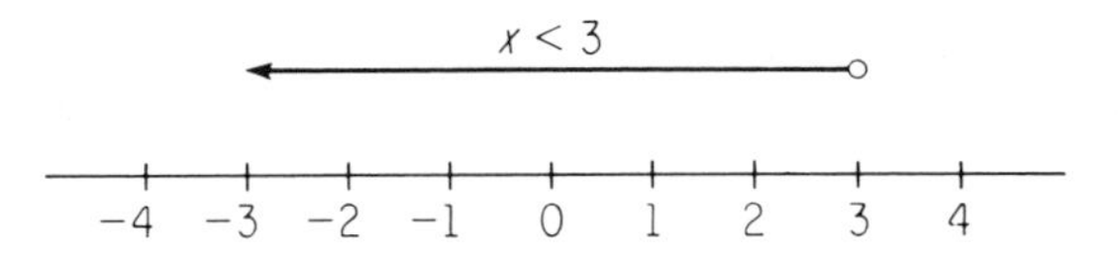

Example 2: Determine the solution set of the inequality $2x + 5 \leqslant 9x + 12$. Sketch the solution set on the real number line.

Solution: (The reader should supply the reason for each step.)

$$
\begin{aligned}
&2x + 5 \leqslant 9x + 12\\
&(2x + 5) + (-5) \leqslant (9x + 12) + (-5)\\
&2x + (5 + (-5)) \leqslant 9x + (12 + (-5))\\
&2x + 0 \leqslant 9x + 7\\
&2x \leqslant 9x + 7\\
&2x + (-9x) \leqslant (-9x) + (9x + 7)\\
&2x + (-9)x \leqslant (-9x + 9x) + 7\\
&-7x \leqslant 0 + 7\\
&-7x \leqslant 7\\
&x \geqslant -1.
\end{aligned}
$$

The solution set of $2x + 5 \leqslant 9x + 12$ is equivalent to the solution set of $x \geqslant -1$, i.e., $\{x \mid x \in R, x \geqslant -1\}$. The geometric representation of the solution set is given in the accompanying diagram.

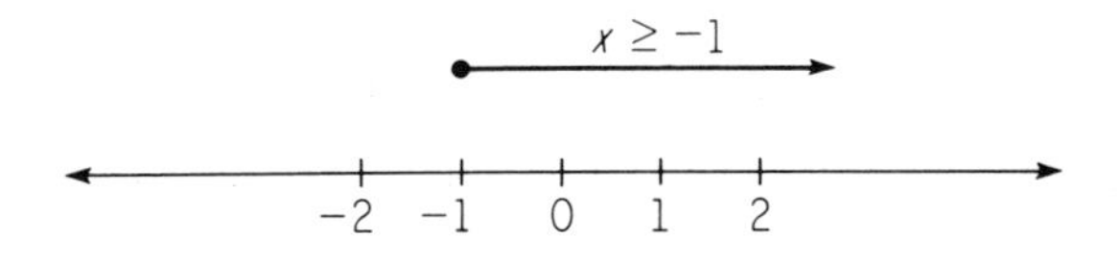

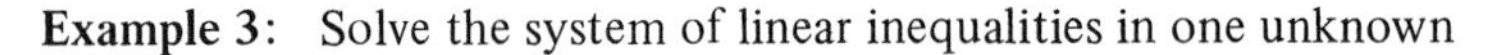

Example 3: Solve the system of linear inequalities in one unknown

$$
\begin{cases} x + 2 \geqslant 0 \\ x - 3 < 1 \end{cases}
$$

and sketch the solution set.

Solution: To obtain the solution of the system we must find the set of real numbers that satisfy each inequality simultaneously. We will first find the solution set of each inequality. The intersection of these sets will be the desired solution set.

The solution set of the inequality $x + 2 \geqslant 0$ is

$$\{x \mid x \in R, x \geqslant -2\}.$$

The solution set for the inequality $x - 3 < 1$ is

$$\{x \mid x \in R, x < 4\}.$$

The intersection of these solution sets $\{x \mid x \in R, x \geqslant -2\} \cap \{x \mid x \in R, x < 4\}$ is

$$\{x \mid x \in R, -2 \leqslant x < 4\}.$$

The geometric diagram as shown is particularly useful in illustrating the solution of system. The diagram shows that the portion of the real-number line common to both

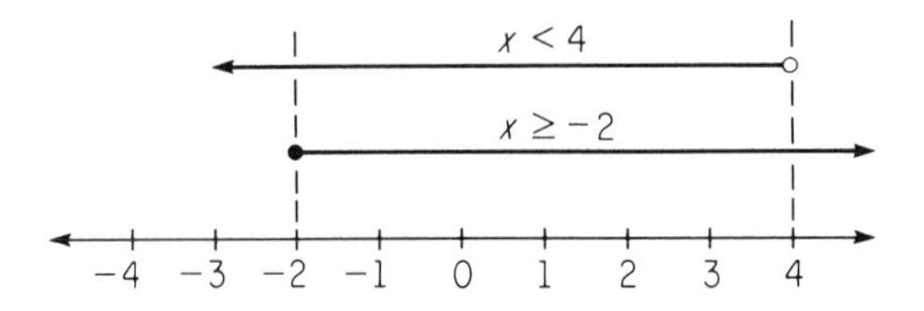

geometric figures is the interval $-2 \leqslant x < 4$. Thus, the solution set may be drawn as shown in the accompanying illustration.

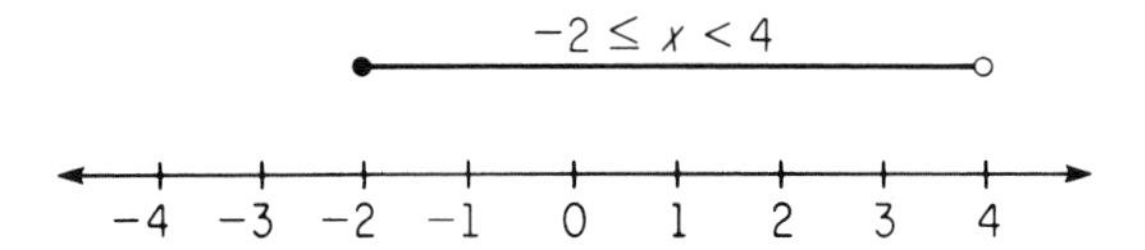

Example 4: Solve

$$\frac{2x - 1}{2 - 3x} > 0$$

and sketch the solution set on the real line.

***Solution*:** The expression $\frac{2x - 1}{2 - 3x}$ is greater than zero if

1. $2x - 1$ is positive (i.e., $2x - 1 > 0$) *and* $2 - 3x$ is positive (i.e., $2 - 3x > 0$), *or*
2. $2x - 1 < 0$ *and* $3 - 2x < 0$.

Thus the solution set of

$$\frac{2x - 1}{2 - 3x} > 0$$

is $\{x \mid 2x - 1 > 0 \text{ and } 2 - 3x > 0\} \cup \{x \mid 2x - 1 < 0 \text{ and } 3 - 2x < 0\}$. Let

$$A = \{x \mid 2x - 1 > 0 \text{ and } 2 - 3x > 0\}$$
$$B = \{x \mid 2x - 1 < 0 \text{ and } 2 - 3x < 0\}.$$

For the sets A, B we must find those real numbers which satisfy the statement form for the sets A, B. To do this we will find the solution set corresponding to each of these statement forms. We first consider the inequalities $2x - 1 > 0$ and $2 - 3x > 0$. We wish to determine those real numbers that simultaneously satisfy both inequalities. Hence, we must find the intersection of the solution sets of $2x - 1 > 0$ and $2 - 3x > 0$. As in the previous example, we will consider each inequality separately and then find the intersection of the solution sets.

Case I: If $2x - 1 > 0$ then $2x > 1$, hence $x > 1/2$. Now consider the inequality $2 - 3x > 0$. For $2 - 3x > 0$, we have $2 > 3x$. Thus $2/3 > x$, which is equivalent to

$x < 2/3$. Thus the solution set to $2x - 1 > 0$ is $\{x \mid x \in R, x > 1/2\}$ and the solution set to $2 - 3x > 0$ is $\{x \mid x \in R, x < 2/3\}$.

To illustrate the intersection of these solution sets, we diagram them as shown on the real line and then determine their intersection from the drawing. From the diagram, the intersection is the set $\{x \mid x \in R, 1/2 < x < 2/3\}$. Thus $A = \{x \mid x \in R, 1/2 < x < 2/3\}$.

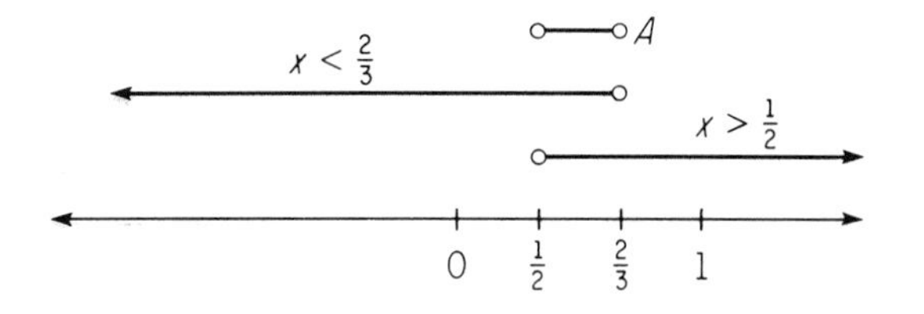

To determine set B we must find the intersection of the solution sets of the inequalities $2x - 1 < 0$, $2 - 3x < 0$.

Case II: If $2x - 1 < 0$ then $2x < 1$; hence $x < 1/2$. Now for $2 - 3x < 0$ we have $2 < 3x$; hence $2/3 < x$. Thus, the solution set to $2x - 1 < 0$ is $\{x \mid x \in R, x < 1/2\}$, and the solution set to $2 - 3x < 0$ is $\{x \mid x \in R, 2/3 < x\}$.

We again diagram each solution set on the real line to determine the intersection of these two solution sets (see the accompanying illustration). From the diagram we observe that the solution set is empty. Thus, $B = \emptyset$.

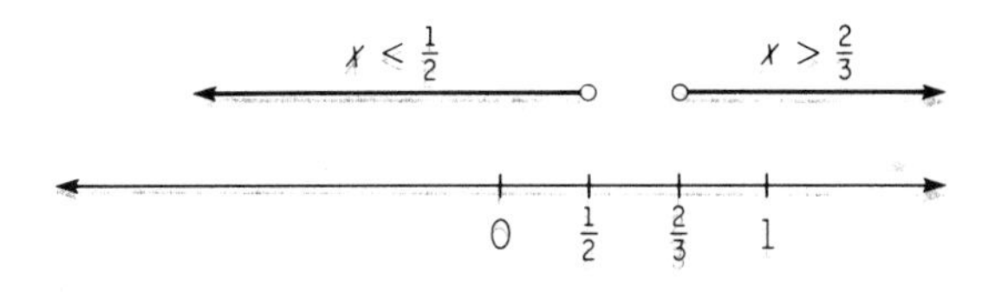

Therefore, the solution set of

$$\frac{2x - 1}{2 - 3x} > 0$$

is $A \cup B = A \cup \emptyset = A = \{x \mid x \in R, 1/2 < x < 2/3\}$.

The basic concepts which have been developed in this section will be extended in subsequent sections to a two-dimensional coordinate system and to systems of linear inequalities in two unknowns.

Problem Set 7.1

A. 1. Find the solution set to the following inequalities and sketch the solution set on the real line.

(a) $\frac{1}{2}x + 3 > -\frac{3}{2}x - 9$. (b) $2x - 3 < 2x - 1$.

(c) $2x + 3 < 2x - 1$. (d) $x + 3 \geqslant 2x + 5$.
(e) $2 - x \leqslant -4 + 2x$.

2. Solve $\dfrac{x}{x - 2} \geqslant 0$ and sketch the solution set on the real line.
3. Solve the following inequalities.
(a) $3 < 2x - 1 < 5$. (b) $-5 \leqslant 3 - 2x < 3$. (c) $-1 \geqslant 2 - 3x \geqslant -7$.
4. Find the solution set to the expression $\dfrac{x}{x - 2} > -5$. (*Hint*: To the given expression, add 5 to both sides of the inequality, and then by algebraic manipulation obtain an expression of the form $\dfrac{ax + b}{cx + d} > 0$. Next, proceed in the same manner as in Prob. 2 of this problem set.)

B. 1. Using the axioms of the reals and the definition of the order relations, prove the following.
(a) Let a, b, c be real numbers and $c < 0$. Prove $a < b$ if and only if $ac > bc$.
(b) Prove $1 > 0$, where 1 is the multiplicative identity for the reals and 0 is the additive identity for the reals.
(c) Prove that if $a > 1$ then $a^2 > a$.
(d) Prove if $0 < a < 1$ then $a^2 < a$.
(e) Prove or disprove that for any real numbers a, b, c, and d that if $a > b$ and $c > d$ then $ac > bd$.

7.2 Linear Inequalities in Two Unknowns

In the preceding section we observed that linear inequalities in one unknown defined over the reals led us to consider solution sets which were subsets of the reals. In contrast to much of our earlier work involving equalities where we frequently encountered unique solutions, most of our work with inequalities in one unknown generated solution sets that were infinite. Interpreted geometrically, we were able to display the solution set (unless the solution set was empty) as some portion of the real-number line.

We now wish to consider the extension of the basic inequality concepts to linear inequalities in two unknowns and their corresponding geometric interpretation.

Our interest will be focused on inequalities of the following general types:

$$Ax + By + C > 0$$
$$Ax + By + C \geqslant 0$$
$$Ax + By + C < 0$$
$$\text{or} \quad Ax + By + C \leqslant 0 \quad \text{where } A, B \text{ are not both zero.}$$

The reader will recall that linear equations of the form $Ax + By + C = 0$, where A, B are not both zero, had an infinite number of ordered pair solutions and the corresponding geometric form was a straight line drawn in the Cartesian plane. Each of the four general types of inequalities in two unknowns will also involve an infinite number of ordered pair solutions. However, the geometric form of the solution will correspond to some portion of the Cartesian plane.

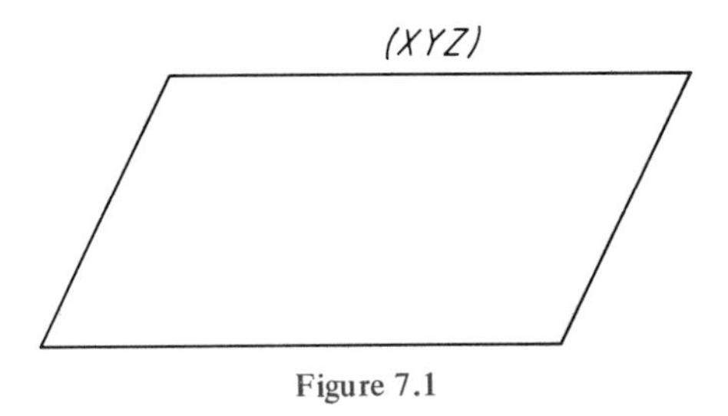

Figure 7.1

Since a plane is infinite in extent, let us first consider how we might conveniently subdivide a plane. Let (XYZ) be a portion of a Euclidean plane drawn pictorially as shown in Figure 7.1.

We will assume that any line L that lies in the plane will divide the plane into two distinct spaces called half-planes. We designate one half-plane as H_1 and the other half-plane as H_2. The line L is the boundary of the two half-planes and does not belong to either half-plane. If the plane is considered as the universe under discussion, then the union of the half-planes and L is the plane, i.e., $H_1 \cup L \cup H_2 = (XYZ)$.

A given point in the plane will either be in half-plane H_1, on the boundary line L, or in half-plane H_2. In Figure 7.2 the point P is on the side of the boundary containing all the points of H_1 and on the opposite side of the boundary from the set of points in H_2.

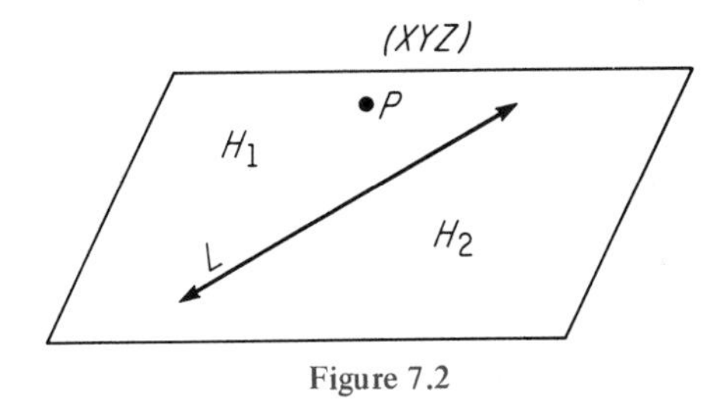

Figure 7.2

Let us now translate these Euclidean plane concepts to a two-dimensional Cartesian coordinate system. (Figure 7.3). Let $Ax + By + C = 0$ be an arbitrary line drawn in the xy-plane. The line will divide the plane into the two half-planes, H_1 and H_2. A dotted line is drawn to indicate it does not belong to either half-plane. The set of points which are on the line are associated with the order pair solutions of the equation $Ax + By + C = 0$, i.e., $\{(x,y) \mid Ax + By + C = 0\}$. The set of points in one half-plane will be associated with the ordered pairs given by $\{(x,y) \mid Ax + By + C > 0\}$ and the set of points in the other half-plane will be associated with $\{(x,y) \mid Ax + By + C < 0\}$. In other words, the inequality statements correspond to the set of points on one side or the other of a given line in the plane. Conversely, the three algebraic forms $Ax + By + C > 0$, $Ax + By + C = 0$, and $Ax + By + C < 0$ may be regarded as open sentences which partition all of the ordered pairs of real numbers into three disjoint sets. Thus, an arbitrarily chosen ordered pair will correspond to one and only one of the algebraic forms such that a true statement is obtained when the ordered pair is substituted into that specific open sentence.

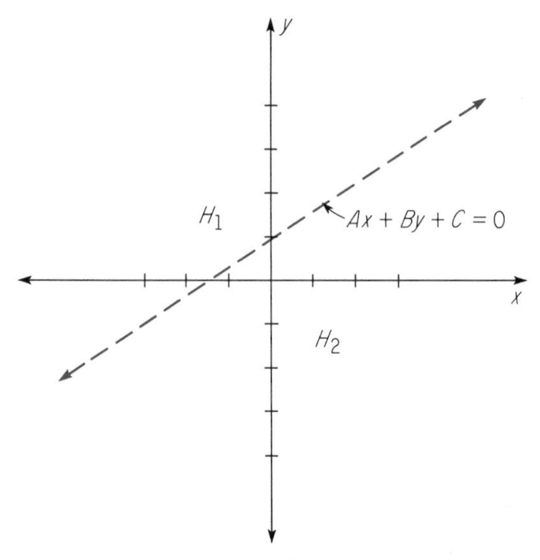

Figure 7.3

For example, consider the three open sentences

1. $3x + 4y > 12.$
2. $3x + 4y = 12.$
3. $3x + 4y < 12.$

The arbitrarily chosen ordered pair (2, 3) corresponds uniquely to the truth set of $3x + 4y > 12$, since

1. $3(2) + 4(3) = 18 > 12$ is true.
2. $3(2) + 4(3) \neq 12.$
3. $3(2) + 4(3) \not< 12.$

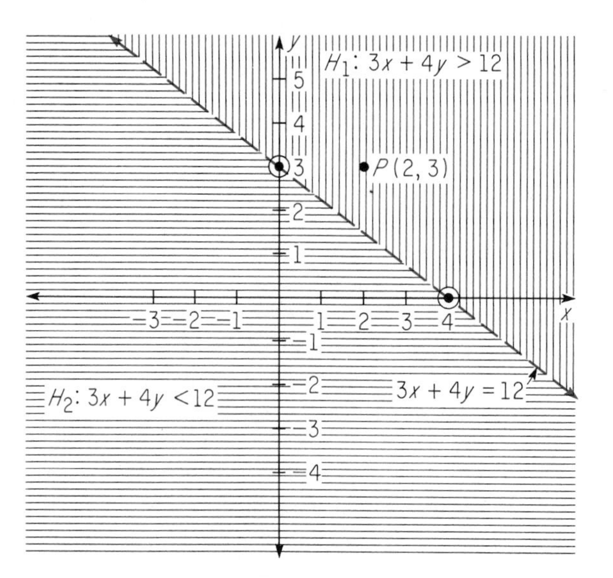

Figure 7.4

The geometric representation of these three disjoint possibilities is shown in Figure 7.4. The line $3x + 4y = 12$ is drawn using the intercepts (0,3) and (4,0). The point $P(2,3)$ lies in half-plane H_1 and thus H_1 corresponds to the inequality $3x + 4y > 12$. Half-plane H_2 corresponds to the inequality $3x + 4y < 12$. To identify which half-plane corresponds to a given inequality, it is sufficient to choose any point that is not on the boundary and substitute the corresponding ordered pair into either inequality. If a true statement results, the half-plane containing the point will be associated with that inequality. If a false statement is obtained, the inequality will be associated with the half-plane on the other side of the boundary line. A convenient point to choose for making this test, providing it is not on the boundary line, is the origin, i.e., (0,0).

Example 1: Sketch the graph of the solution set of the inequality $3x - 2y - 6 < 0$ in the xy-plane.

Solution: First we sketch the graph of the line $3x - 2y - 6 = 0$ by using the intercepts (0,−3) and (2,0) (Figure 7.5). Now we use the origin (0,0), to test the

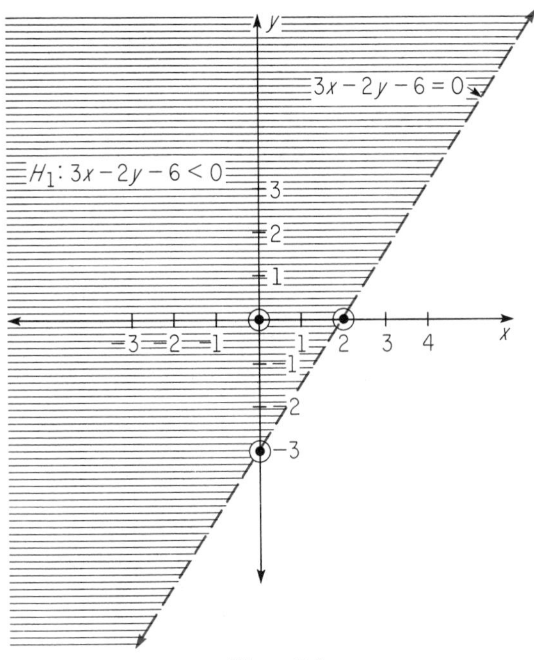

Figure 7.5

given inequality $3x - 2y - 6 < 0$. Thus, $3(0) - 2(0) - 6 < 0$ gives us $-6 < 0$ which is a true statement. Hence, we associate the inequality $3x - 2y - 6 < 0$ with the shaded half-plane H_1 containing the point (0,0).

In the presentation thus far, we have limited our discussion to those cases in which the boundary line did not belong to either half-plane. This was convenient for introducing the material, but it is not a necessary restriction. We have also chosen examples where the coefficients of both the x and y terms were nonzero. However, it is frequently important to sketch, in the plane, inequalities in which the coefficient of one variable is zero and the coefficient of the other variable is nonzero. We will complete this section by considering an example which will illustrate both of these points.

Example 2: Sketch the graph of the solution set of the inequality $-3y \leqslant 6$.

Solution: First, we sketch the graph of the line $-3y = 6$, i.e., $y = -2$ in the xy-plane. Note that the line is now drawn as a solid line (Figure 7.6) indicating that the line

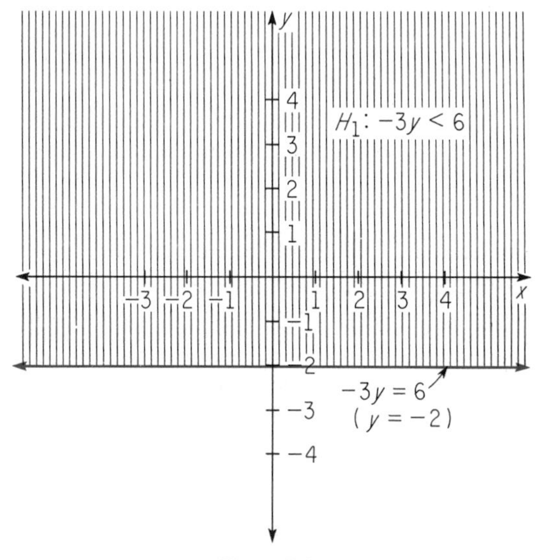

Figure 7.6

will be part of the solution set. Using the origin as a test point in the given inequality $-3y \leqslant 6$ we have

$$-3(0) \leqslant 6 \qquad \text{or} \qquad 0 \leqslant 6$$

which is a true statement. Thus, the shaded half-plane $-3y < 6$ which contains the origin (0,0) together with the line $-3y = 6$ ($y = -2$) represent the entire solution set.

Problem Set 7.2

A. 1. Sketch the graph of the solution set of $-3x - 2y < -6$.

2. For the inequality $-2x \geqslant 8$, sketch the solution set in the xy-coordinate plane.

3. Sketch the solution set of the following inequalities in a two-dimensional coordinate system.

   (a) $2x + y < 2x + y - 3$.
   (b) $2(x - 3y) > 2x - 18$.
   (c) $x > y - 1$.
   (d) $3x - y < 4x - 2y + 3$.
   (e) $4y - 2x \geqslant 7$.

4. Sketch the solution set of $x + 2y \geqslant 4$ in the xy-plane when $x \in A, y \in A$, and $A = \{0,1,2,3,4,5,6,7,8,9\}$.
5. Let $X = Y = \{0,1,2,3,4,5,6,7,8,9\}$. Find the solution set for the inequality $3y + 2x < 18$ when $x \in X$ and $y \in Y$.
6. Sketch the solution set of the inequalities in the x_1x_2-coordinate plane where the horizontal axis is the x_1-axis and the x_2-axis is the vertical axis.
   (a) $x_1 - x_2 \geqslant -3$.
   (b) $2x_1 - 3x_2 - 9 \leqslant 0$.
   (c) $4x_1 + 3x_2 \geqslant 2x_1 + 4x_2$.
   (d) $5x_2 + 5 - x_1 \leqslant 4x_1 + 4x_2 - 5$.
   (e) $3 \geqslant 2x_2 + 3x_1$.
7. (a) Graph on the same set of axes the system of inequalities

$$\begin{cases} x_1 + 2x_2 \leqslant 6 \\ x_1 \geqslant 0 \\ x_2 \geqslant 0. \end{cases}$$

   (b) Is the intersection of all the inequalities in (a) bounded?
   (c) Superimpose on the graph of (a) the graphs of the straight lines whose equations are $-3x_1 + x_2 = 3$, $-3x_1 + x_2 = 0$, and $-3x_1 + x_2 = -18$.
   (d) Is there an ordered pair solution to (a) which would also lie on a straight line of the form $-3x_1 + x_2 = k$ and which will make k (1) a minimum, (2) a maximum? If so, identify any such ordered pairs.
8. A company makes sweaters and blouses. These products are manufactured in two different factories, say A and B. Factory A makes 2 sweaters and 2 blouses every hour. Factory B makes 3 sweaters and 1 blouse every hour. The manufacturer must quickly produce at least 18 sweaters and 10 blouses for a given store. Graph the inequalities representing the various hour combinations at which the company could operate both factories simultaneously. What hour combination will minimize the total production time? (*Hint*: Let x_1 represent the number of hours factory A operates and x_2 represent the number of hours factory B operates.)

7.3 Systems of Linear Inequalities in Two Unknowns

In Chapter 6 we discussed in considerable detail systems of linear equations. We observed that systems involving only two unknowns could be solved by using rather standard algebraic techniques such as the substitution method as well as the matrix theory approach which was based on certain elementary row operations. The corresponding geometric interpretation for systems of equations involving two variables gave rise to a consideration of the straight lines representing each equation and whether or not they had a single point in common, no points in common, or an infinite number of points in common.

Although it is possible to consider algebraic methods for obtaining solutions to systems of linear inequalities involving two unknowns, any such procedure becomes extremely cumbersome. The reasons for this become evident if we recall, for example, that multiplication of an inequality by a negative real number changes the direction of an inequality and that other algebraic operations which apply to systems of equations may not be indiscriminately applied to systems of inequalities. As a consequence, we will rely on a geometric interpretation of the system of linear inequalities in two unknowns.

Based on the graphs which may be drawn, a solution, if it exists, can be determined. This procedure can best be illustrated by considering a few typical examples.

Example 1: Find the solution set to the system of linear inequalities

$$\begin{cases} 4x + 3y \geqslant 12 \\ 3x - 2y < 6. \end{cases}$$

Solution: In a coordinate system (Figure 7.7) first sketch the graph of the lines $4x + 3y = 12$ and $3x - 2y = 6$ using the x- and y-intercepts. Observe that the

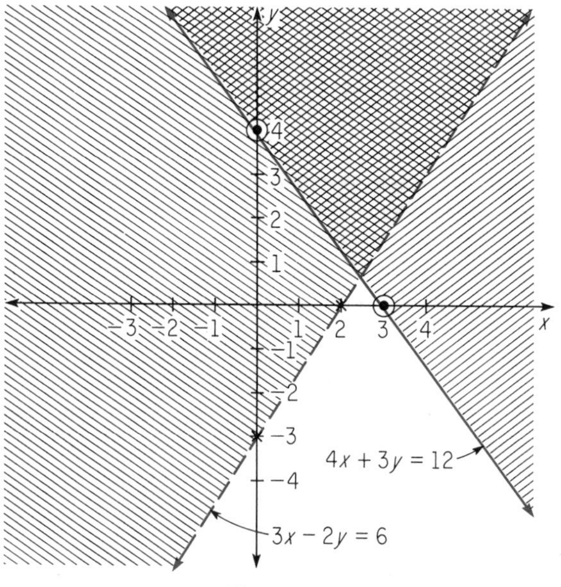

Figure 7.7

line $4x + 3y = 12$ is drawn as a solid line indicating the line is part of the solution set to the inequality; the line $3x - 2y = 6$ is drawn as a dotted line indicating that the line is not part of the solution set of the corresponding inequality. Using the origin (0,0) as a test point for each inequality, we can determine the appropriate half-plane and shade the appropriate region as shown. The cross-hatched region, ▦ , represents the portion of the plane common to both inequalities and represents the solution set to the system of inequalities.

In set form, the solution set is the intersection of the two inequalities, i.e.,

$$\{(x,y) \mid 4x + 3y \geqslant 12\} \cap \{(x,y) \mid 3x - 2y < 6\}.$$

Example 2: Find the solution set to the system of inequalities

$$\begin{cases} 2x - 4y \geqslant 6 \\ \ x - 2y \leqslant -1. \end{cases}$$

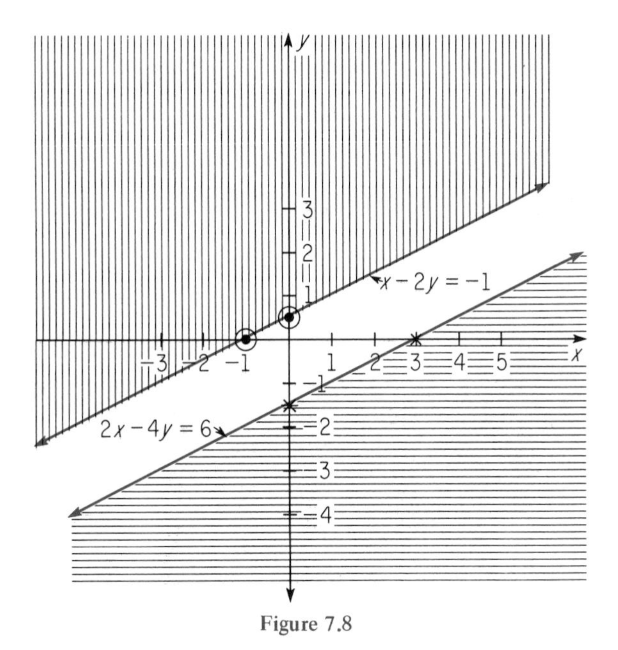

Figure 7.8

Solution: First sketch the graph of each line $2x - 4y = 6$ and $x - 2y = -1$ using the x- and y-intercepts (Figure 7.8). The origin (0,0) is used as a test point and the appropriate region is shaded. There is no overlap in the shaded regions and the lines representing the respective boundaries are parallel. Thus, the solution set is empty.

In set form, we have

$$\{(x,y) \mid 2x - 4y \geqslant 6\} \cap \{(x,y) \mid x - 2y \leqslant -1\} = \emptyset.$$

Certain situations require that we consider a system of linear inequalities in two unknowns which will involve more than two inequalities.

Example 3: Find the solution set to the system of inequalities

$$\begin{cases} x + 2y \leqslant 8 \\ x + 2y \leqslant 4 \\ -2x + 3y < 12 \\ y > 0. \end{cases}$$

Solution: As a matter of convenience we proceed in a step-by-step pattern to obtain the common solution set. The graphs of $x + 2y \leqslant 8$ and $x + 2y \leqslant 4$ are shown in Figure 7.9. The cross-hatched region common to both inequalities lies below the line (and includes the line) $x + 2y = 4$.

Figure 7.10 displays the intersection of the inequality $x + 2y \leqslant 4$ and the inequality $-2x + 3y < 12$. The cross-hatched region common to both lies below the lines $x + 2y = 4$ and $-2x + 3y = 12$. In the one case the boundary is included and in the other case the boundary is not included.

Finally, we impose the additional condition that $y > 0$ onto the geometric con-

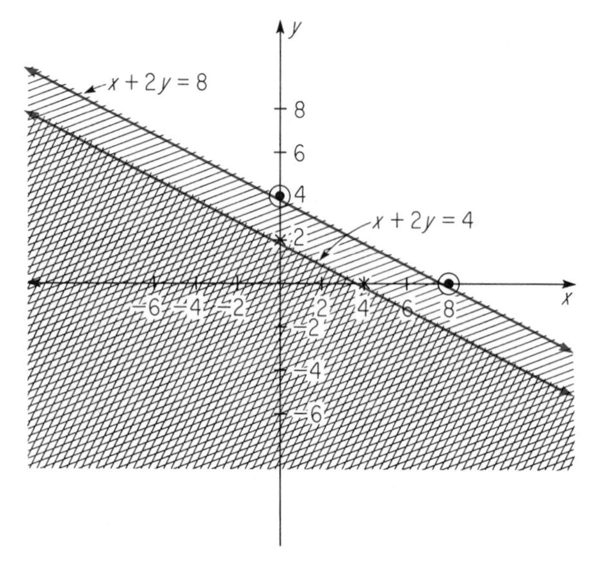

Figure 7.9

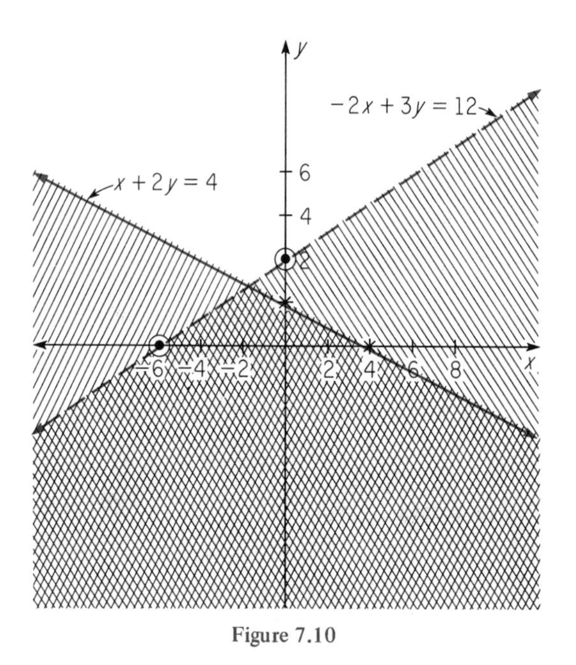

Figure 7.10

figuration shown in Figure 7.10 to obtain the triangular region shown in Figure 7.11 which represents the solution set to the system of inequalities.

In the preceding example, one of the given inequalities, $x + 2y < 8$, had no direct bearing on the solution set of the other three inequalities, since the solution set of

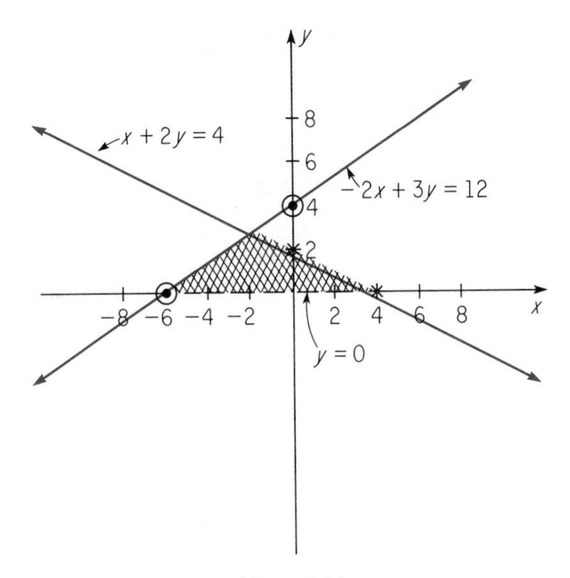

Figure 7.11

$x + 2y < 4$ was a subset of the solution set of $x + 2y < 8$. In practice our work can be shortened if we can observe such relationships at the outset. Note also that the condition $y > 0$ imposes the condition that our solution must consist only of regions above the x-axis and could be used as an initial condition in sketching the solution of each inequality.

The triangular shaped region which we obtained as the solution set in Example 3 consists of an infinite number of points. However, we regard the solution set as **bounded,** since the area of the region is a finite number.

Example 4: Find the solution set to the system of linear inequalities

$$\begin{cases} x + \ \ y \geqslant 1 \\ 3x + 3y \leqslant 9. \end{cases}$$

Solution: The graph of each inequality is shown in Figure 7.12. The cross-hatched region is an infinite strip that lies between the two parallel lines and includes the lines in the solution set.

However, since the area of the strip is infinite, we regard the solution set as **unbounded**.

The geometric techniques which we have discussed for solving systems of linear inequalities may be generalized to systems involving more than two unknowns. In the next section we will use these same methods to introduce the basic concepts of linear programming.

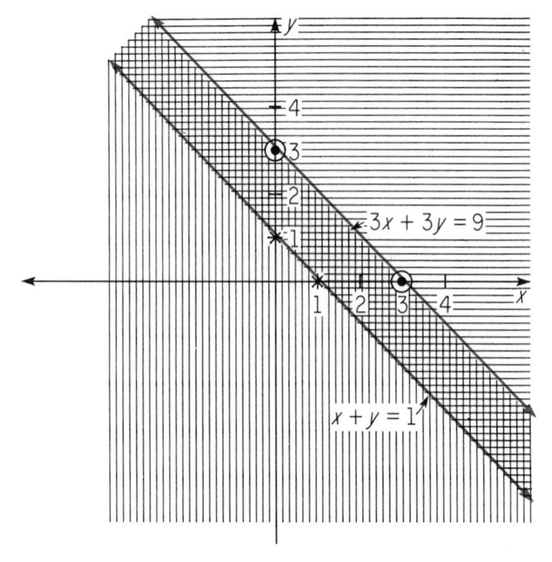

Figure 7.12

Problem Set 7.3

A. 1. Sketch the solution set in the xy-plane for each of the following systems of linear inequalities:

(a) $\begin{cases} x + y \leqslant 2 \\ y \geqslant 0 \\ x \geqslant 0. \end{cases}$ (b) $\begin{cases} x - y > 2 \\ x + y < 5 \\ y > -2. \end{cases}$ (c) $\begin{cases} x > 0 \\ y > 0. \end{cases}$

2. Find the solution set to the following system of linear inequalities:

(a) $\begin{cases} -4x + 3y \leqslant 12 \\ -2x + 3/2y \geqslant 6. \end{cases}$ (b) $\begin{cases} 2x + 3y < 6 \\ 3x + 2y > 12 \\ y > 0. \end{cases}$ (c) $\begin{cases} y > x \\ x < -1 \\ y < 0 \\ x > -5. \end{cases}$

3. Sketch the solution set in the xy-plane for each of the following systems of linear inequalities when the variables x, y can only assume integer values.

(a) $\begin{cases} y + 2x > 2 \\ y > 0 \\ x > 0 \\ 3y + 4x < 24. \end{cases}$ (b) $\begin{cases} y < 0 \\ x < 3 \\ x > -2 \\ y > -5. \end{cases}$

4. Sketch the solution set for the following system of linear inequalities and determine whether the solution set is bounded or unbounded.

(a) $\begin{cases} 4x + 3y > 12 \\ -3x + 2y > -6. \end{cases}$ (b) $\begin{cases} y < 4 \\ -3x + 2y > -6 \\ 4x + 3y > 12. \end{cases}$ (c) $\begin{cases} y \geqslant 0 \\ x \geqslant 0 \\ 4x + 3y < 12. \end{cases}$

(d) $\begin{cases} x \geqslant 0 \\ -3x + 2y \geqslant -6 \\ y \leqslant 0. \end{cases}$ (e) $\begin{cases} -3x + 2y \leqslant -6 \\ 4x + 3y \leqslant 12 \\ y \leqslant 0. \end{cases}$

5. Find a system of linear inequalities that represent each of the following solution sets shaded in the xy-plane.

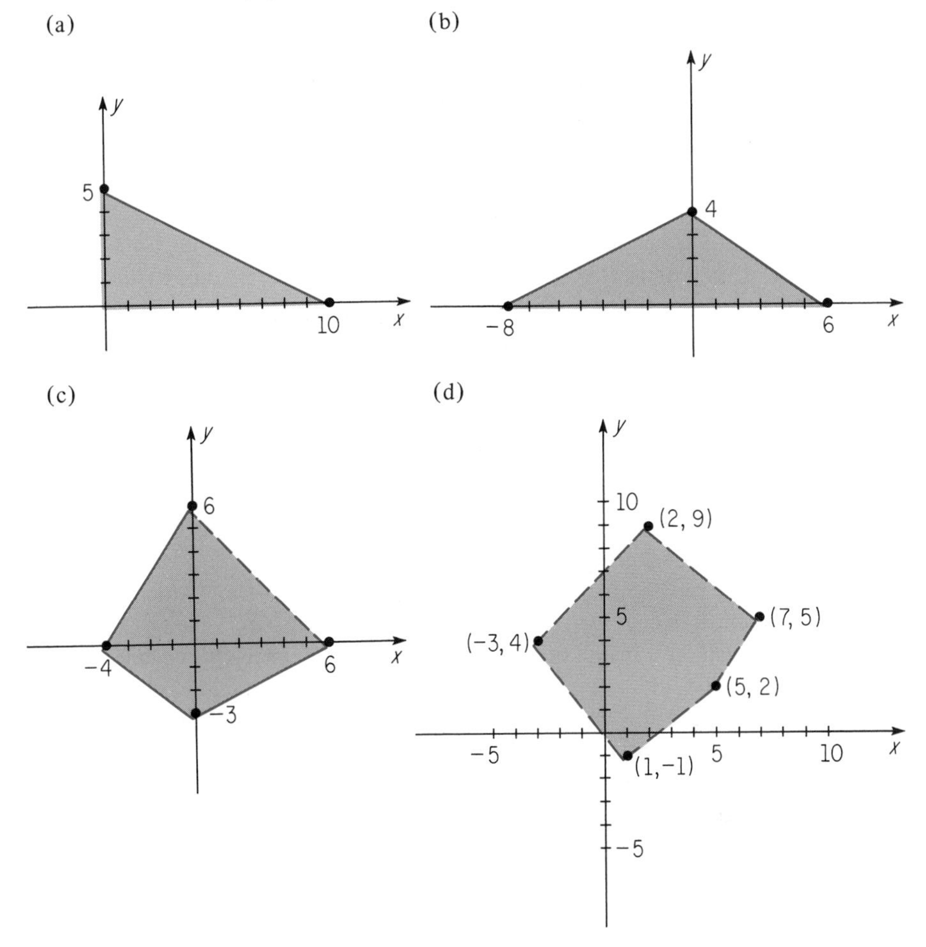

6. A company manufactures two models of radios, super sound, and modern sound. It costs \$5 to make one super sound radio and \$10 to make one modern sound radio. The company can spend at most \$500 in one day. State the system of linear inequalities that apply to the given data and graph the solution set to this system of linear inequalities.

7.4 Fundamentals of Linear Programming

Linear programming is one of the most recent and one of the most important branches of modern mathematics. Although some of the basic concepts of linear programming were used by both J. Von Neumann and W. Leontief prior to 1940, most experts trace the beginning of linear programming as a separate field of activity to the work undertaken by the mathematicians F. L. Hitchcock, L. Kantorovitch, T. C. Koopmans, and G. B. Dantzig in the 1940's.

In essence, a linear program problem may be described as one in which an effort is made to either maximize or minimize a linear function, generally of two or more variables, where the variables are subject to specified constraints.

Prior to 1947 linear programming problems were generally solved by trial and error techniques. Frequently the solutions obtained were only approximations of the desired exact solutions. In 1947 G. B. Dantzig developed a general method, called the simplex method, for solving linear programming problems. In the past 25 years the simplex method has undergone considerable refinement and other techniques for handling linear programming problems have been developed.

The field of linear programming is continuing to expand and is now being widely used in such diverse areas as economics, agriculture, engineering, and medical technology. Contributing to the rapid increase in the use of linear programming techniques has been the parallel development of digital computers. The procedures used for solving a given problem whether it be by the simplex method or other specialized methods, frequently involve complex computations which must be repeated over and over again. A modern computer can be programmed to automatically perform the necessary computations and produce a desired solution. We, of course, cannot in this text present the necessary computer programming information that would enable the reader to solve a problem by the use of a computer. However, we will present the basic theory and indicate the solutions to some typical although simplified problems.

Most linear programming problems are expressed in a form which is related to systems of inequalities which were discussed in the preceding section. The solution set to a system of inequalities consists of all ordered n-tuples that satisfy each and every inequality. Geometrically, the solution set is the intersection of the graphs of the various inequalities. In 2-space the solution set, if it exists, will be some portion of the plane and may be bounded or unbounded. We wish to focus our attention on a particular type of set called a convex set which will play an important role in determining the solution set of a linear programming problem.

DEFINITION: A set of points in n-space is said to be a **convex set** if for every pair of points P, Q in the set, the line segment joining these points is also in the set.

Note:

(i) The line segment from b to a in n-space is a set of points α which may be represented as $\alpha = ra + (1 - r)b$ where $0 \leqslant r \leqslant 1$. For example, if $r = 0$ then $\alpha = b$, if $r = 1$ then $\alpha = a$, and if $r = 1/2$ then $\alpha = \dfrac{a+b}{2}$ which is the midpoint of the line segment from b to a.

(ii) A set of points will be considered convex if it contains less than two elements.

Example 1: The shaded drawings below indicate various geometric configurations of sets of points in two-space. Indicate whether or not the sets are convex.

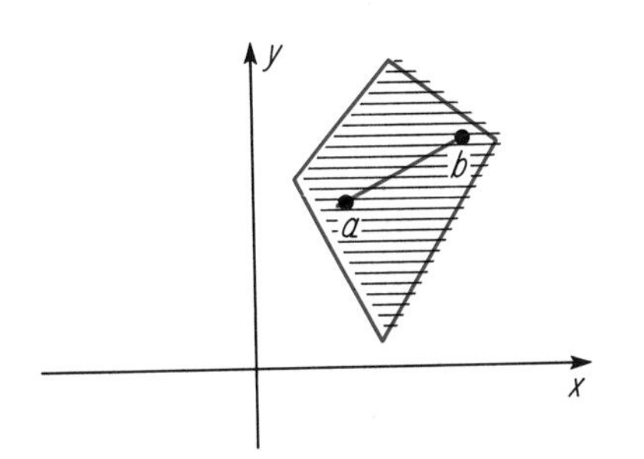

Figure 7.13

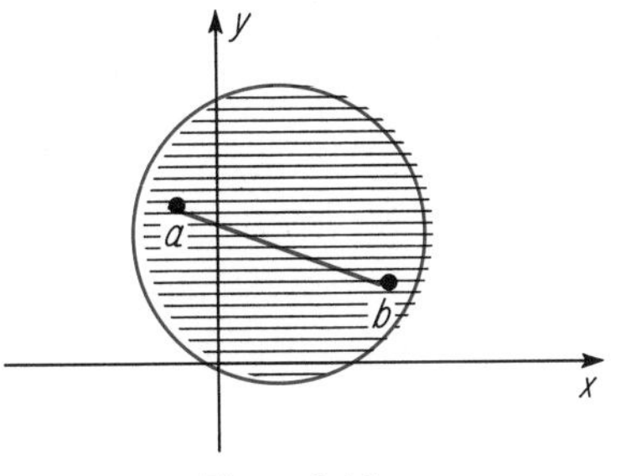

Figure 7.14

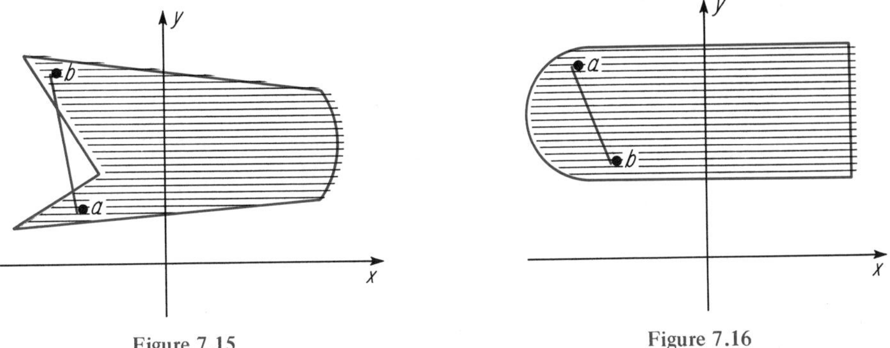

Figure 7.15

Figure 7.16

Solution: In each of the above drawings, a typical line segment from a point a to a point b is drawn. The reader can visualize other possible line segments that could be drawn in each figure. Line segments whose end points are in the set as drawn in Figure 7.13, 7.14, and 7.16 would all remain within the set of points illustrated and thus these figures represent convex sets. However, the line segment shown in Figure 7.15 does not entirely lie within the set, thus this set is not convex.

As a consequence of the definition of a convex set, many important theorems regarding convex sets can be established. For our immediate purposes we are interested only in two-space and the intersection of convex sets. We will list, therefore, certain basic properties which could be formally proven but which we believe the reader will find intuitively easy to accept.

1. The intersection of two or more convex sets is another convex set.
2. The Cartesian plane (two-space) is a convex set.
3. A Cartesian half-plane is a convex set.
4. A Cartesian half-plane including the boundary line, called a **closed half-plane**, is a convex set.
5. The intersection of a finite number of closed half-planes forms a convex set, called a **polyhedral convex set**. The intersection may be bounded, i.e., of finite area or the intersection may be unbounded, i.e., infinite in extent.

Example 2: Sketch the convex set in the xy-coordinate plane that is determined by the system of linear inequalities.

$$\begin{cases} 2x - y \leqslant 2 \\ x + y \geqslant 3 \\ y \leqslant 5 \\ y \geqslant 2\,. \end{cases}$$

Solution: Each of the inequalities is drawn as a closed half-plane (Figure 7.17). The intersection of all of these convex sets is the bounded polyhedral convex set which is shown as the shaded region in Figure 7.17.

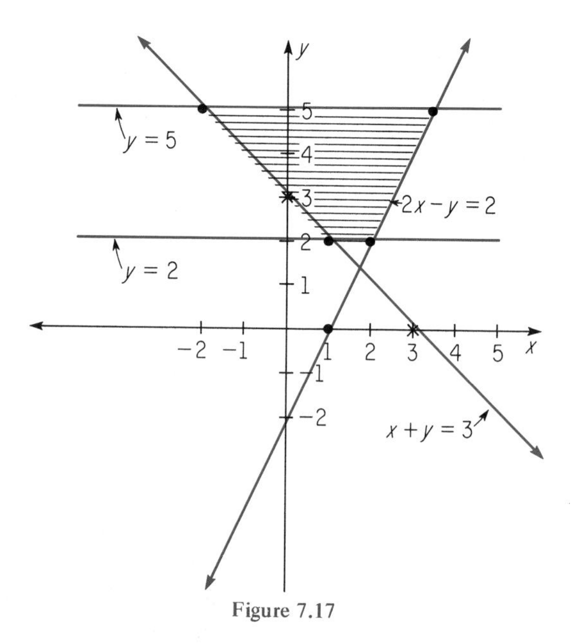

Figure 7.17

Example 3: Sketch the convex set in the xy-coordinate plane that is determined by the system of linear inequalities.

$$\begin{cases} x + y \geqslant 0 \\ 2x - 3y \leqslant 6 \\ -2x + y \geqslant -8\,. \end{cases}$$

Solution: The intersection of all the closed half-planes is shown as the shaded region in Figure 7.18.

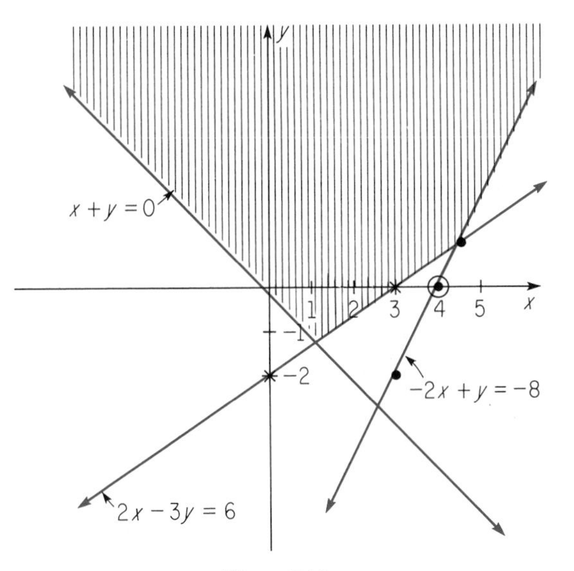

Figure 7.18

The essential distinction between the type of solution set obtained in Example 2 as compared to the solution set obtained in Example 3 is that although there are an infinite number of ordered pair solutions to each system the solution set in Example 2 is a bounded polyhedral convex set and the solution set in Example 3 is an unbounded polyhedral convex set.

The concept of boundedness in two-space can be related to area and in three space boundedness corresponds to volume. However, in n-space where n is greater than three, we cannot relate boundedness to some intuitively appealing geometric model. It is interesting to mention nevertheless, that precise mathematical formulations have been given by mathematicians to boundedness in n-space which are devoid of geometric interpretations. Our goal at this point is not aimed at building such an abstract mathematical system and we will be content to rely on the intuitive concept of boundedness as it relates to the solution of specific problems.

Let us now state in a general way the form of a typical linear programming problem. For a given **objective function** f where

1. $f = b_1x_1 + b_2x_2 + \ldots + b_nx_n$, maximize (minimize) the function f subject to the given constraints

2. $$\begin{cases} a_{11}x_1 + a_{12}x_2 + \ldots + a_{1n}x_n \leqslant c_1 \\ a_{21}x_1 + a_{22}x_2 + \ldots + a_{2n}x_n \leqslant c_2 \\ \quad \cdot \quad \cdot \quad \cdot \\ a_{m1}x_1 + a_{m2}x_2 + \ldots + a_{mn}x_n \leqslant c_m, \end{cases}$$

3. $$\begin{cases} x_1 \geqslant 0 \\ x_2 \geqslant 0 \\ x_3 \geqslant 0 \\ \vdots \\ x_n \geqslant 0\,. \end{cases}$$

Note: The a_{ij}'s, b_i's, and the c_i's are fixed real numbers and x_i's represent the variables.

The set of points (ordered n-tuples) that satisfy the constraints 2 and 3 is called the **feasible set**. The elements of the feasible set are called **feasible points**. A feasible point which also satisfies the given objective function is a **solution** to the stated linear programming problem.

The following examples will help to clarify the basic characteristics of a linear programming problem and will illustrate how we can obtain a solution, (if one exists).

Example 4: Let $f = 2x_1 + 3x_2$ be the objective function and the constraints be given as

$$\begin{cases} x_1 + x_2 \leqslant 8 \\ x_1 \geqslant 0 \\ x_2 \geqslant 0\,. \end{cases}$$

Determine geometrically feasible points (x_1, x_2) for which (1) f is a minimum (2) f is a maximum.

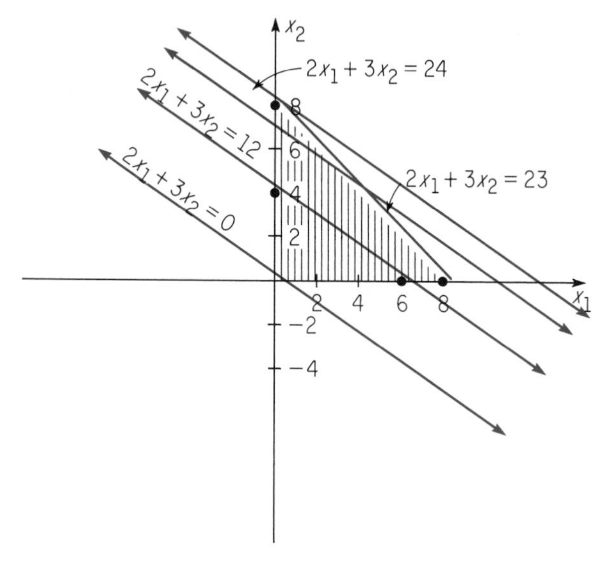

Figure 7.19

Solution: The graph of the constraint set is shown as the shaded region in Figure 7.19. Now let the objective function f assume different values. In each case we obtain a linear equation. A family of parallel straight lines each having a slope of $-3/2$ will be generated. We are interested particularly in those straight lines which intersect the constraint set. A few such typical lines are shown in Figure 7.19.

As the drawing indicates, f will be a minimum when $x_1 = 0$ and $x_2 = 0$. Hence (0,0) which is a feaxible point is also a solution since (0,0) minimizes the function f. To determine if any feasible point will maximize f, we can again refer to the drawing for some insight. When f takes on the value 24, i.e., $2x_1 + 3x_2 = 24$, the corresponding straight line appears to intersect the feasible set at a point; namely, (0,8), that will maximize f. For further confirmation we could compute some functional values of f for corresponding values of x_1 and x_2 (Table 7.1). The feasible point (0,8) maximizes f and is a solution to the given problem.

Table 7.1

x_1	x_2	$f = 2x_1 + 3x_2$
0	0	0
1	1	5
0	4	12
8	0	16
1	6	20
1	7	23
0	8	24

In the previous example the solution that made the objective function f a minimum was the ordered pair corresponding to a vertex of a polyhedral convex set and the solution that made f a maximum was an ordered pair that was also a vertex of a polyhedral convex set. This example simply reflects a very general situation which is expressed by the following theorem.

THEOREM 1: A linear function $f = a_1x_1 + a_2x_2$ defined at every point of a convex set has its maximum (minimum) value, if one exists, at a vertex on the boundary of a polyhedral convex set.

If f is defined at every point of a convex set and if the convex set is bounded, then both a maximum and a minimum will exist for f. If the convex set is unbounded, then either a maximum or a minimum for f may not exist. However, if a maximum or minimum does exist, it will occur at a vertex.

Example 5: Let $f = x_1 + x_2$ and the constraints be given as

$$\begin{cases} 5x_1 + 9x_2 \geqslant 48 \\ -6x_1 + 5x_2 \geqslant -26 \\ x_1 \geqslant 0 \\ x_2 \geqslant 0\,. \end{cases}$$

Find the maximum and minimum of f, if either exists.

Solution: The constraint set is shown as the shaded region in Figure 7.20. Observe that the straight lines $5x_1 + 9x_2 = 48$ and $-6x_1 + 5x_2 = -26$ intersect at the point whose coordinates are (6,2). **Theorem 1** guarantees that the vertex (6,2) or the vertex (0,5) are the only possible points which could maximize or minimize f.

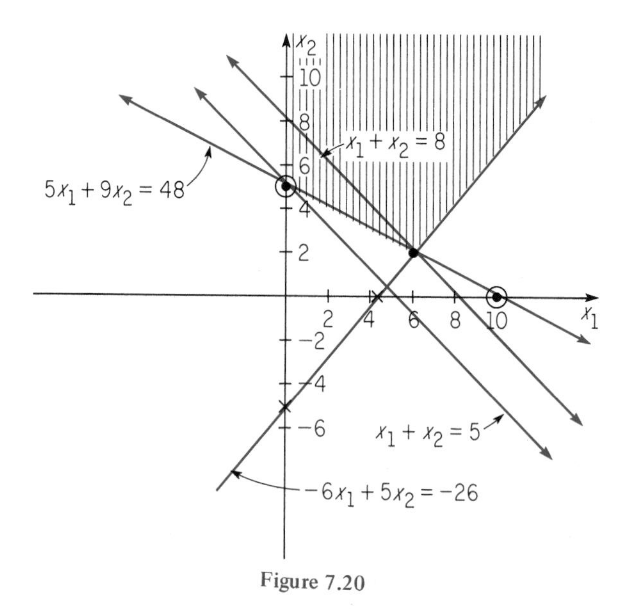

Figure 7.20

If we substitute the ordered pair (0,5) in the linear function f we obtain $f = 5$. If we substitute the ordered pair (6,2) into the function f we obtain $f = 8$. Both the line $x_1 + x_2 = 5$ and $x_1 + x_2 = 8$ are shown in Figure 7.20. These lines are two of the family of parallel lines which have a slope of -1 generated by letting f assume different numerical values. The vertex (0,5) represents the ordered pair which minimizes the objective function f. Since the constraint set is unbounded above, an infinite number of ordered pairs will give values of the objective function larger than the **8** obtained using the coordinates of the vertex (6,2). For example, the ordered pair (8,10) which is an element of the solution set assigns a value of **18** to the objective function f.

In this section we have introduced some of the basic theory of linear programming.

In the last section of this chapter we will consider the application of the theory to a wide variety of applied problems.

Problem Set 7.4

A. 1. Determine which of the following shaded regions are convex sets.

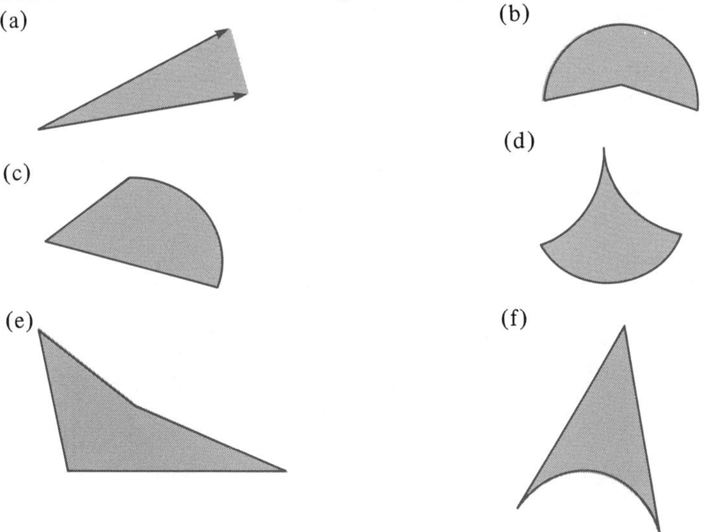

2. Sketch the solution set to the following systems of linear inequalities. Is each solution set convex? bounded?

(a) $\begin{cases} x \geqslant 0 \\ y \geqslant 0 \\ x + y \geqslant 1. \end{cases}$

(b) $\begin{cases} x \geqslant 0 \\ y \geqslant 0 \\ 2x + y \geqslant 8 \\ x + 2y \geqslant 10. \end{cases}$

(c) $\begin{cases} x \geqslant 0 \\ y \geqslant 0 \\ x + 2y \leqslant 10 \\ 2x + y \leqslant 8. \end{cases}$

(d) $\begin{cases} 2x + y \geqslant 8 \\ x + 2y \geqslant 10 \\ x + 3y \leqslant 24 \\ 5x + 2y \leqslant 50. \end{cases}$

3. Let $g = 5x_1 + 5x_2$ be the objective function and the constraint set be determined by the following system of linear inequalities,

$$\begin{cases} x_1 \geqslant 0 \\ x_2 \geqslant 0 \\ x_1 + x_2 \geqslant 6 \\ x_1 + 3x_2 \geqslant 10\,. \end{cases}$$

Sketch the constraint set with x_1 the horizontal axis and x_2 the vertical axis and find the point (x_1, x_2) for which g is a minimum.

4. Let $f = 3x_1 + 5x_2$ be an objective function and let the constraint set be determined by $x_1 + x_2 \geqslant 6$, $x_1 + 3x_2 \geqslant 10$, $3x_1 + 4x_2 \leqslant 24$ and $6x_1 + x_2 \leqslant 60$. Determine the feasible points that will make the objective function f (a) minimum, (b) maximum. Sketch the constraint set in an x_1x_2-plane.
5. Let the constraint set be determined by $x_1 \geqslant 0$, $x_1 + 12x_2 \leqslant 24$, $x_1 \leqslant 12$, $-2x_1 + 3x_2 \leqslant 6$, and $x_1 + 3x_2 \leqslant 24$ and let the objective function be $f = x_1 - x_2$. Draw the constraint set in an x_1x_2-plane and find feasible points for which the objective function f is a (a) maximum, (b) minimum.
6. Solve each of the following systems of linear equalities by graphing. Determine the polyhedral convex set that represents the solution set, and find the coordinates of the vertices.

   (a) $\begin{cases} x_1 \geqslant 3 \\ x_2 \geqslant 1 \\ 5x_1 + 4x_2 \leqslant 40. \end{cases}$ (b) $\begin{cases} x_2 \leqslant 7 \\ x_1 \leqslant 5 \\ 5x_1 + 4x_2 \geqslant 33. \end{cases}$ (c) $\begin{cases} -x_1 + x_2 \leqslant 4 \\ 5x_1 + 4x_2 \leqslant 61 \\ -x_1 + 5x_2 \geqslant 24 \\ 2x_1 + 3x_2 \geqslant 30. \end{cases}$
7. A school is heated by two heating systems, one oil and the other gas. For a given day, at most 30 units of oil and 40 units of gas are used. To maintain a normal temperature, the sum of the number of units of gas and two-thirds the number of units of oil is at least 30. If oil costs \$3 a unit and gas costs \$5 a unit, how many units should each system consume in order to minimize the cost of fuel?

7.5 Problems in Linear Programming

In the preceding section we introduced the necessary background information that will now permit us to use the basic linear programming techniques to solve applied problems. Since no new theory is involved, three detailed examples have been chosen to illustrate the wide variety of fields in which linear programming methods are applicable.

Example 1: A metal part for an automobile engine requires two operations on a piece of metal. A metal worker must shape the piece after which a machinist must put holes and threads in the shaped metal. The shaper requires at least one hour and the machinist requires at least one-half the shaper's time plus half an hour for the given part. The manufacturer wants the part completed in five hours or less. The manufacturer pays \$4 per hour to the shaper and \$6 per hour to the machinist. The manufacturer would like to minimize the cost of producing the part with respect to the wages paid to the workers. How much time should each worker spend on the part to minimize the cost?

Solution: Let x_1 represent the number of hours the shaper spends on the part and let x_2 represent the number of hours the machinist spends on the part. From the statements in the problem we have the following inequalities:

$$\begin{cases} x_1 \geqslant 1 \\ x_2 \geqslant \dfrac{x_1}{2} + \dfrac{1}{2} \\ x_1 + x_2 \leqslant 5\,. \end{cases}$$

The objective function is $C = 4x_1 + 6x_2$, where C represents that part of the cost of the automobile part directly related to the labor in making the part. The manufacturer wishes to minimize C.

By Theorem 1 in Sec. 7.4, the minimum of the objective function will occur at a vertex of the constraint set. Thus we need only to graph the constraint set and check the vertices of the constraint set in the objective function. From Figure 7.21

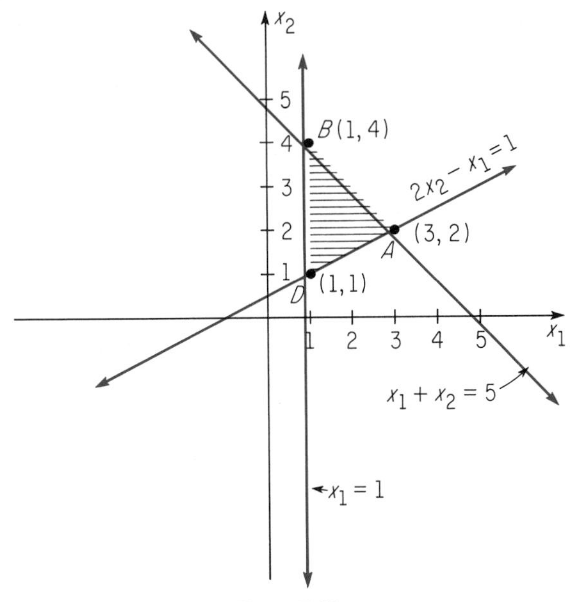

Figure 7.21

we see that the constraint set has three vertices. Substituting the coordinates of each vertex in the objective function, we obtain the results shown in Table 7.2.

Table 7.2

Vertex	Coordinates (x_1, x_2)	Objective Function $C = 4x_1 + 6x_2$
A	(3, 2)	24
B	(1, 4)	28
D	(1, 1)	10

The coordinates associated with vertex D will make the cost function C a minimum.

Hence, the practical solution to the problem is that the shaper should spend one hour on the part and the machinist should also spend one hour on the part.

Let us now consider an example in which we wish to maximize an objective function.

Example 2: A dressmaker has the following materials available: 18 sq yds of cotton, 20 sq yds of rayon, and 20 sq yds of wool. A pants suit requires 1 sq yd of cotton, 1-1/2 sq yds of rayon and 2 sq yds of wool. An evening dress requires 2 sq yds of cotton, 2 sq yds of rayon, and 1 sq yd of wool. A suit sells for $40 and a gown sells for $55. How many pants suits and how many gowns should the dressmaker manufacturer in order to obtain the maximum amount of income?

Solution: Let x_1 represent the number of pants suits and let x_2 represent the number of gowns. We wish to maximize the objective function $P = 40x_1 + 55x_2$ subject to the given constraints in the problem which are:

$$\begin{cases} x_1 + 2x_2 \leqslant 18 \\ \frac{3}{2}x_1 + 2x_2 \leqslant 20 \\ 2x_1 + x_2 \leqslant 20 \\ x_1 \geqslant 0 \\ x_2 \geqslant 0\,. \end{cases}$$

The intersection of the graphs of each inequality will give us the constraint set in Figure 7.22. The vertices of the constraint set are the points labeled *A, B, C, D, E* whose coordinates are respectively (0,9), (4,7), (8,4), (10,0) and (0,0).

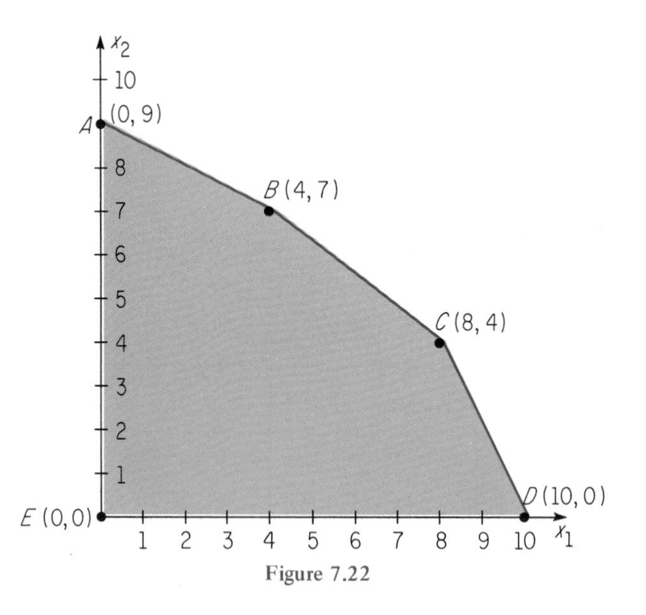

Figure 7.22

The maximum of the objective function *P* will occur at one of these vertices. Table 7.3 lists a value of the objective function obtained by substituting the coordinates of each vertex into the objective function.

Table 7.3

Vertex	Coordinates of Vertex (x_1,x_2)	Value of Objective Function $P = 40x_1 + 55x_2$
A	(0, 9)	495
B	(4, 7)	545
C	(8, 4)	540
D	(10, 0)	400
E	(0, 0)	0

From Table 7-3, we observe that the coordinates of vertex B will maximize the objective function. Thus, in order for the dressmaker to obtain the maximum amount of income he should make four pants suits and seven gowns.

The last example which we will consider in this section is taken from the food nutrition field.

Example 3: A nutritionist in a large hospital wishes to serve two meals that will provide the necessary nutritional requirements to each patient and at the same time minimize the weight of the food consumed. The first meal contains 2 units of carbohydrates, 1 unit of fat, and 2 units of protein. The second meal has 6 units of carbohydrates, 1 unit of fat, and 3 units of protein. The weight of the first meal is 25 oz, and the weight of the second meal is 30 oz. The minimum nutritional requirements are: carbohydrates, 24 units; fats, 7 units; and protein, 18 units. Determine the minimum weight of the food that can be consumed and that will also satisfy the minimum nutritional requirements.

Solution: Let x_1 represent the number of first meals and x_2 represent the number of second meals. We wish to minimize $W = 25x_1 + 30x_2$ subject to the following

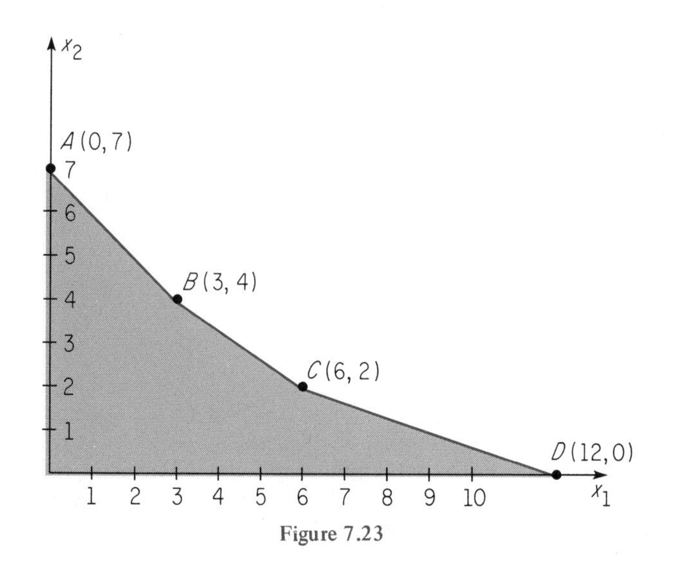

Figure 7.23

constraints: $2x_1 + 6x_2 \geqslant 24$, $x_1 + x_2 \geqslant 7$, $2x_1 + 3x_2 \geqslant 18$, $x_1 \geqslant 0$, $x_2 \geqslant 0$. The constraint set which is obtained by graphing the intersection of the given inequalities is shown in Figure 7.23. The vertices of the constraint set are the points labeled *A, B, C, D* with the respective coordinates (0,7), (3,4), (6,2) and (12,0).

In order to determine which vertex will make the objective function, $W = 25x_1 + 30x_2$ a minimum, we substitute the coordinates of each vertex in the objective function. The functional values obtained are shown in Table 7.4.

Table 7.4

Vertex	Coordinates of Vertex	Value of Objective Function $W = 25x_1 + 30x_2$
A	(0, 7)	210
B	(3, 4)	195
C	(6, 2)	210
D	(12, 0)	300

The coordinates of vertex *B* will minimize the objective function. Thus, the combination of three of the first meal and four of the second meal will minimize the weight and provide the minimum nutritional requirements.

As we indicated in our initial introduction, it was our intent to introduce some of the basic concepts of linear programming. The reader should be aware, however, that the field of linear programming is much broader than the brief introduction to the topic we have provided.

For example, the concepts which we have discussed in two-space can be generalized to three-space and *n*-space. However, in handling problems beyond three-space we can no longer rely on geometric models and we must resort to rather complex algebraic methods.

In many applied problems the variables may assume only certain specific values. For example, in the pants suit-gown problem discussed in this section, the variables may only assume integer values. Introducing such restrictions as constraints on an objective function can complicate matters considerably and frequently are best handled by a computer analysis.

The interested reader will find at the end of this section a Suggested Reading List which may be consulted for more detailed information on such items as slack variables, the tableau technique, and the general simplex method for finding solutions to more complex linear programming problems.

Problem Set 7.5

A. 1. A real estate firm has 72 acres of tillable land and 72 acres of nonfertile land. An industrial plant requires 6 acres of tillable land and 10 acres of nonfertile land. A small estate in an exclusive development requires 12 acres of tillable land and 8 acres of nonfertile land. How should the real estate firm apportion the land to maximize income if an (a) industrial plant and a small estate each sell for

\$60,000, (b) industrial plant sells for \$75,000 and a small estate sells for \$50,000, and (c) industrial plant sells for \$40,000 and a small estate sells for \$60,000?

2. An oil company has three grades of oil that are obtained from two different locations, say A and B. For distribution, the company needs 288 gals of top grade oil, 570 gals of medium grade oil, and 400 gals of low grade oil. Location A produces 21 gals of top grade oil, 15 gals of medium grade oil, and 20 gals of low grade oil each day. Location B produces 12 gals of top grade oil, 45 gals of medium grade oil, and 15 gals of low grade oil. How many days should each location operate if it costs (a) \$500 per day to operate each location, (b) \$300 per day to operate location A and \$600 to operate location B, (c) \$700 per day to operate location A and \$400 per day to operate location B?
3. A nut company has 400 lb of cashews and 200 lb of pecans. It can sell the cashews at 60¢ per lb and the pecans at 85¢ per lb. It can also sell a mixture of nuts consisting of one part pecans and three parts cashews for 65¢ per lb or a second mixture consisting of two parts pecans and two parts cashews for 75¢ per lb. How many pounds of each mixture should it produce to maximize income?
4. A publishing company publishes books in both paperback and hardbound copies. Two factories, say A and B, manufacture these books for the publishing company. Each day factory A produces 200 paperback and 100 hardbound copies and factory B produces 100 paperback and 200 hardbound books. The publishing company has an order for 1700 paperback and 1600 hardbound copies. What is the minimum number of days each factory should operate to produce enough books for the order?
5. A farmer has a 200-acre farm and can plant tomatoes or corn in any combination of acreage. Tomatoes require 10 hours of labor and \$10 operating expenses for each acre planted. Corn requires 30 hours of labor and \$18 operating expenses for each acre planted. The net revenue per acre for the corn is \$50. The farmer has available \$2,250 for the planting season and enough laborers to do 3,000 hours of labor.
   (a) Let x_1 represent the number of acres of tomatoes planted and let x_2 represent the number of acres of corn planted; find the inequalities that represent the constraint set.
   (b) Write an expression that represents the objective function.
   (c) Find the number of acres of tomatoes and corn the farmer should plant to provide a maximum income.

Review Exercises

7.1 State definitions for the following:
(a) The real number c is less than the real number d.
(b) Closed interval between the real numbers c and d.
(c) Half-plane.
(d) Convex set.
(e) Constraint set.
(f) Objective function.
(d) Feasible set.
(e) Solution to a linear programming problem.

7.2 Solve the inequality $2x + 7 < 4x + 11$ and justify each step by reference to an axiom, definition, or theorem. Diagram the solution set on the real-number line.

7.3 Solve

$$\frac{3 - x}{x - 1} < 0$$

and sketch the solution set on the real line.

7.4 Sketch the graph of the solution set of the inequality $-y - 3x + 6 < 0$ in the xy-plane.

7.5 Sketch the graph of the solution set of the inequality $2x - y + 5 \leqslant 3x - 2y + 4$ in the xy-plane.

7.6 Find the solution set to the system of linear inequalities

$$\begin{cases} -4x + 3y \leqslant 12 \\ 2y - 3x > 6 \,. \end{cases}$$

7.7 For the inequalities

$$\begin{cases} x_1 + x_2 \geqslant 0 \\ x_2 \leqslant 0 \\ 4x_2 - 5x_1 \geqslant -40 \end{cases}$$

find the solution set and sketch in the $x_1 x_2$-plane.

7.8 Determine the solution set to the system of linear inequalities

$$\begin{cases} 2x_1 + 3x_2 \geqslant 6 \\ x_1 + \frac{3}{2}x_2 < 0 \,. \end{cases}$$

7.9 Let $f = 3x_1 + 2x_2$ be an objective function and let the constraints be given by $x_1 + 6x_2 \leqslant 51$, $x_2 - 7x_1 \leqslant -13$, $x_2 - 2x_1 \geqslant -11$, and $5x_2 - 2x_1 \geqslant 1$. Determine geometrically feasible points (x_1, x_2) for which f is a (a) maximum (b) minimum.

7.10 Let the coordinates of the vertices of a polyhedral convex set be given as (5,2), (2,6), (7,9) and (10,4). Determine a system of linear inequalities that forms this convex set.

7.11 Let the constraint set be determined by $2x_1 + 5x_2 \geqslant 24$, $x_2 \leqslant 10$, $x_1 \geqslant 10$, $x_1 \geqslant 2$, and $3x_1 - 8x_1 \geqslant -50$, and let the objective function be $g = 4x_1 + 3x_2$. Draw the constraint set in an $x_1 x_2$-plane and find feasible points for which the objective function g is a (a) maximum, (b) minimum.

7.12 A toy manufacturer has 114, 140 and 98 lbs of steel, wood, and plastic respectively. A boy's toy requires 3, 4, and 3 lbs of steel, wood and plastic, respectively; and a girl's toy requires 3, 3, and 2 pounds of steel, wood, and plastic, respectively. If the boy's toy sells for \$7 and the girl's toy sells for \$5, how many of each toy should be made to obtain maximum income?

Suggested Reading

Bush, Grace A., and Young, John E., *Foundations of Mathematics with Applications to the Social and Management Sciences*. McGraw-Hill, New York, 1968.

Campbell, Hugh G., *An Introduction to Matrices, Vectors and Linear Programming.* Appleton-Century-Crofts, New York, 1965.

Kemeny, John G., Snell, J. Laurie, and Thompson, Gerald L., *Introduction to Finite Mathematics*. Prentice-Hall, Englewood Cliffs, N. J., 1966.

Owen, Guillermo, *Finite Mathematics.* W. B. Saunders, Philadelphia, Pa., 1970.

APPENDIX 1

Relations and Functions

In Sec. 2.7 the topic of the Cartesian product of two or more sets was discussed. In order to define Cartesian product of two sets we had to define first the idea of an ordered pair. We considered an ordered pair to be a pair of elements in which the order is fixed. We write the ordered pair of elements x and y as (x,y), where x is called the **first component** and y is called the **second component**, or member, of the ordered pair (x,y). The Cartesian product between two sets, A and B, denoted $A \times B$, was defined as the set of all ordered pairs where first elements are from set A and whose second elements are from the set B.

We wish now to consider certain types of subsets of the Cartesian product of two sets. Many of these subsets are of particular importance and play a significant role in a more advanced study of mathematics.

An important concept which one encounters not only in mathematics but in many other disciplines is the concept of a **relation**. Before we make a formal mathematical definition of a relation, let us consider some specific examples of relations.

Example 1:

1. Mr. Trite is the instructor of students taking journalism at Whileaway College.
2. x is less than y, where x, y are counting numbers less than ten.
3. American cars are not built to last as long as a foreign made car.
4. Speed depends upon time and distance.
5. Mathematics courses are more difficult than any other courses at Whileaway College.

In each of the five examples we are relating elements from one set to those of another (or same) set or elements of one set to elements of more than one set. (Statement 4 relates speed to both time and distance.) A general association between sets which we designate as a relation is defined as follows:

DEFINITION: A relation R from set A to set B is a subset of $A \times B$. If the ordered pair (a,b) is in R, i.e., $(a,b) \in R$, then we say a is related to b and we

write aRb. If $A = B$, then R is said to be a **relation on A** (i.e., $R \subset A \times A$).
Note: *aRb if and only if $(a,b) \in R$.*

Frequently sets are specified by the descriptive method. Similarly, a relation may be specified by a descriptive method. In the statement form for a relation, the set or sets on which the relation is defined are often implied.

Example 2: Let R be the relation, "x is less than y," where x and y are counting numbers less than six. What is set A; set B? Write R by (1) set-builder notation, and (2) by listing the elements in the set.

Solution:

$$A = B = \{1,2,3,4,5\}.$$
$$R = \{(x,y) \mid x \in A, y \in B \text{ and } x \text{ is less than } y\}.$$
$$R = \{(1,2),(1,3),(1,4),(1,5),(2,3),(2,4),(2,5),(3,4),(3,5),(4,5)\}.$$

Note: *$(x,y) \in R$ iff* x is less than y.*

In many situations we may be interested in the set of all first components of the relation or the set of all second components of a relation. We identify these sets in accordance with the following definition.

DEFINITION: Let R be a relation from set A to set B. The set of all elements of A that occur as first elements of ordered pairs in R is called the **domain of R**, denoted D_R; and the set of all elements of B that occur as second elements of ordered pairs in R is called the **range of R**, denoted R_R. Symbolically, we write

$$D_R = \{a \in A \mid (a,b) \in R\}$$
$$R_R = \{b \in B \mid (a,b) \in R\}.$$

Example 3: Let $A = \{1,2,3,4,5\}$, $B = \{1,2,3,4,5,6\}$, and let $R = \{(x,y) \mid x \in A, y \in B$ and x is greater than $y\}$. List the elements in R. Find the domain of R and the range of R.

Solution:

$$R = \{(2,1),(3,1),(3,2),(4,1),(4,2),(4,3),(5,1),(5,2),(5,3),(5,4)\}$$
$$D_R = \{2,3,4,5\} \text{ and } R_R = \{1,2,3,4\}.$$

Suppose we now define a relation T by interchanging the first and second components of each ordered pair in the relation R of Example 3. We then have that $T = \{(1,2),(1,3),(2,3),(1,4),(2,4),(3,4),(1,5),(2,5),(3,5),(4,5)\}$. Note that $T \subset B \times A$ and also that $D_T = \{1,2,3,4\}$ and $R_T = \{2,3,4,5\}$.

*Means "if and only if."

When two relations exhibit this association we call T the inverse relation of the relation R and denote it as R^{-1}. Formally stated, the inverse relation is defined as follows.

DEFINITION: Let R be a relation from set A to set B. The **inverse relation of** R, denoted R^{-1}, is a relation from B to A defined by

$$R^{-1} = \{(y,x) \mid y \in B, x \in A, (x,y) \in R\}.$$

Example 4: Let $A = \{2,4,6,8,10\}$, $B = \{1,2,3,4,5,6,7,8,9,10,11\}$, and $R = \{(x,y) \mid x \in A, y \in B \text{ and } y = x + 1\}$. Write out R and R^{-1} in the ordered pair form.

Solution: R consists of all ordered pairs where the first element is selected from A and the second element is one more than the first element and selected from B. Thus,

$$R = \{(2,3),(4,5),(6,7),(8,9),(10,11)\}.$$

R^{-1} is defined as $R^{-1} = \{(y,x) \mid y \in B, x \in A, (x,y) \in R\}$. Thus $R^{-1} = \{(3,2),(5,4),(7,6),(9,8),(11,10)\}$.

A very important relation on any set A is the **equality relation.** Certain properties of this relation are:

(i) for any $a \in A$, $a = a$

(ii) if $a, b \in A$ and $a = b$, then $b = a$

(iii) if $a, b, c \in A$ and $a = b$ and $b = c$, then $a = c$.

There are other relations which satisfy the above three properties. We now define this important class of relations.

DEFINITION: Let R be a relation on a set A such that

(i) aRa for every $a \in A$ **(reflexive property)**

(ii) if $a, b \in A$ such that aRb, then bRa **(symmetric property)**

(iii) if $a, b, c \in A$ such that aRb and bRc, then aRc **(transitive property).**

Then we say R is an **equivalence relation on** A.

Example 5: Show that the similarity relation of triangles in Euclidean geometry is an equivalence relation.

Solution:

1. Let T_1 represent a triangle and R represent the similarity relation. Then T_1 is similar to itself so that T_1RT_1 and the relation is **reflexive**.
2. Let T_1, T_2 represent triangles. If T_1 is similar to T_2, T_2 similar to T_1. Thus if T_1RT_2 then T_2RT_1 and the relation is **symmetric**.
3. Let T_1, T_2, T_3 represent triangles. If T_1 is similar to T_2 and T_2 in turn is similar to T_3, then T_1 is similar to T_3. Thus if T_1RT_2 and T_2RT_3, then T_1RT_3 and the relation is **transitive**.

Since all three properties are satisfied, the similarity relation is an equivalence relation on the set of all triangles in Euclidean geometry.

We now consider a special type of relation between sets that is of fundamental importance and basic in the study of mathematics. This relation which we now define is that of **function**.

DEFINITION: Let f be a relation from set A into set B. f is a function (or mapping) from set A into set B, denoted $f: A \longrightarrow B$, if and only if
(i) $D_f = A$, and
(ii) if $(a,b_1) \in f$ and $(a,b_2) \in f$ then $b_1 = b_2$.

An equivalent statement for the above definition is: A function (or mapping) from set A into set B, denoted $f: A \longrightarrow B$, is any rule of correspondence that assigns to each element of set A a unique element of set B. Further, since $f \subset A \times B$, we call this collection of ordered pairs the **graph** of the function. Furthermore, since f is a relation, it has a domain and range with the domain of the function the entire set on which it is defined. The range of a function is a subset of the second set.

There are various ways of displaying a function. But in order to define a function we *must* specify its domain and indicate (define) a rule of correspondence. Some of the more important ways of defining a function are by: (1) table, (2) graph, and (3) equations which give a rule of correspondence.

Example 6:
(a) Let $A = \{1,2,3\}$, $B = \{a,b,c,d\}$. Let $f: A \longrightarrow B$ such that $f = \{(1,d),(2,c),(3,b)\}$. We could indicate this in tabular form as follows:

(a,b)	
1	d
2	c
3	b

(b) Let $A = \{1,2,3,4\}$, $B = \{1,2,3,4,5,6,7,8,9\}$. Let $g: B \longrightarrow A$ such that g is defined by the following table:

b	1	2	3	4	5	6	7	8	9
a	1	2	1	2	1	2	1	2	1

DEFINITION: If x is any element of the domain of a function $f: X \longrightarrow Y$, then the unique element y in Y that corresponds to x is called the **image of x** and is denoted $f(x)$, i.e., $(y = f(x))$. The image of a set $A, A \subset X$ is the set of all images of the elements of A and is written $f(A)$. Hence the range of a function may be defined as $f(X)$ and we may further note that $f(X) = R_f \subset Y$. Symbolically for any $A \subset X, f(A) = \{f(a) \mid a \in A\}$.

Example 7: Let

$$X = \{1,2,3,4,5\},\ Y = \{1,2,3, \ldots, 10\}$$
$$f = \{(x,y) \mid y = 2x, x \in X\}.$$

The image of 1 is $f(1) = 2$ and image of 3 is $f(3) = 6$. The set of all images denoted $f(X)$ is given by

$$f(X) = \{2,4,6,8,10\}.$$

In context, the symbol $f(x)$ is read as the value of the function f at x.

Many elements in the domain of a function may have the same image. The set of elements which all have the same image are denoted as a set of preimages and defined as follows:

DEFINITION: If b is any range element of a function, $f\colon X \longrightarrow Y$, then any domain element whose image is b is called a **preimage of b**. The preimage of a set $B \subset Y$ is the set of all preimages of elements of B and is written $f^{-1}(B)$ and is read, "The set of preimages of set B." Symbolically,

$$f^{-1}(B) = \{x \in X \mid f(x) \in B\}.$$

Example 8:

(a) Let $x = \{1,2,3,4\}$, $Y = \{1,2,3,4,5,6,7,8,9,10\}$. Let f be defined by the following table:

x	1	2	3	6	7	8	9	10
$f(x)$	1	2	2	4	2	6	10	10

(b) Let $B = \{1,2,10\} \subset Y$; then $f^{-1}(B) = \{1,2,3,7,9,10\}$.

Since a function is a special relation, it is reasonable to ask if the inverse relation of a function is also a function. For example, if f is a function from set A into set B, then how do we define, if it is possible, a relation which will be the inverse function for the function f? The next two definitions are necessary before we can precisely define an inverse function.

DEFINITION: A function $f\colon A \longrightarrow B$ is **onto** if and only if every element of B is the image of some element in set A. When a function is onto we may write, $f\colon A \underset{\text{onto}}{\longrightarrow} B$.

DEFINITION: A function $f\colon A \longrightarrow B$ is **one-to-one**, 1-1, if and only if every image in B has an unique preimage in set A. To denote that a function is 1-1, we write $f\colon A \xrightarrow{1\text{-}1} B$.

Some alternate forms of the definition of a 1-1 function are as follows:
(a) A function $f: A \longrightarrow B$ is one-to-one if and only if,

$$\text{if } f(a_1) = f(a_2), \text{ then } a_1 = a_2$$

or
(b) a function $f: A \longrightarrow B$ is one-to-one if and only if,

$$\text{if } a_1 \neq a_2, \text{ then } f(a_1) \neq f(a_2).$$

Example 9:
(a) Let $f = \{(a,1),(b,1),(c,2),(d,3)\}$ where $A = \{a,b,c,d\}$, $B = \{1,2,3\}$. The function f, $f: A \longrightarrow B$ is onto set B, but it is not one-to-one, since both a and b are mapped to 1.
(b) Let $A = \{a,b,c\}$, $B = \{1,2,3,4,5\}$ and $g = (a,1),(b,3),(c,2)$. The function g, $g: A \longrightarrow B$, is 1-1, but it is not onto set B since not all of the elements of B have preimages.

We are now prepared to give the definition of an inverse function of a function f from a set A into a set B.

DEFINITION: Let f be a function from set A into set B such that f is 1-1 and onto. Then the function from set B into set A that assigns to each element of B its unique preimage under f is called the **inverse function** of f and is denoted f^{-1}.

If $f: A \underset{\text{onto}}{\overset{1\text{-}1}{\longrightarrow}} B$ such that $f(x) = y$, then $f^{-1}: B \underset{\text{onto}}{\overset{1\text{-}1}{\longrightarrow}} A$ such that $f^{-1}(y) = x$.

Example 10: Let $A = \{x,y,z,a,b\}$, $B = \{1,3,5,2,4\}$ and let $f = \{(x,1),(y,3),(z,5),(a,2),(b,4)\}$. Then $f: A \underset{\text{onto}}{\overset{1\text{-}1}{\longrightarrow}} B$. Thus by definition $f^{-1} = \{(1,x),(3,y),(5,z),(2,a),(4,b)\}$ where $f^{-1}: B \underset{\text{onto}}{\overset{1\text{-}1}{\longrightarrow}} A$.

Functions are sets of ordered pairs and it is possible to have sets which represent equal functions. Equality of functions is defined as follows:

DEFINITION: Two functions $f: A \longrightarrow B$ and $g: A \longrightarrow B$ are said to be equal, denoted $f = g$ if and only if $f(a) = g(a)$ for all $a \in A$.

Example 11: Let $A = \{1,2,3,4,5\}$, $B = \{1,4,9,16,25\}$ and $f = \{(1,1),(2,4),(3,9),(4,16),(5,25)\}$. If g is defined as $\{(x,y)\, x \in A, y \in B \text{ and } x^2 = y\}$ then $f = g$.

We now consider an operation on functions which is basic to the further study of functions. The operation is that of composition of functions.

DEFINITION: The **composition of a function f with a function g**, symbolized $f \circ g$, is the function h given by:
(i) $D_h = \{x | x \in D_g \text{ and } g(x) \in D_f\}$
(ii) $h(x) = (f \circ g)(x) = f(g(x))$ for every $x \in D_h$.

Example 12: Let $A = \{1,2,3,4\}$, $B = \{1,2,3,4,5,6,7,8\}$, $C = \{1,2,3, \ldots, 8,9,10\}$, $D = \{1,2,3, \ldots, 23,24,25\}$, $f = \{(x,y) | x \in A, y \in B, y = 2x\}$, and $g = \{x,y) | x \in C, y \in D, y = 3x\}$. Then $g \circ f: A \longrightarrow D$ such that

$$\begin{aligned}(g \circ f)(x) &= g(f(x)) \\ &= g(2x) \\ &= 6x.\end{aligned}$$

Exercises

1 Let $A = \{a \mid a \in N$ and a is less than or equal to ten$\}$, and $S = \{(a,b) \mid a \in N, b \in N$ and $b = 2a - 1\}$. Find:
(a) domain of S.
(b) range of S.
(c) S^{-1}.
(d) domain of S^{-1}.
(e) range of S^{-1}.

2 Determine whether the following relations are (i) reflexive, (ii) symmetric, and (iii) transitive:
(a) person a is shorter than person e.
(b) person c is married to person r.
(c) the counting number a is not equal to the counting number b.
(d) statement a implies statement b.
(e) set A is a subset of set B.
(f) statement a is equivalent to statement b.

3 Let h be a function from C into D, where $C = \{1,2,3, \ldots, 10\}$, $D = \{1,2,3, \ldots, 20\}$ and $h = \{(p,q) \mid p \in C, q \in D$ and $q = 2p - 1\}$.
(a) Find range of h.
(b) Is h a one-to-one function?
(c) Is h onto D?
(d) If $B = \{1,9,17,19\} \subset R_h$, find $h^{-1}(B)$.
(e) If $G = \{1,3,5,7\} \subset C$, find $h(G)$.

4 Let $g = \{(q,t) \mid q \in R, t \in U$, and $t = 3q\}$, $R = \{1,2,3,4,5\}$, and $U = \{3,6,9,12,15\}$.
(a) Is g a function that is one-to-one and onto U?
(b) If g is 1-1 and onto U, then find g^{-1}.
(c) Find the function $g \circ g^{-1}$. State its domain and range.
(d) Find the function $g^{-1} \circ g$. State its domain and range.

APPENDIX **2**

Multinomial Expansions

The multinomial theorem may be considered as a natural extension of the binomial theorem. The multinomial theorem will afford us a means of expanding an expression of the form $(x_1 + x_2 + \cdots + x_k)^n$ where n is a nonnegative integer and k is greater than or equal to three.

In the binomial expansion, the numerical coefficient of each term is related to a two-celled ordered partition. For example, in the term $\binom{n}{r,n\text{-}r}x^r y^{n-r}$ we choose r x's from n of the factors and n-r y's from the n factors. The number of ways this can be done is given by $\binom{n}{r,n\text{-}r}$ and represents the numerical coefficient of the term.

Now let us consider a multinomial that is raised to a positive integer power. In particular, let the multinomial be $(a + b + c)^2$, and let us determine the expansion of the given multinomial. First, by definition $(a + b + c)^2 = (a + b + c)\,(a + b + c)$. In the expansion we choose either an a or b or c from each factor. A tree diagram will be helpful in displaying these choices.

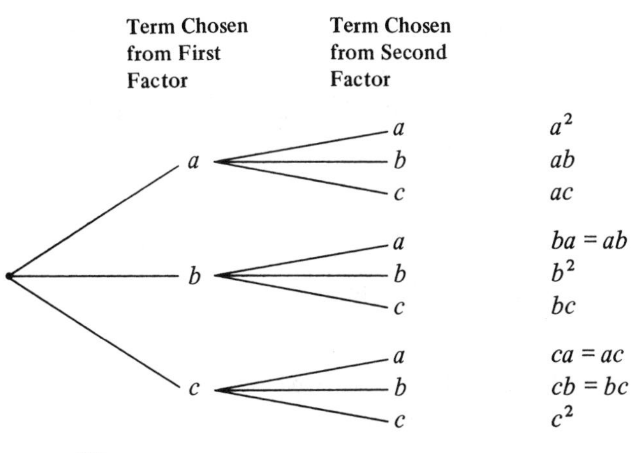

At the end of the branch the product of the letters is listed. If we now add up these products (adding like terms together) we obtain the expansion for $(a + b + c)^2$; namely,

$$(a + b + c)^2 = a^2 + \underline{ab} + \underline{\underline{ac}} + \underline{ab} + b^2 + \underline{\underline{\underline{bc}}} + \underline{\underline{ac}} + \underline{\underline{\underline{bc}}} + c^2$$
$$= a^2 + 2ab + 2ac + 2bc + b^2 + c^2.$$

The choice of one term from each factor may be represented by a partition. For this particular example, the partition will have three cells, since there are three terms; namely, a, b, and c in the multinomial. But the expression is being squared, so that at least one of the cells of the partition will always be empty, since there are at most two distinct terms which can be chosen; i.e., one from each factor. In the first cell of the partition there will be a set of numbers indicating from what factor or factors, if any, the first term in the expansion was chosen. In a similar manner, the numbers appearing in the second and third cells will specify from what factor the term is chosen. The number of partitions that have the same number of terms in each cell (though the terms came from different factors), is the coefficient of that term in the expansion of the multinomial. For example, the coefficient of the term containing ab in the expansion of $(a + b + c)^2$ is 2, since the only possible partitions are $[\{1\}, \{2\}, \{0\}]$ and $[\{2\}, \{1\}, \{0\}]$ indicating a came from the first factor and b from the second, or vice versa. The number of partitions can be found without listing each partition and is given by $\binom{2}{1,1,0}$, which is the familiar notation for the arrangement of n objects into k cells where not all objects are distinguishable. Thus, $\binom{2}{1,1,0} = \dfrac{2!}{(1!)(1!)(0!)} = 2$. Each number in the symbol $\binom{2}{1,1,0}$ has meaning in relation to a term in the expansion of a multinomial as given below.

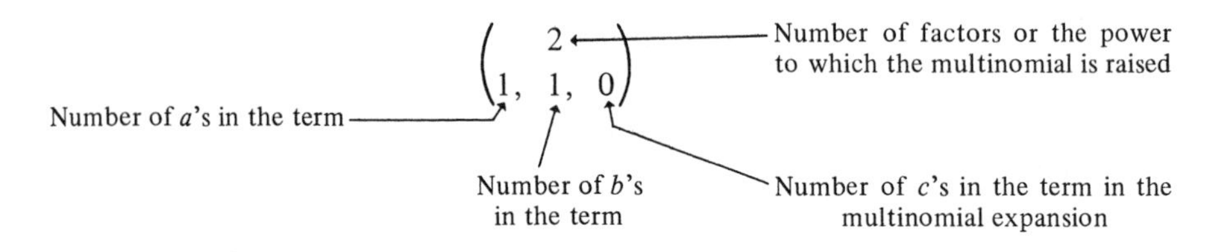

We can now write out the expansion of $(a + b + c)^2$ in the following manner:

$$(a + b + c)^2 = \binom{2}{2,0,0}a^2 + \binom{2}{0,2,0}b^2 + \binom{2}{0,0,2}c^2 + \binom{2}{1,1,0}ab + \binom{2}{1,0,1}ac + \binom{2}{0,1,1}bc$$
$$= a^2 + b^2 + c^2 + 2ab + 2ac + 2bc.$$

The above example may be generalized to the following theorem:

MULTINOMIAL THEOREM: The expansion of $(x_1 + x_2 + \cdots + x_k)^n$ is found by adding all terms of the form

$$\binom{n}{n_1, n_2, \ldots, n_k} x_1^{n_1} x_2^{n_2}, \cdots, x_k^{n_k}$$

where $n_1 + n_2 + \cdots + n_k = n$.

Another important part in the multinomial expansion is to know how many terms there are in the expansion so that we can be sure we have the complete list of all the terms. To this end we state the following:

The total number of terms in the multinomial expansions of $(x_1 + x_2 + \cdots + x_k)^n$ *where n is a positive integer and k is the number of terms in the multinomial, is given by the combination symbol formula:*

$$\binom{n + k - 1}{k - 1, n}.$$

The following example illustrates how we may apply the multinomial theorem and the preceding counting formula to a particular problem.

Example 1: Expand $(a + b + c)^4$ by using the multinomial theorem.

Solution: The number of terms in the expansion is

$$\begin{aligned}\binom{4 + 3 - 1}{3 - 1, 4} &= \binom{6}{2,4} \\ &= \frac{6 \cdot 5 \cdot 4!}{2 \cdot 4!} \\ &= 15.\end{aligned}$$

Thus there are fifteen partitions which we now list and relate to the corresponding literal factors in the expansion:

[{4}, {0}, {0}]	a^4
[{0}, {4}, {0}]	b^4
[{0}, {0}, {4}]	c^4
[{3}, {1}, {0}]	a^3b
[{3}, {0}, {1}]	a^3c
[{1}, {3}, {0}]	ab^3
[{1}, {0}, {3}]	ac^3
[{0}, {3}, {1}]	b^3c
[{0}, {1}, {3}]	bc^3
[{2}, {2}, {0}]	a^2b^2
[{2}, {0}, {2}]	a^2c^2
[{0}, {2}, {2}]	b^2c^2
[{2}, {1}, {1}]	a^2bc
[{1}, {2}, {1}]	ab^2c
[{1}, {1}, {2}]	abc^2

We may now write out the expansion of $(a + b + c)^4$.

$$\begin{aligned}(a+b+c)^4 = {} & \binom{4}{4,0,0}a^4 + \binom{4}{0,4,0}b^4 + \binom{4}{0,0,4}c^4 + \binom{4}{3,1,0}a^3b + \binom{4}{3,0,1}a^3c \\ & + \binom{4}{1,3,0}ab^3 + \binom{4}{1,0,3}ac^3 + \binom{4}{0,3,1}b^3c + \binom{4}{0,1,3}bc^3 \\ & + \binom{4}{2,2,0}a^2b^2 + \binom{4}{2,0,2}a^2c^2 + \binom{4}{0,2,2}b^2c^2 + \binom{4}{2,1,1}a^2bc \\ & + \binom{4}{1,2,1}ab^2c + \binom{4}{1,1,2}abc^2 \\ = {} & a^4 + b^4 + c^4 + 4a^3b + 4a^3c + 4ab^3 + 4ac^3 + 4b^3c + 4bc^3 \\ & + 6a^2b^2 + 6a^2c^2 + 6b^2c^2 + 12a^2bc + 12ab^2c + 12abc^2\end{aligned}$$

The multinomial theorem though completely general does unfortunately give rise to some tedious computations even for problems of apparent simplicity. For example, the expansion of $(a + b + c + d + e)^7$ by the counting formula will require the listing of 330 separate and distinct terms computed as follows:

$$\binom{7+5-1}{5-1, 7} = \binom{11}{4,7} = 330.$$

For obvious reasons, we will not continue the discussion beyond this brief introduction to the topic.

Exercises

1 Expand $(x + y + z)^3$.

2 How many terms are in the expansion of $(p + q + r + s)^9$?

3 Expand $(a + b + c + d + e + f)^2$.

4 Use the multinomial theorem to calculate $(1153)^3$. (*Hint*: $1153 = 1000 + 100 + 50 + 3$.)

5 Calculate $(235)^4$.

APPENDIX **3**

Tables

Table 3.A
Squares and Square Roots

n	n^2	$\sqrt{n}$	$\sqrt{10n}$	n	n^2	$\sqrt{n}$	$\sqrt{10n}$	n	n^2	$\sqrt{n}$	$\sqrt{10n}$
1.0	1.00	1.000	3.162	4.0	16.00	2.000	6.325	7.0	49.00	2.646	8.367
1.1	1.21	1.049	3.317	4.1	16.81	2.025	6.403	7.1	50.41	2.665	8.426
1.2	1.44	1.095	3.464	4.2	17.64	2.049	6.481	7.2	51.84	2.683	8.485
1.3	1.69	1.140	3.606	4.3	18.49	2.074	6.557	7.3	53.29	2.702	8.544
1.4	1.96	1.183	3.742	4.4	19.36	2.098	6.633	7.4	54.76	2.720	8.602
1.5	2.25	1.225	3.873	4.5	20.25	2.121	6.708	7.5	56.25	2.739	8.660
1.6	2.56	1.265	4.000	4.6	21.16	2.145	6.782	7.6	57.76	2.757	8.718
1.7	2.89	1.304	4.123	4.7	22.09	2.168	6.856	7.7	59.29	2.775	8.775
1.8	3.24	1.342	4.243	4.8	23.04	2.191	6.928	7.8	60.84	2.793	8.832
1.9	3.61	1.378	4.359	4.9	24.01	2.214	7.000	7.9	62.41	2.811	8.888
2.0	4.00	1.414	4.472	5.0	25.00	2.236	7.071	8.0	64.00	2.828	8.944
2.1	4.41	1.449	4.583	5.1	26.01	2.258	7.141	8.1	65.61	2.846	9.000
2.2	4.84	1.483	4.690	5.2	27.04	2.280	7.211	8.2	67.24	2.864	9.055
2.3	5.29	1.517	4.796	5.3	28.09	2.302	7.280	8.3	68.89	2.881	9.110
2.4	5.76	1.549	4.899	5.4	29.16	2.324	7.348	8.4	70.56	2.898	9.165
2.5	6.25	1.581	5.000	5.5	30.25	2.345	7.416	8.5	72.25	2.915	9.220
2.6	6.76	1.612	5.099	5.6	31.36	2.366	7.483	8.6	73.96	2.933	9.274
2.7	7.29	1.643	5.196	5.7	32.49	2.387	7.550	8.7	75.69	2.950	9.327
2.8	7.84	1.673	5.292	5.8	33.64	2.408	7.616	8.8	77.44	2.966	9.381
2.9	8.41	1.703	5.385	5.9	34.81	2.429	7.681	8.9	79.21	2.983	9.434
3.0	9.00	1.732	5.477	6.0	36.00	2.449	7.746	9.0	81.00	3.000	9.487
3.1	9.61	1.761	5.568	6.1	37.21	2.470	7.810	9.1	82.81	3.017	9.539
3.2	10.24	1.789	5.657	6.2	38.44	2.490	7.874	9.2	84.64	3.033	9.592
3.3	10.89	1.817	5.745	6.3	39.69	2.510	7.937	9.3	86.49	3.050	9.644
3.4	11.56	1.844	5.831	6.4	40.96	2.530	8.000	9.4	88.36	3.066	9.695
3.5	12.25	1.871	5.916	6.5	42.25	2.550	8.062	9.5	90.25	3.082	9.747
3.6	12.96	1.897	6.000	6.6	43.56	2.569	8.124	9.6	92.16	3.098	9.798
3.7	13.69	1.924	6.083	6.7	44.89	2.588	8.185	9.7	94.09	3.114	9.849
3.8	14.44	1.949	6.164	6.8	46.24	2.608	8.246	9.8	96.04	3.130	9.899
3.9	15.21	1.975	6.245	6.9	47.61	2.627	8.307	9.9	98.01	3.146	9.950

Table 3.B
Areas Under the Normal Curve

z	0.00	0.01	0.02	0.03	0.04	0.05	0.06	0.07	0.08	0.09
-3.4	0.0003	0.0003	0.0003	0.0003	0.0003	0.0003	0.0003	0.0003	0.0003	0.0002
-3.3	0.0005	0.0005	0.0005	0.0004	0.0004	0.0004	0.0004	0.0004	0.0004	0.0003
-3.2	0.0007	0.0007	0.0006	0.0006	0.0006	0.0006	0.0006	0.0005	0.0005	0.0005
-3.1	0.0010	0.0009	0.0009	0.0009	0.0008	0.0008	0.0008	0.0008	0.0007	0.0007
-3.0	0.0013	0.0013	0.0013	0.0012	0.0012	0.0011	0.0011	0.0011	0.0010	0.0010
-2.9	0.0019	0.0018	0.0017	0.0017	0.0016	0.0016	0.0015	0.0015	0.0014	0.0014
-2.8	0.0026	0.0025	0.0024	0.0023	0.0023	0.0022	0.0021	0.0021	0.0020	0.0019
-2.7	0.0035	0.0034	0.0033	0.0032	0.0031	0.0030	0.0029	0.0028	0.0027	0.0026
-2.6	0.0047	0.0045	0.0044	0.0043	0.0041	0.0040	0.0039	0.0038	0.0037	0.0036
-2.5	0.0062	0.0060	0.0059	0.0057	0.0055	0.0054	0.0052	0.0051	0.0049	0.0048
-2.4	0.0082	0.0080	0.0078	0.0075	0.0073	0.0071	0.0069	0.0068	0.0066	0.0064
-2.3	0.0107	0.0104	0.0102	0.0099	0.0096	0.0094	0.0091	0.0089	0.0087	0.0084
-2.2	0.0139	0.0136	0.0132	0.0129	0.0125	0.0122	0.0119	0.0116	0.0113	0.0110
-2.1	0.0179	0.0174	0.0170	0.0166	0.0162	0.0158	0.0154	0.0150	0.0146	0.0143
-2.0	0.0228	0.0222	0.0217	0.0212	0.0207	0.0202	0.0197	0.0192	0.0188	0.0183
-1.9	0.0287	0.0281	0.0274	0.0268	0.0262	0.0256	0.0250	0.0244	0.0239	0.0233
-1.8	0.0359	0.0352	0.0344	0.0336	0.0329	0.0322	0.0314	0.0307	0.0301	0.0294
-1.7	0.0446	0.0436	0.0427	0.0418	0.0409	0.0401	0.0392	0.0384	0.0375	0.0367
-1.6	0.0548	0.0537	0.0526	0.0516	0.0505	0.0495	0.0485	0.0475	0.0465	0.0455
-1.5	0.0668	0.0655	0.0643	0.0630	0.0618	0.0606	0.0594	0.0582	0.0571	0.0559
-1.4	0.0808	0.0793	0.0778	0.0764	0.0749	0.0735	0.0722	0.0708	0.0694	0.0681
-1.3	0.0968	0.0951	0.0934	0.0918	0.0901	0.0885	0.0869	0.0853	0.0838	0.0823
-1.2	0.1151	0.1131	0.1112	0.1093	0.1075	0.1056	0.1038	0.1020	0.1003	0.0985
-1.1	0.1357	0.1335	0.1314	0.1292	0.1271	0.1251	0.1230	0.1210	0.1190	0.1170
-1.0	0.1587	0.1562	0.1539	0.1515	0.1492	0.1469	0.1446	0.1423	0.1401	0.1379
-0.9	0.1841	0.1814	0.1788	0.1762	0.1736	0.1711	0.1685	0.1660	0.1635	0.1611
-0.8	0.2119	0.2090	0.2061	0.2033	0.2005	0.1977	0.1949	0.1922	0.1894	0.1867
-0.7	0.2420	0.2389	0.2358	0.2327	0.2296	0.2266	0.2236	0.2206	0.2177	0.2148
-0.6	0.2743	0.2709	0.2676	0.2643	0.2611	0.2578	0.2546	0.2514	0.2483	0.2451
-0.5	0.3085	0.3050	0.3015	0.2981	0.2946	0.2912	0.2877	0.2843	0.2810	0.2776
-0.4	0.3446	0.3409	0.3372	0.3336	0.3300	0.3264	0.3228	0.3192	0.3156	0.3121
-0.3	0.3821	0.3783	0.3745	0.3707	0.3669	0.3632	0.3594	0.3557	0.3520	0.3483
-0.2	0.4207	0.4168	0.4129	0.4090	0.4052	0.4013	0.3974	0.3936	0.3897	0.3859
-0.1	0.4602	0.4562	0.4522	0.4483	0.4443	0.4404	0.4364	0.4325	0.4286	0.4247
-0.0	0.5000	0.4960	0.4920	0.4880	0.4840	0.4801	0.4761	0.4721	0.4681	0.4641
0.0	0.5000	0.5040	0.5080	0.5120	0.5160	0.5199	0.5239	0.5279	0.5319	0.5359
0.1	0.5398	0.5438	0.5478	0.5517	0.5557	0.5596	0.5636	0.5675	0.5714	0.5753
0.2	0.5793	0.5832	0.5871	0.5910	0.5948	0.5987	0.6026	0.6064	0.6103	0.6141
0.3	0.6179	0.6217	0.6244	0.6293	0.6331	0.6368	0.6406	0.6443	0.6480	0.6517
0.4	0.6554	0.6591	0.6628	0.6664	0.6700	0.6736	0.6772	0.6808	0.6844	0.6879
0.5	0.6915	0.6950	0.6985	0.7019	0.7054	0.7088	0.7123	0.7157	0.7190	0.7224
0.6	0.7257	0.7291	0.7324	0.7357	0.7389	0.7422	0.7454	0.7486	0.7517	0.7549
0.7	0.7580	0.7611	0.7642	0.7673	0.7704	0.7734	0.7764	0.7794	0.7823	0.7852
0.8	0.7881	0.7910	0.7939	0.7967	0.7995	0.8023	0.8051	0.8078	0.8106	0.8133
0.9	0.8159	0.8186	0.8212	0.8238	0.8264	0.8289	0.8315	0.8340	0.8365	0.8389
1.0	0.8413	0.8438	0.8461	0.8485	0.8508	0.8531	0.8554	0.8577	0.8599	0.8621
1.1	0.8643	0.8665	0.8686	0.8708	0.8729	0.8749	0.8770	0.8790	0.8810	0.8830
1.2	0.8849	0.8869	0.8888	0.8907	0.8925	0.8944	0.8962	0.8980	0.8997	0.9015
1.3	0.9032	0.9049	0.9066	0.9082	0.9099	0.9115	0.9131	0.9147	0.9162	0.9177
1.4	0.9192	0.9207	0.9222	0.9236	0.9251	0.9265	0.9278	0.9292	0.9306	0.9319
1.5	0.9332	0.9345	0.9357	0.9370	0.9382	0.9394	0.9406	0.9418	0.9429	0.9441
1.6	0.9452	0.9463	0.9474	0.9484	0.9495	0.9505	0.9515	0.9525	0.9535	0.9545
1.7	0.9554	0.9564	0.9573	0.9582	0.9591	0.9599	0.9608	0.9616	0.9625	0.9633
1.8	0.9641	0.9649	0.9656	0.9664	0.9671	0.9678	0.9686	0.9693	0.9699	0.9706
1.9	0.9713	0.9719	0.9726	0.9732	0.9738	0.9744	0.9750	0.9756	0.9761	0.9767
2.0	0.9772	0.9778	0.9783	0.9788	0.9793	0.9798	0.9803	0.9808	0.9812	0.9817
2.1	0.9821	0.9826	0.9830	0.9834	0.9838	0.9842	0.9846	0.9850	0.9854	0.9857
2.2	0.9861	0.9864	0.9868	0.9871	0.9875	0.9878	0.9881	0.9884	0.9887	0.9890
2.3	0.9893	0.9896	0.9898	0.9901	0.9904	0.9906	0.9909	0.9911	0.9913	0.9916
2.4	0.9918	0.9920	0.9922	0.9925	0.9927	0.9929	0.9931	0.9932	0.9934	0.9936
2.5	0.9938	0.9940	0.9941	0.9943	0.9945	0.9946	0.9948	0.9949	0.9951	0.9952
2.6	0.9953	0.9955	0.9956	0.9957	0.9959	0.9960	0.9961	0.9962	0.9963	0.9964
2.7	0.9965	0.9966	0.9967	0.9968	0.9969	0.9970	0.9971	0.9972	0.9973	0.9974
2.8	0.9974	0.9975	0.9976	0.9977	0.9977	0.9978	0.9979	0.9979	0.9980	0.9981
2.9	0.9981	0.9982	0.9982	0.9983	0.9984	0.9984	0.9985	0.9985	0.9986	0.9986
3.0	0.9987	0.9987	0.9987	0.9988	0.9988	0.9989	0.9989	0.9989	0.9990	0.9990
3.1	0.9990	0.9991	0.9991	0.9991	0.9992	0.9992	0.9992	0.9992	0.9993	0.9993
3.2	0.9993	0.9993	0.9994	0.9994	0.9994	0.9994	0.9994	0.9995	0.9995	0.9995
3.3	0.9995	0.9995	0.9995	0.9996	0.9996	0.9996	0.9996	0.9996	0.9996	0.9997
3.4	0.9997	0.9997	0.9997	0.9997	0.9997	0.9997	0.9997	0.9997	0.9997	0.9998

APPENDIX **4**

Solutions and Answers to Selected Problems

Sec. 1.1

A. 1. False. 2. False. 3. False. 4. True. 5. True. 6. False. 7. False.

B. 1. Inductive. Conclusion does not follow, since 9 is odd and yet 9 is not prime.

Sec. 1.2

A. 1. F = set of fish
S = set of swimmers

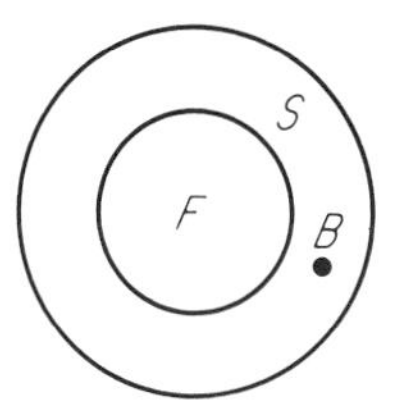

Conclusion does not follow.

2. F = set of fish
S = set of swimmers
P = set of perch

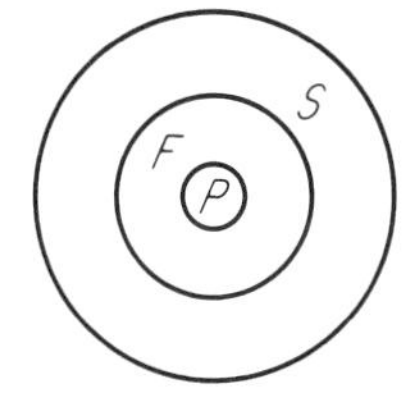

Conclusion does follow.

3. Conclusion does not follow.
4. Conclusion does not follow.

B. 1. Conclusion does not follow.
2. Conclusion does follow.
3. Conclusion does not follow.
4. Conclusion does not follow.
5. Conclusion does follow.

Sec. 1.3

A. 1. True. 2. False. 3. True. 4. False. 5. False.

B. 1. Statement, compound, conditional.
2. Not a statement, since it is not a declarative sentence.
3. Statement, compound, conditional.
4. Not a statement.
5. Statement, simple.
6. Statement, compound, conjunction.
7. Statement, compound, conjunction. (See Prob. Set 1.2, A.4.)
8. Not a statement.
9. Statement, compound, disjunction.
10. Not a statement.

C. (a) $\sim p \vee (q \longleftrightarrow p) \longrightarrow q$ or $(\sim p \vee q \longleftrightarrow p) \longrightarrow q$. (b) $\sim p \vee (q \longleftrightarrow p \longrightarrow q)$.
(c) $\sim(p \vee q \longleftrightarrow p \longrightarrow q)$.

D. 1. $\sim r$. 2. $\sim(p \vee q)$. 3. $\sim q \longrightarrow r$. 4. $(p \longrightarrow q) \wedge \sim p$.
5. $p \longrightarrow q \longleftrightarrow \sim p \vee q$.

E. 1. I do not actively discuss all issues.
2. If I am involved and I actively discuss all issues then I know the answers to all issues.
3. If I actively discuss all issues then I know the answers to all issues.
4. I am involved if and only if I actively discuss all issues.
5. If I am involved then I actively discuss all issues and I know the answers to all issues.
6. I do not actively discuss all issues or I know the answers to all issues.
7. It is not the case that I actively discuss all issues or I know the answers to all issues.
8. I do not actively discuss all issues and I do not know the answers to all issues.
9. I actively discuss all issues or I am involved, and I know the answers to all issues.
10. I actively discuss all issues, or I am involved and I know the answers to all issues.

Sec. 1.4

A. 1. True. 2. False. 3. True. 4. True. 5. True. 6. False.
7. False. 8. False. 9. False. 10. True.

B. 1.

$\sim$	$\sim$	p
T	F	T
F	T	F

2.

$\sim$	$(\sim$	p	$\vee$	$q)$
F	F	T	T	T
T	F	T	F	F
F	T	F	T	T
F	T	F	T	F
(5)	(3)	(1)	(4)	(2)

3.

$\sim(p \longrightarrow q)$
F
T
F
F

4. $p \longrightarrow q \longleftrightarrow \sim q \vee \sim p$

F
F
T
T

5. $(p \longrightarrow q) \wedge (q \longrightarrow r) \longrightarrow (p \longrightarrow r)$

T
T
T
T
T
T
T
T

6.

p	$\longrightarrow$	q	$\wedge$	r
T	T	T	T	T
T	F	T	F	F
T	F	F	F	T
T	F	F	F	F
F	T	T	T	T
F	T	T	F	F
F	T	F	F	T
F	T	F	F	F
①	⑤	②	④	③

7. $p \longrightarrow q \vee r$

T
T
T
F
T
T
T
T

8. $p \wedge q \longrightarrow p$

T
T
T
T

9. $(p \vee q) \wedge \sim p \longrightarrow q$

T
T
T
T

10. $(p \longrightarrow q) \wedge p \longrightarrow q$

T
T
T
T

11.

p	$\wedge$	$(q$	$\vee$	$r)$
T	T	T	T	T
T	T	T	T	F
T	T	F	T	T
T	F	F	F	F
F	F	T	T	T
F	F	T	T	F
F	F	F	T	T
F	F	F	F	F
①	⑤	②	④	③

12. $(p \wedge q) \vee r$

T
T
T
F
T
F
T
F

C. $2^5 = 32$.

Sec. 1.5

A. 1. True. 2. False. 3. True. 4. False. 5. False.

B. 1.

P	P	$\vee$	$\sim$	Q
T	T	T	F	T
T	T	T	T	F
F	F	F	F	T
F	F	T	T	F

P, $P \vee \sim Q$ are not equivalent, $P \not\Leftrightarrow (P \vee \sim Q)$, since in the truth table, the truth values row-by-row are not the same. $P \Rightarrow P \vee \sim Q$, since the definition of implication is satisfied, i.e., there is no case when P is true and $P \vee \sim Q$ is false. $P \vee \sim Q \not\Rightarrow P$, since the definition of implication is not satisfied in the fourth row.

2. $(\sim P \vee Q) \Leftrightarrow (P \rightarrow Q)$
$(\sim P \vee Q) \Rightarrow (P \rightarrow Q)$
$(P \rightarrow Q) \Rightarrow (\sim P \vee Q)$.

3. $\sim(P \vee Q) \Leftrightarrow (\sim P \wedge \sim Q)$
$\sim(P \vee Q) \Rightarrow (\sim P \wedge \sim Q)$
$(\sim P \wedge \sim Q) \Rightarrow \sim(P \vee Q)$.

4. $(P \rightarrow Q) \wedge (Q \rightarrow R) \not\Leftrightarrow (P \rightarrow R)$
$(P \rightarrow Q) \wedge (Q \rightarrow R) \Rightarrow (P \rightarrow R)$
$(P \rightarrow R) \not\Rightarrow (P \rightarrow Q) \wedge (Q \rightarrow R)$.

5. $P \not\Leftrightarrow (P \wedge \sim Q)$
$P \not\Rightarrow (P \wedge \sim Q)$
$(P \wedge \sim Q) \Rightarrow P$.

C. **Converse:** If I pass this course then I study.
Inverse: If I do not study then I do not pass this course.
Contrapositive: If I do not pass this course then I do not study.

D.
1. If the person is a student then he does not study.
2. If the person is a woman, then the person is fickle.
3. If the book is a mathematics book then it cannot be read easily.
4. If you go then he will not go.
5. If you exercise regularly then you are physically fit.

E. 1. $(p \leftrightarrow r) \rightarrow [(u \wedge r) \rightarrow (p \vee q)]$.
F F T (T) T T T F F F F
The expression is true, when p is logically false, q is logically false, r is logically true, and u is logically false.
2. $\sim p \wedge (q \vee u) \wedge \sim(p \leftrightarrow q)$ is false.
3. $(q \rightarrow u) \leftrightarrow (\sim q \vee r) \wedge r$ is true.
4. $(\sim p \vee r) \rightarrow (u \rightarrow p) \leftrightarrow \sim(p \vee r)$ is true.
5. $p \wedge (p \vee r) \rightarrow u \leftrightarrow \sim(p \vee q) \rightarrow \sim r$ is false.

F. 1. $\sim(\sim p \vee \sim q)$. Use the equivalence, $\sim(p \wedge q) \Leftrightarrow \sim p \vee \sim q$. Negate each side of the equivalence to obtain $\sim\sim(p \wedge q) \Leftrightarrow \sim(\sim p \vee \sim q)$. But $(p \wedge q) \Leftrightarrow \sim\sim \cdot (p \wedge q)$ by the law of double negation. Then by substitution we have $(p \vee q) \Leftrightarrow \sim(\sim p \vee \sim q)$, which is the desired result.
2. $\sim(\sim p \vee q) \vee r$.
3. $(\sim p \vee \sim q) \vee \sim(\sim p \vee \sim r)$.
4. $\sim(\sim p \vee \sim q) \vee \sim[\sim(\sim r \vee s) \vee \sim(\sim s \vee r)]$.
5. $\sim\{\sim[\sim(\sim p \vee q) \vee \sim(\sim q \vee p)]\} \vee \{\sim[\sim(\sim r \vee s) \vee \sim(\sim s \vee r)]\}$.

G. Let p represent "Income is available"
q represent "Expenses will rise to meet it"
r represent "Time is available"
s represent "Work will expand to fill it"

$(p \rightarrow q) \wedge (r \rightarrow s)$ implies $(p \wedge r) \rightarrow (q \wedge s)$ which can be shown by use of the definition of implication. When p and r are assumed true then q and s must be true in order that the premises be true. Thus the conjunction of the premises implies the conclusion.

Sec. 1.6

A. 1. $p \wedge (q \vee r) \Longleftrightarrow (p \wedge q) \vee (p \wedge r)$.

2. $p \longrightarrow (q \wedge r) \Longleftrightarrow (p \longrightarrow q) \wedge (p \longrightarrow r)$.

3. To determine if the expressions are equivalent we use the definition.

$(q$	$\wedge$	$r)$	$\longrightarrow$	p	$(q \longrightarrow p)$	$\wedge$	$(r \longrightarrow p)$
T	T	T	T	T	T	T	T
T	F	F	T	T	T	T	T
F	F	T	T	T	T	T	T
F	F	F	T	T	T	T	T
T	T	T	F	F	F	F	F
T	F	F	(T)	F	F	(F)	T
F	F	T	(T)	F	T	(F)	F
F	F	F	T	F	T	T	T

Since the two expressions are not the same row-by-row (see circled letters in truth table), under the major connective, the expressions are not equivalent.

4. $p \longleftrightarrow (q \longrightarrow r) \not\Longleftrightarrow (p \longleftrightarrow r) \longrightarrow (p \longleftrightarrow r)$.

5. The expressions $p \longleftrightarrow (q \wedge r)$, $(p \longleftrightarrow q) \wedge (p \longleftrightarrow r)$ are not equivalent.

6. The conditional is not commutative. 9. Yes.

Sec. 1.7

A. 1. Let r represent "Mathematics is easy," and t "You should study logic." The argument symbolized is

$$\begin{array}{l} t \longrightarrow r \\ \sim r \\ \hline \therefore \sim t \end{array}.$$

$(t \longrightarrow r)$			$\wedge$	$\sim r$	$\Rightarrow \sim t$
T	T	T	F	F	F
F	T	T	F	F	T
T	F	F	F	T	F
F	T	F	T	T	T

The conjunction of the premises does imply the conclusion. Thus we have an argument that is **valid**.

2. Valid. 3. Valid.
4. Let a represent "The sun is shining," and b "It is raining."
The argument symbolized is

$$\begin{array}{l} a \vee b \\ \underline{b} \\ \therefore a \end{array}$$

$(a \vee b)$	$\wedge$	b	$\nRightarrow a$
T T T	T	T	T
T T F	F	F	T
F T T	(T)	T	(F)
F F F	F	F	F

Row 3 of the truth table shows that the conjunction of the premises does not imply the conclusion. Thus the argument is invalid.

5. Invalid. 6. Valid. 7. Valid. 8. Valid. 9. Valid. 10. Invalid
11. Invalid. 12. Invalid. 13. Invalid.

B. 1. Valid. 2. Valid. 3. Valid. 4. Valid. 5. Valid.
6. Invalid. 7. Invalid. 8. Invalid. 9. Valid. 10. Invalid.

C. 1. Let c represent "The room is cold," and d represent "Tomorrow is Friday."
The argument in symbolic form is

$$\begin{array}{l} \underline{c} \\ \therefore c \vee d. \end{array}$$

Let the conclusion be false. The disjunction is false when both components are false., i.e., c is false and d is false. The argument is **valid** since we cannot consistently assign false to the conclusion and true to the conjunction of the premises. (Any instance in which the conclusion is false, the conjunction of the premises will also be false.)

2. Invalid.
3. $\underline{p}$
$\therefore p \wedge q$. Three cases must be considered as shown in the following table:

	p	$\wedge$	q
Case 1:	F	F	F
Case 2:	F	F	T
Case 3:	T	F	F

Case 1: The conclusion is false when both p and q are false. In this instance the argument seems valid BUT we must show the argument is valid, if it is, in all instances.
Case 2: The conclusion is false when p is false and q is true. Again in this instance the argument seems valid.
Case 3: The conclusion is false when p is true and q is false. But now the premise has a true truth value and the conclusion is false. For this case, the premise does not imply the conclusion. Therefore, the argument is *invalid.*

4. Invalid. 5. Valid. 6. Valid.

D. 1. (1) P.
(2) $\sim P \vee Q$.
(3) $Q \longrightarrow R$.
$\therefore R$

(4) Q lines 1,2, Law of disjunctive syllogism
$\therefore R$ lines 3,4, Modus ponens.

2. (1) $P \longrightarrow (Q \vee R)$.
(2) $\sim R \wedge \sim Q$.
(3) $R \longrightarrow P$
$\therefore \sim R$.

(4) $\sim R$, line 2, Simplification.

3. (1) $P \longrightarrow Q \vee R$.
(2) $\sim(S \vee T) \longrightarrow P$.
(3) $T \longrightarrow \sim U$.
(4) $U \wedge \sim Q$.
(5) $U \longrightarrow \sim S$.
$\therefore R$.

(6) U.
(7) $\sim S$.
(8) $\sim T$.
(9) $\sim S \wedge \sim T$.
(10) $\sim (S \vee T)$.
(11) P.
(12) $Q \vee R$.
(13) $\sim Q$
$\therefore R$.

4. Let p represent "I take physics," c represent "I take calculus," w represent "My time will be well spent," and y represent "I will take psychology." The argument is symbolized as given below:
(1) $p \vee c \longrightarrow w$.
(2) p
$\therefore \sim y \vee w$.

(3) $p \vee c$, line 2, Addition.
(4) w, line 1, 3, Modus ponens
$\therefore \sim y \vee w$, line 4, Addition.

5. (1) $(h \longrightarrow b). \wedge (o \longrightarrow d)$.
(2) $h \wedge o$.
$\therefore d \wedge b$.

(3) $h \longrightarrow b$.
(4) $o \longrightarrow d$.
(5) h.
(6) o.
(7) b.
(8) d.
$\therefore d \wedge b$.

6. (1) $\sim r \vee a$.
(2) $s \longrightarrow \sim a$
$\therefore r \longrightarrow \sim s$.

(3) $r \longrightarrow a$.
(4) $a \longrightarrow \sim s$
$\therefore r \longrightarrow \sim s$.

7. (1) $p \longrightarrow q$
p
$\therefore q$

p represents "x is an element of set A."
q represents "x is an element of set B."
The argument is valid since it has the Modus ponens valid argument form.

8. Let w represent "I work," p represent "I am prosperous," and t represent "I have a good time." The argument symbolized is:
(1) $w \longrightarrow p$.
(2) $\sim w \longrightarrow t$
$\therefore p \vee t$.

The argument is **valid.**

10. (1) $P \vee (R \longrightarrow Q)$.
(2) $(P \longrightarrow R) \vee \sim S$.
(3) $S \wedge \sim R$.
$\therefore R \longrightarrow Q$.

(4) S.
(5) $P \longrightarrow R$.
(6) $\sim R$.
(7) $\sim P$.
$\therefore R \longrightarrow Q$.

Sec. 1.8

A. 1. All integers are not odd numbers.
2. There is a set which is not a subset of itself.
3. All rationals are repeating decimals.
4. All elements of set A are elements of set B.
5. All natural numbers do not have a multiplicative inverse.

B. 1. Let D represent the set of students, P represent the set of stupid people, and let j represent the person Jones.

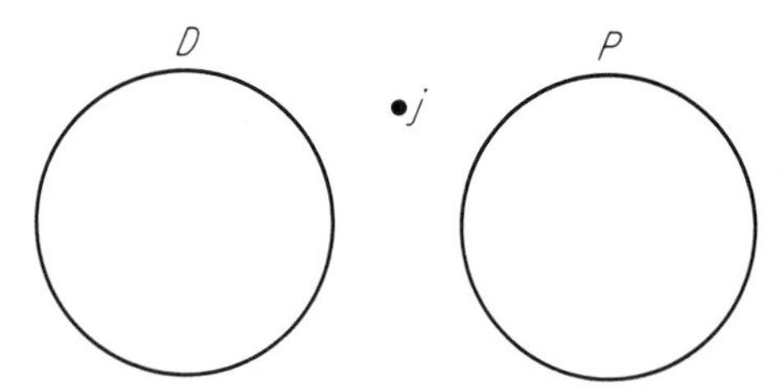

Argument is invalid, since the conclusion does not necessarily follow from the given premises. (It would seem to if we placed j in the region determined by P, but no premise necessitates that it be in the P-region.)
2. Valid. 3. Invalid. 4. Valid. 5. Valid.

Sec. 1.9

A. 1. Let a, b, c be nonzero integers. By hypothesis a is a factor of b and also b is a factor of c. Since a is a factor of b we have $a \cdot p = b$ for some integer p. Also, since b is a factor of c we have $b \cdot q = c$ for some integer q. Since $(a \cdot p)q = b \cdot q$ we have $(a \cdot p) \cdot q = c$, so that $a(p \cdot q) = c$ where $p \cdot q$ is an integer. Thus a is a factor of c.
2. Let a, b, c be real numbers. By hypothesis $a < b$ so by definition of less than there is a positive real number, say r, such that $a + r = b$. Thus we have $(a + r)c = bc$ or $ac + rc = bc$. Adding $-rc$ to both sides of the last equation, we obtain $ac = bc + (-rc)$. But rc is negative so $-rc$ is positive and again by definition we have $bc < ac$.
3. Let a, b, c be real numbers. By hypothesis $a = b$. The sum of a and c is a real number. By the reflexive property of real numbers we have $a + c = a + c$. Then by substitution we have $a + c = b + c$, the desired conclusion.
4. By hypothesis, x^2 is even. Using a proof by contradiction, assume x is not an even integer. Then x is an odd integer so that $x = 2k + 1$ for some integer k.

Now $x^2 = (2k+1)^2$ so that $x^2 = 4k^2 + 4k + 1$ or $x^2 = 2(2k^2 + 2k) + 1$ where $2k^2 + 2k$ is an integer. But then x^2 is an odd integer by definition. Since this contradicts the hypothesis our assumption that x is not even must be false; therefore, x is even.

6. Let $x = 2$, then $x^2 + x - 5 = 1$ (the property of an integer being prime is generally restricted to positive integers, thus we start with $x = 2$). Let $x = 3$, then $x^2 + x - 5 = 7$, which is prime. Let $x = 4$, then $x^2 + x - 5 = 15$, but 15 is not prime. Thus $x^2 + x - 5$ does not generate only primes.
7. Two triangles may be similar and still not be congruent. As a counterexample, consider two equilateral triangles, one triangle having each side 3 units in length and the other triangle with each side 19 units in length. These triangles are similar but not congruent.

Sec. 2.2

A. 1. False. 2. False. 3. True. 4. False. 5. False.

B. 1. $2^5 = 32$.

2. (a) $\{0,1,2,3,4,5,6,7,8,9\}$. (b) $\{3,4\}$.
   (c) $\{1,2,3,4,5\}$. (d) $\{\emptyset, \{1\}, \{2\}, \{1,2\}\}$.
3. (a) $\{x \mid x \in N$ and x is an odd number less than $10\}$.
   (b) $\{y \mid y$ is one of the first five letters of the alphabet$\}$.
   (c) $\{z \mid z \in N$ and z is an even number less than $10\}$.
   (d) $\{\alpha \mid \alpha$ is one of the first three presidents of the United States$\}$.
4. $P(B) = \{\emptyset, \{\emptyset\}, \{\{1\}\}, \{\{1,2\}\}, \{\emptyset, \{1\}\}, \{\emptyset, \{1,2\}\}, \{\{1\}, \{1,2\}\}, \{\emptyset, \{1\}, \{1,2\}\}\}$.
5. (a) $\emptyset$ (empty set). (b) $\{1,2,3,4,5,6,7,8,9\}$. (c) $\{2,4\}$. (d) $\{2\}$.
6. Subsets of C are: $\emptyset, \{w\}, \{x\}, \{\{y,z\}\}, \{w,x\}, \{w, \{y,z\}\}, \{w,x, \{y,z\}\}, \{x, \{y,z\}\}$.

C. 1. Assume the empty set is not a subset of some set A. So there is an element, say x_1, such that $x_1 \in \emptyset$ and $x_1 \notin A$. But $x_1 \in \emptyset$ is not possible since this states that the empty set contains an element, but this contradicts the definition of the empty set. Thus our assumption is false and the empty set is a subset of any set.

2. By hypotheses $N \subset I$, $I \subset Q$. Since $N \subset I$, we have for any element x, if $x \in N$, then $x \in I$. Similarly, for any x, if $x \in I$ then $x \in Q$. By a rule of logic, we obtain "if $x \in N$ then $x \in Q$." Then by the definition of subset we have $N \subset Q$.

Sec. 2.3

A. 1. False. 2. False. 3. False. 4. False. 5. False.

B. 1. (a) $P - Q = \{3,4,8\}$. (b) $Q - P = \emptyset$.
   (c) $Q \cup P = \{3,4,5,6,7,8\}$. (d) $P \cap Q = \{5,6,7\}$.

2. (a) $\tilde{C} = \{a,e,i,o,u\}$. (b) $D \cap \tilde{C} = \{a,e,i,o,u\}$. (c) $D - C = D$.
   (d) $(\widetilde{C \cup D}) = \emptyset$. (e) $\tilde{C} \cap \tilde{D} = \emptyset$.
3. (a) $X \cup Y = \{2,3,4,5,6,8\}$. (b) $X \cap Y = \{4\}$. (c) $Y - X = \{3,5\}$.

C. 1. $A - \mathcal{U} = \{x \mid x \in A$ and $x \notin \mathcal{U}\}$. $A - \mathcal{U} = \emptyset$, since an element in a set, which is a subset of $\mathcal{U}$, will also be an element of $\mathcal{U}$.

2. Let $\mathcal{U} = \{1,2,3\}$, $A = \{2\}$, then $\tilde{A} = \{1,3\}$ and $\mathcal{U} - A = \{1,3\}$. $\mathcal{U} - A$ does

equal $\tilde{A}$. By definition of set complement, $A = \{x \mid x \in \mathcal{U}$ and $x \notin A\}$ and the two sets are equal.

3. (a) $\widetilde{(A - C)} = \{a, c, e, f, g, h, i\}$. (b) $\tilde{A} \cap C = \emptyset$.
   (c) $A - \mathcal{U} = \emptyset$. (d) $\widetilde{(C \cap B)} = \{a, b, d, e, f, g, h, i\}$.
   (e) $\mathcal{U} - (C \cap B) = \{a, b, d, e, f, g, h, i\}$.

Sec. 2.4.1

A. 1.

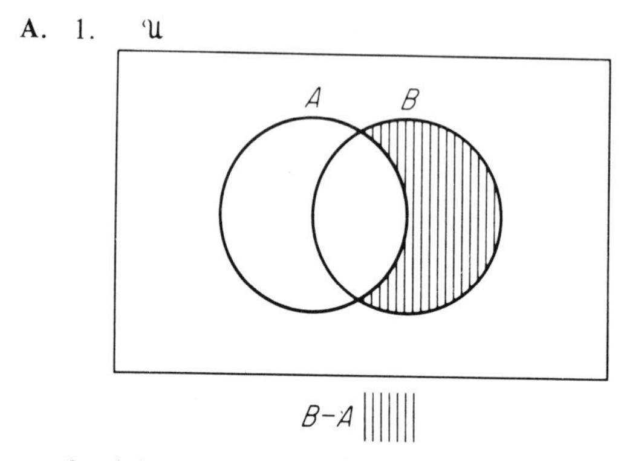

2. (1) $C \subset D$. To show that $C \cap D = C$ by Venn diagrams. When $C \subset D$, $C \cap D$ and

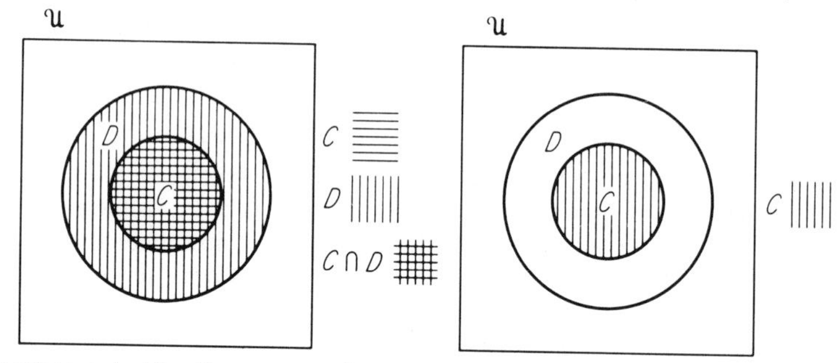

C are represented by the same region, hence $C \cap D = C$.

(2) $C \subset D$. To show that $C \cup D = D$ by Venn diagrams. When $C \subset D$, $C \cup D$ and

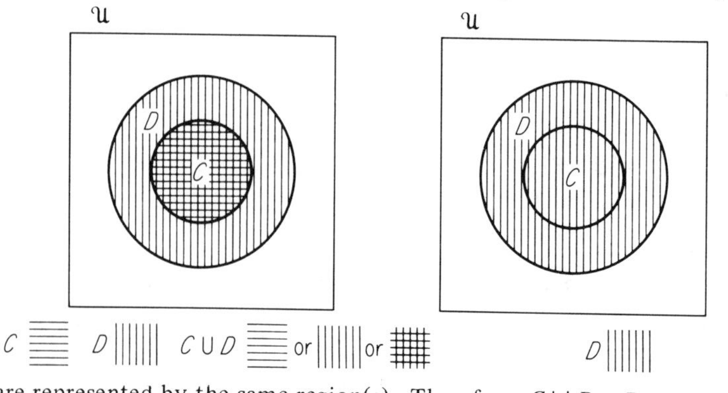

D are represented by the same region(s). Therefore, $C \cup D = D$.

3. $A \cup (B \cap C) = (A \cup B) \cap (A \cup C)$

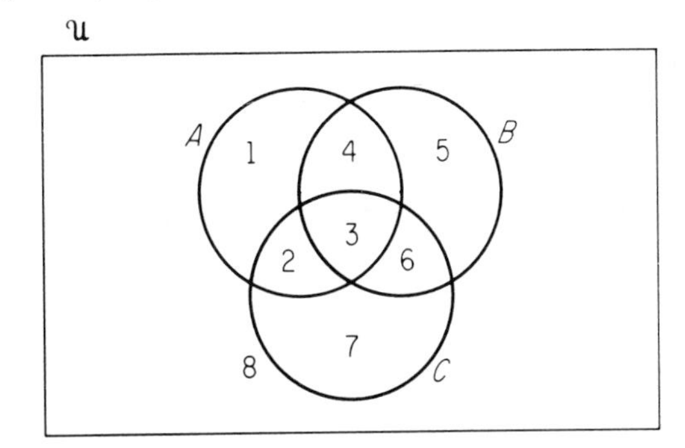

$A = \{1,2,3,4)$ $B = \{3,4,5,6\}$ $C = \{2,3,6,7\}$
$B \cap C = \{3,6\}$ $A \cup B = \{1,2,3,4,5,6\}$
$A \cup (B \cap C) = \{1,2,3,4,6\}$ $A \cup C = \{1,2,3,4,6,7\}$
$(A \cup B) \cap (A \cup C) = \{1,2,3,4,6\}$.
Since the sets, $A \cup (B \cap C)$ and $(A \cup B) \cap (A \cup C)$ are represented by the same regions, the sets are equal.

4. To show $\tilde{\tilde{A}} = A$ by Venn diagram.

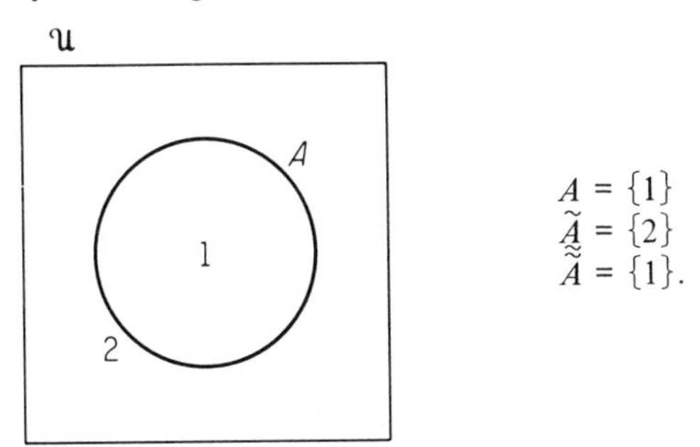

$A = \{1\}$
$\tilde{A} = \{2\}$
$\tilde{\tilde{A}} = \{1\}$.

Since A and $\tilde{A}$ are represented by the same region, the sets are equal.

5. To show $\tilde{A} \cap (B \cup \tilde{C}) = (\tilde{A} \cap B) \cup (\tilde{A} \cap \tilde{C})$ by Venn diagram and numbered regions. Consider the following diagram

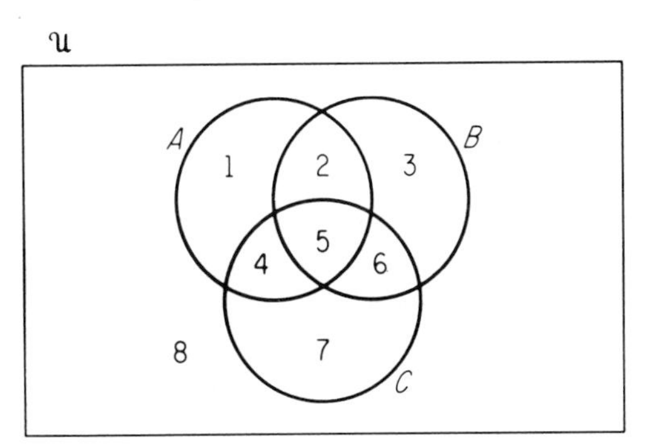

$A = \{1,2,4,5\}$ $B = \{2,3,5,6\}$ $C = \{4,5,6,7\}$ $\tilde{A} = \{3,6,7,8\}$
$\tilde{C} = \{1,2,3,8\}$ $B \cup \tilde{C} = \{1,2,3,5,6,8\}$ $\tilde{A} \cap B = \{3,6\}$ $\tilde{A} \cap (B \cup \tilde{C}) = \{3,6,8\}$
$\tilde{A} \cap \tilde{C} = \{3,8\}$ $(\tilde{A} \cap B) \cup (\tilde{A} \cap \tilde{C}) = \{3,6,8\}$.

Since the sets $\tilde{A} \cap (B \cup C)$ and $(\tilde{A} \cap B) \cup (\tilde{A} \cap \tilde{C})$ are represented by the same regions, the sets are equal.

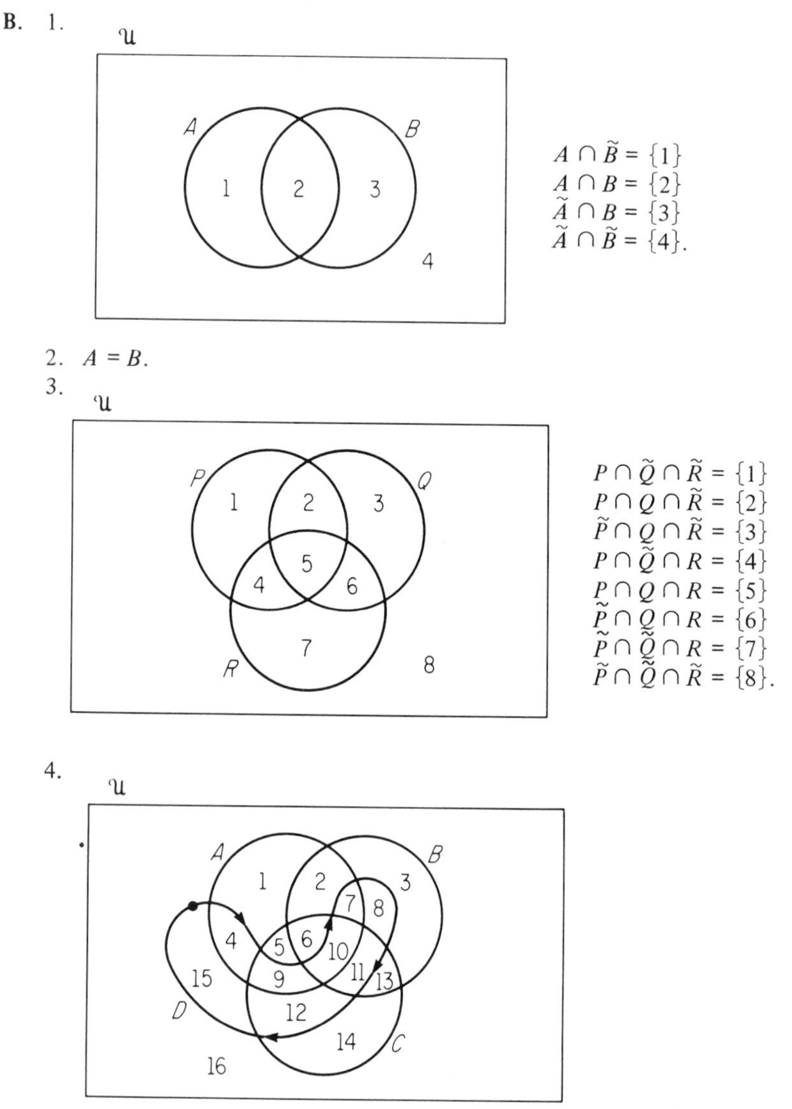

$A \cap \tilde{B} = \{1\}$
$A \cap B = \{2\}$
$\tilde{A} \cap B = \{3\}$
$\tilde{A} \cap \tilde{B} = \{4\}$.

$P \cap \tilde{Q} \cap \tilde{R} = \{1\}$
$P \cap Q \cap \tilde{R} = \{2\}$
$\tilde{P} \cap Q \cap \tilde{R} = \{3\}$
$P \cap \tilde{Q} \cap R = \{4\}$
$P \cap Q \cap R = \{5\}$
$\tilde{P} \cap Q \cap R = \{6\}$
$\tilde{P} \cap \tilde{Q} \cap R = \{7\}$
$\tilde{P} \cap \tilde{Q} \cap \tilde{R} = \{8\}$.

Divide each of the eight disjoint regions determined by A, B, C into two disjoint regions, thus giving sixteen disjoint regions.

Sec. 2.4.2

A. 1. (a) Let Q be the truth set for the statement q. The statement form $q \wedge \sim q$ may be translated to set language as $Q \cap \tilde{Q}$. Since by a Venn diagram $Q \cap \tilde{Q} = \emptyset$. The statement form $q \wedge \sim q$ is logically false.

(b) Logically true. (c) Logically true. (d) Neither.

(e) $p \longrightarrow (\sim p \longrightarrow q) \Longleftrightarrow p \longrightarrow (p \vee q)$
$\Longleftrightarrow \sim p \vee (p \vee q)$.

Let P represent the truth set for p and let Q represent the truth set for the statement q. Then the symbolic expression $\sim p \vee (p \vee q)$ may be translated into set language as $\tilde{P} \cup (P \cup Q)$. By the Venn diagram shown we have

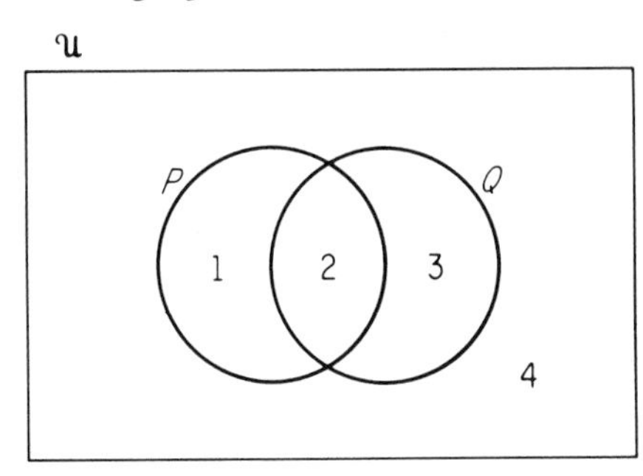

$P = \{1,2\}$ $Q = \{2,3\}$
$\tilde{P} = \{3,4\}$ $P \cup Q = \{1,2,3\}$
$\tilde{P} \cup (P \cup Q) = \{1,2,3,4\}$.

Since $\tilde{P} \cup (P \cup Q)$ represents the universe, the expression $p \longrightarrow (\sim p \longrightarrow q)$ is logically true.

2. (a) $A \cap B \subset A$.
(b) $B \subset A \cup B$.
$A - B \subset A$ since there is no case in which there is an element in $A - B$ and not in A.
(c)

A	B	$\tilde{B}$	$A - B$
$\in$	$\in$	$\notin$	$\notin$
$\in$	$\notin$	$\in$	$\in$
$\notin$	$\in$	$\notin$	$\notin$
$\notin$	$\notin$	$\in$	$\notin$

3. (a) $\widetilde{(P \cup Q)} = \tilde{P} \cap \tilde{Q}$. (b) $P \subset P \cup Q$. (c) $P \cap Q \subset Q$. (d) $\tilde{\tilde{P}} = P$.

4. Let P be the truth set for statement p and let Q be the truth set for statement q. The symbolic expression $(p \wedge \sim q) \wedge (\sim p \wedge q)$ can be translated into set language as $(P \cap \tilde{Q}) \cap (\tilde{P} \cap Q)$. By Venn diagram as shown we have

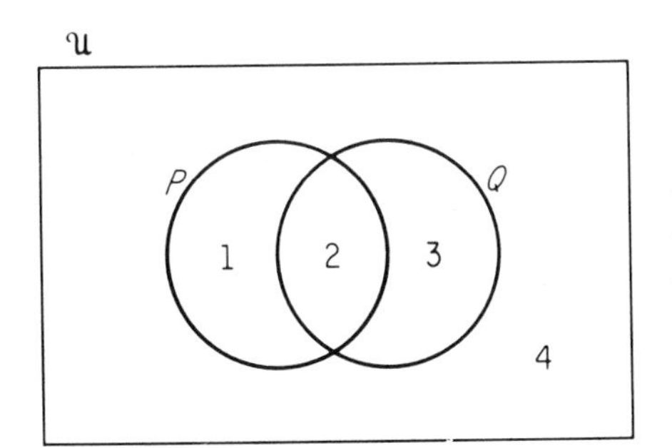

$P = \{1,2\}$ $Q = \{2,3\}$
$\tilde{P} = \{3,4\}$ $\tilde{Q} = \{1,4\}$
$P \cap \tilde{Q} = \{1\}$ $\tilde{P} \cap Q = \{3\}$
$(P \cap \tilde{Q}) \cap (\tilde{P} \cap Q) = \emptyset$.

Therefore the expression $(p \wedge \sim q) \wedge (\sim p \wedge q)$ is logically false.

B. 1. (a) To show $\widetilde{(\tilde{A} \cap B)} = A \cup \tilde{B}$ by membership tables.

A	B	$\tilde{A}$	$\tilde{B}$	$\tilde{A} \cap B$	$\widetilde{(\tilde{A} \cap B)}$	$A \cup \tilde{B}$
$\in$	$\in$	$\notin$	$\notin$	$\notin$	$\in$	$\in$
$\in$	$\notin$	$\notin$	$\in$	$\notin$	$\in$	$\in$
$\notin$	$\in$	$\in$	$\notin$	$\in$	$\notin$	$\notin$
$\notin$	$\notin$	$\in$	$\in$	$\notin$	$\in$	$\in$

$\widetilde{(\tilde{A} \cap B)} = A \cup \tilde{B}$, since in the table the two expressions have the same set membership row-by-row.

2.

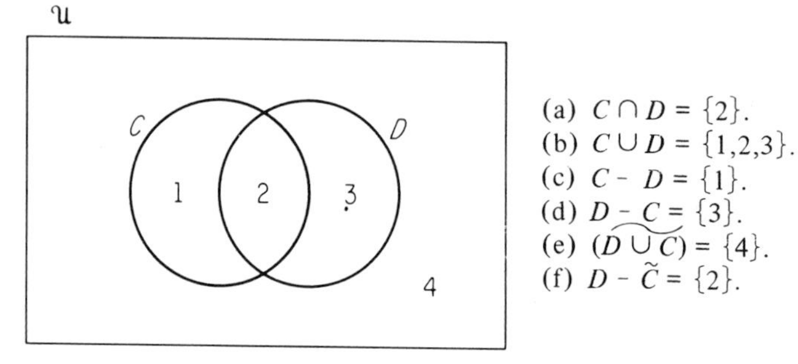

(a) $C \cap D = \{2\}$.
(b) $C \cup D = \{1,2,3\}$.
(c) $C - D = \{1\}$.
(d) $D - C = \{3\}$.
(e) $\widetilde{(D \cup C)} = \{4\}$.
(f) $D - \tilde{C} = \{2\}$.

3. Let a be the statement that has set A for its truth set. Then the set expression $A \cap \tilde{A}$ may be translated to symbolic language as $a \wedge \sim a$ which is logically false. An expression which is logically false has as its truth set the empty set. Therefore $A \cap \tilde{A} = \emptyset$.

4.

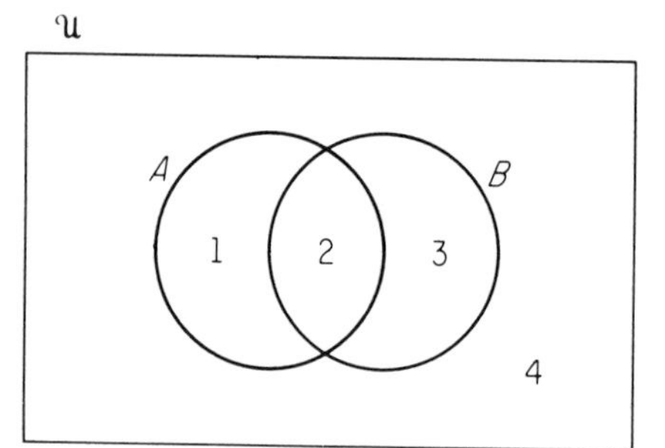

$A \Delta B = (A - B) \cup (B - A)$, $A - B = \{1\}$ $B - A = \{3\}$, $(A - B) \cup (B - A) = \{1,3\}$. Thus $A \Delta B = \{1,3\}$. Yes, $A \Delta B = B \Delta A$.

5. (a) $\{2\}$. (b) $\{1,2,3\}$. (c) $\{1\}$. (d) $\{3\}$. (e) $\{4\}$.

Sec. 2.5

A. 1. False. 2. True. 3. True. 4. False. 5. False.

B. 1. *b*. 2. *d*. 3. *c*. 4. *d*. 5. *c*.

C. 1. To show $A \cup (B \cap C) = (A \cup B) \cap (A \cup C)$ we must show (1) $A \cup (B \cap C) \subset (A \cup B) \cap (A \cup C)$, and (2) $(A \cup B) \cap (A \cup C) \subset A \cup (B \cap C)$.

To prove (1), let $x \in A \cup (B \cap C)$. By definition of union $x \in A$ or $x \in B \cap C$. Then by definition of intersection we have $x \in A$, or $x \in B$ and $x \in C$. But disjunction distributes over conjunction so $x \in A$ or $x \in B$, and $x \in A$ or $x \in C$. By definition of union, $x \in A \cup B$ and $x \in A \cup C$. By definition of intersection, $x \in (A \cup B) \cap (A \cup C)$. Thus we have shown if $x \in A \cup (B \cap C)$ then $x \in (A \cup B) \cap (A \cup C)$. Hence $A \cup (B \cap C) \subset (A \cup B) \cap (A \cup C)$.

The proof of (2) is done similarly, by letting $x \in (A \cup B) \cap (A \cup C)$ and deducing that $x \in A \cup (B \cap C)$.

2. By hypothesis, $A \subset B$. Assume $\tilde{B} \not\subset \tilde{A}$. Thus there is an element, say x_1, such that $x_1 \in \tilde{B}$ and $x_1 \notin \tilde{A}$. Since $x_1 \notin \tilde{A}$, then $x_1 \in A$. But $A \subset B$ so $x_1 \in B$. Thus we have an element, namely x_1, which is in B and $\tilde{B}$. But this is impossible. The assumption $\tilde{B} \not\subset \tilde{A}$ must be false, hence $\tilde{B} \subset \tilde{A}$ must be true. Thus, if $A \subset B$ then $\tilde{B} \subset \tilde{A}$.

6. (a) $B \cup (\tilde{B} \cap C) = (B \cup \tilde{B}) \cap (B \cup C)$ Union distributes over intersection

$= \mathfrak{U} \cap (B \cup C)$ $B \cup \tilde{B} = \mathfrak{U}$ (Complement law)

$= B \cup C$ $A \cap \mathfrak{U} = A$ (Identity law).

(b) $(A \cap B) \cup (A \cap \tilde{B}) = A \cap (B \cup \tilde{B})$ Intersection distributes over union

$= A \cap \mathfrak{U}$ $B \cup \tilde{B} = \mathfrak{U}$ (Complement law)

$= A$ $A \cap \mathfrak{U} = A$ (Identity law).

(c) $((A \cap B) \cup (A \cap \tilde{B})) \cup (\tilde{A} \cap B)$
$= (A \cap (B \cup \tilde{B})) \cup (\tilde{A} \cap B)$
$= (A \cap \mathfrak{U}) \cup (\tilde{A} \cap B)$
$= A \cup (\tilde{A} \cap B)$
$= (A \cup \tilde{A}) \cap (A \cup B)$
$= \mathfrak{U} \cap (A \cup B)$
$= A \cup B.$

(d) $A \cap (A \cup B) = (A \cup \emptyset) \cap (A \cup B)$
$= A \cup (\emptyset \cap B)$
$= A \cup \emptyset$
$= A.$

8. (a) $B \cap (\tilde{B} \cup C) = B \cap C$
$A = (A \cup B) \cap (A \cup \tilde{B})$
$A \cap B = ((A \cup B) \cap (A \cup \tilde{B})) \cap (\tilde{A} \cup B)$
$A \cup (A \cap B) = A.$

(b) $(A \cap \mathfrak{U}) \cup (\tilde{A} \cap \emptyset) = (\tilde{A} \cup \mathfrak{U}) \cap (\emptyset \cup A).$

Sec. 2.6.1

A. 1. False. 2. False. 3. True. 4. True. 5. False.

B. 1. (a)

A	A'	$(A')'$
1	0	1
0	1	0

$A = (A')'$, since the values under A and $(A')'$ are the same row-by-row.

(b)

A	B	A'	B'	$A \cdot B$	$(A \cdot B)'$	$A' + B'$
1	1	0	0	1	0	0
1	0	0	1	0	1	1
0	1	1	0	0	1	1
0	0	1	1	0	1	1

$(A \cdot B)' = A' + B'$, since the values under $(A \cdot B)'$ and $A' + B'$ are the same row-by-row.

(c)

A	A'	$A + A'$
1	0	1
0	1	1

Since $A + A'$ is 1 in each row, $A + A' = 1$.

(d) Since A, $A + A$ are the same row-by-row, they are equal.

A	$A + A$
1	1
0	0

2. (a) $A + (A \cdot B') = (A \cdot 1) + (A \cdot B')$ $\quad A \cdot 1 = A$ Identity law
$= A \cdot (1 + B')$ Distributive law
$= A \cdot 1$ $\quad A + 1 = 1$ Identity law
$= A$ $\quad A \cdot 1 = A$ Identity law.

(b) $A \cdot (B + B') = A \cdot 1$
$= A.$

(c) $(A \cdot B) \cdot C = [(A \cdot A) \cdot (B \cdot B)] \cdot (C \cdot C)$ Idempotent law
$= [A \cdot (A \cdot B) \cdot B] \cdot (C \cdot C)$
$= [(A \cdot B) \cdot A \cdot B] \cdot (C \cdot C)$
$= [(A \cdot B) \cdot A] \cdot [B \cdot (C \cdot C)]$
$= [(A \cdot B) \cdot A] \cdot [(B \cdot C) \cdot C]$
$= [(A \cdot B) \cdot A] \cdot [C \cdot (B \cdot C)]$
$= [(A \cdot B) \cdot A \cdot C] \cdot (B \cdot C)$
$= (A \cdot B) \cdot (A \cdot C) \cdot (B \cdot C).$

(d) $A + (B \cdot C) = (A + B) \cdot (A + C)$ Distributive law.

(e) $A + (A \cdot B) = (A \cdot A) + (A \cdot B)$ Idempotent law
$= A \cdot (A + B)$ Distributive law.

3. (a) $(A' \cdot B')'$. (b) $A \cdot (B' \cdot C')'$. (c) $(A' \cdot B' \cdot C')'$. (d) $(A \cdot B)'$.

Sec. 2.6.2

A. 1.

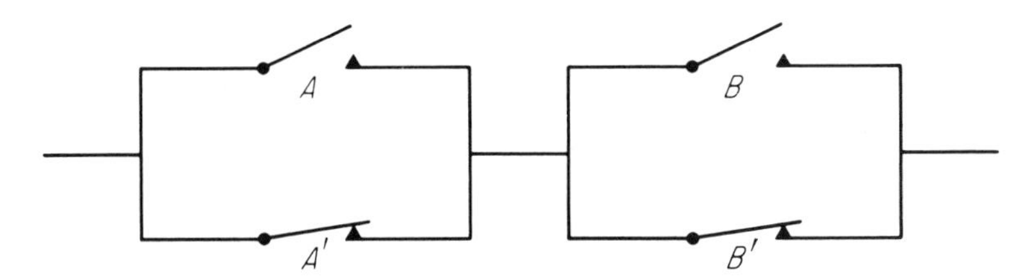

The state of the network expression is always on.

2. (a)

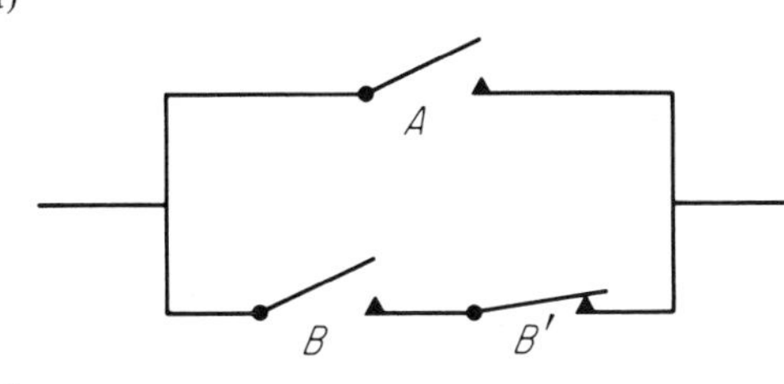

(b)

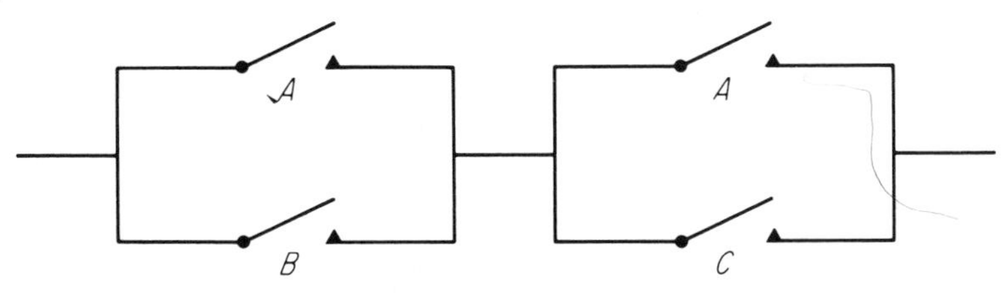

(c)

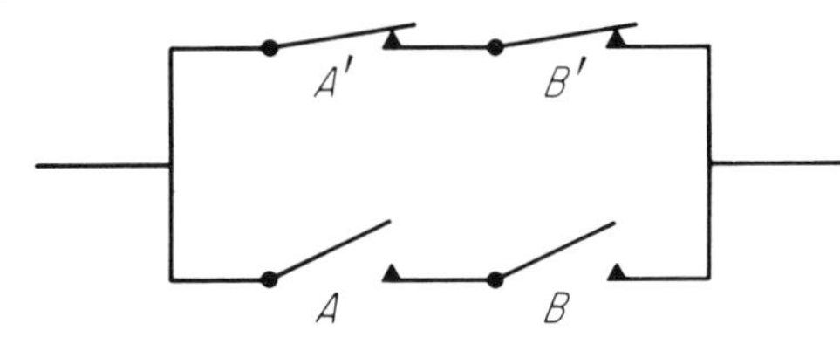

(d)

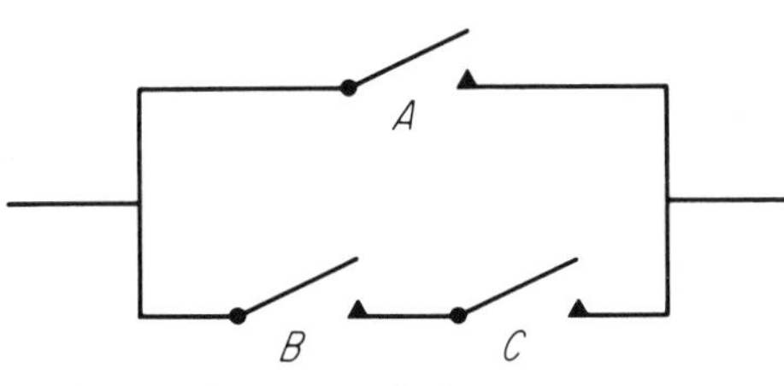

Networks *b* and *d* are equivalent.

3. (a) $(A \cdot A' \cdot B) + (A' \cdot B')$. (b) $(A + B') \cdot ((B \cdot B' \cdot C') + (A' \cdot C))$.
   (c) $(A \cdot B') + (A \cdot B) + (A' \cdot B)$. (d) $(A \cdot B) + A' + B'$.

4. (a) $A' \cdot A$ (b) $A' + B'$. (c) $A' \cdot (B' + C')$. (d) $A \cdot B$.
5. (a)

(b)

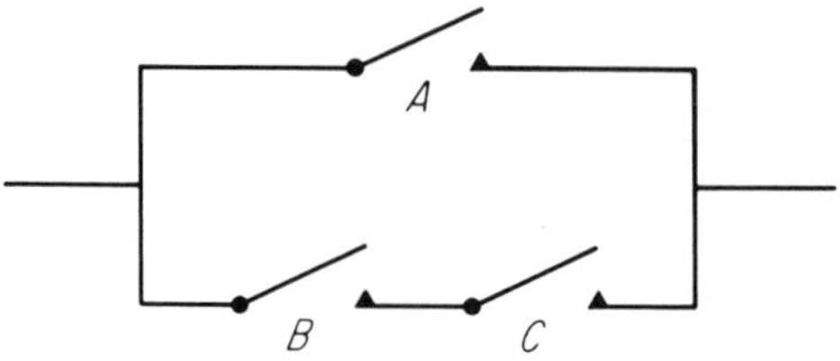

(c)

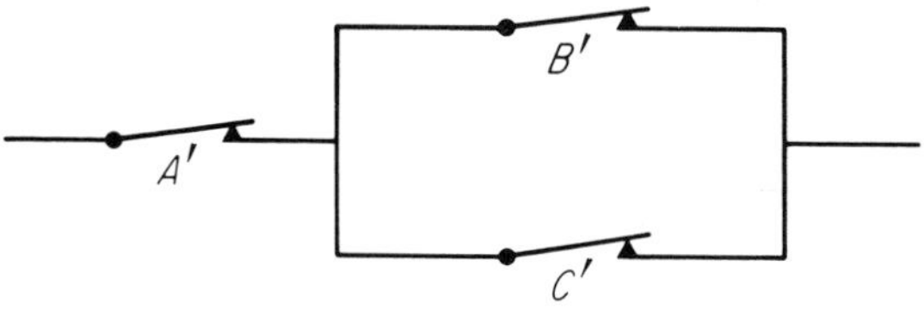

(d)

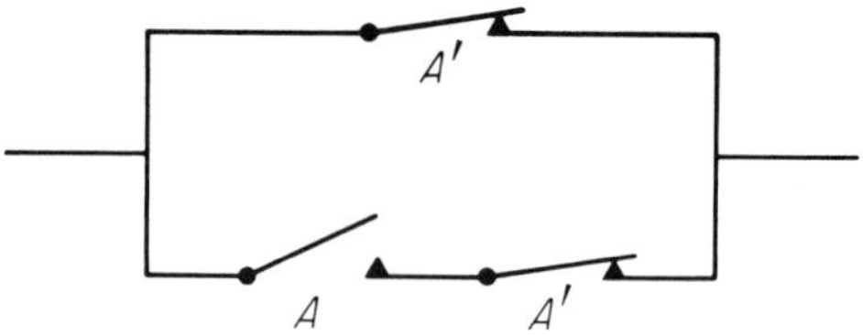

(e)

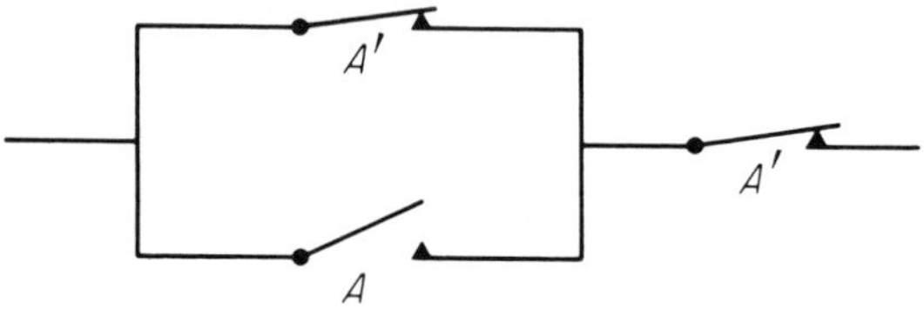

B. 1. a, b, c are equivalent.
2. (a) $(A \cdot B) + (A \cdot C) = A \cdot (B + C)$.

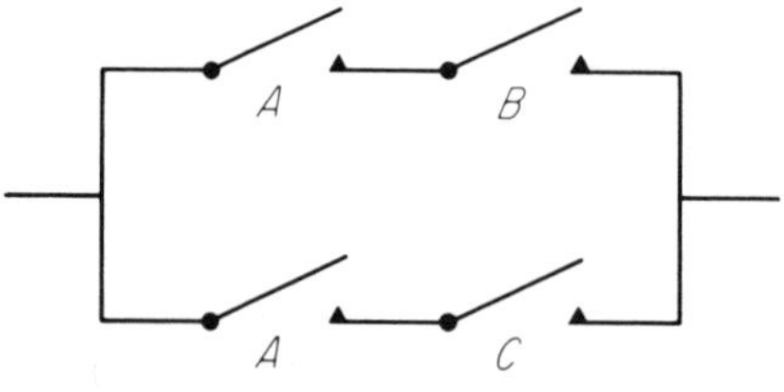

$(A \cdot B) + (A \cdot C)$ 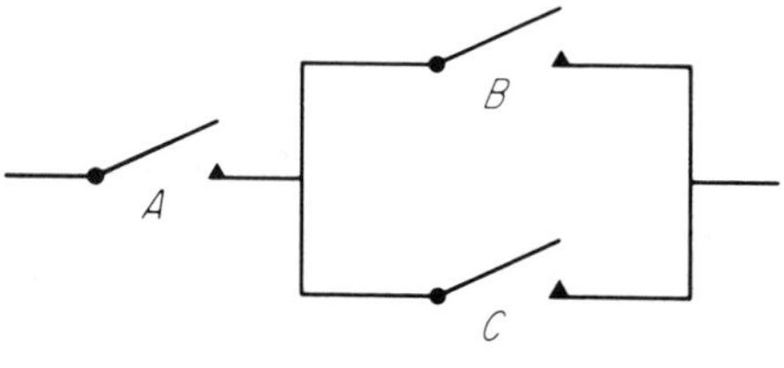 $A \cdot (B + C)$

(b) $B \cdot (A + A') = B \cdot 1.$
$= B$

(c) $(A \cdot B) + (A' \cdot B) = (A + A') \cdot B$
$= 1 \cdot B$
$= B.$

(d) $A + A' = 1.$

(e) $A + A = A.$

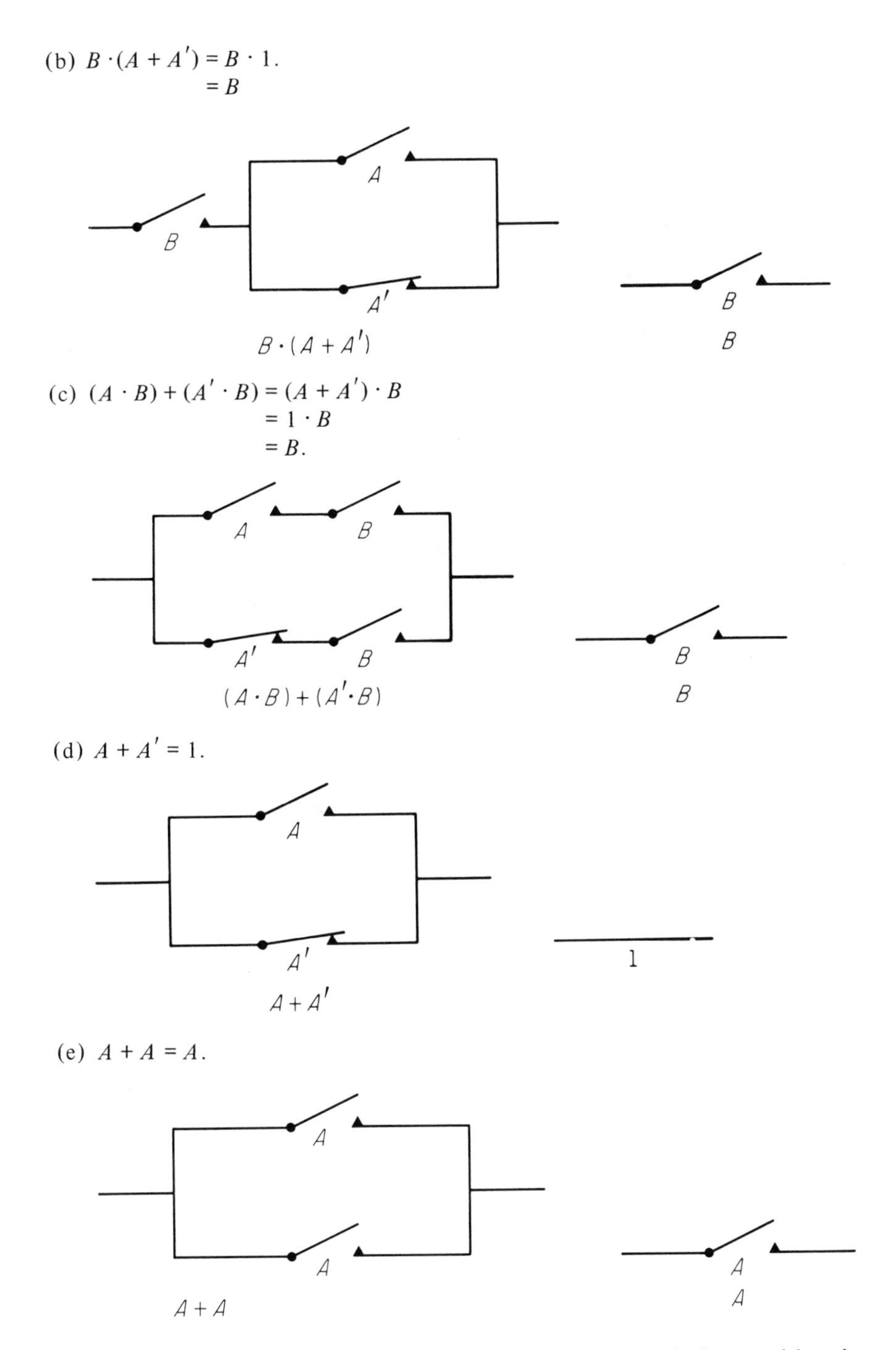

3. The circuit diagrams in a and d are equivalent, f and h are equivalent, and b and e are equivalent.

Sec. 2.7

A. 1. True. 2. False. 3. True. 4. True. 5. True.

B. 1.

$$D \times D = \left\{ \begin{array}{l} (1,1),(1,2),(1,3),(1,4),(1,5),(1,6), \\ (2,1),(2,2),(2,3),(2,4),(2,5),(2,6), \\ (3,1),(3,2),(3,3),(3,4),(3,5),(3,6), \\ (4,1),(4,2),(4,3),(4,4),(4,5),(4,6), \\ (5,1),(5,2),(5,3),(5,4),(5,5),(5,6), \\ (6,1),(6,2),(6,3),(6,4),(6,5),(6,6) \end{array} \right\}.$$

There are 36 elements in $D \times D$.

2. $A \times B = \{(1,1),(1,2),(2,1),(2,2),(3,1),(3,2)\}$
   $B \times A = \{(1,1),(2,1),(1,2),(2,2),(1,3),(2,3)\}$.
3. (a) $A \times A = \{(x,x),(x,y),(y,x),(y,y)\}$.
   (b) $A \times A \times A = \{(x,x,x),(x,x,y),(x,y,x),(x,y,y),(y,x,x),$
   $(y,x,y),(y,y,x),(y,y,y)\}$.
   There are four elements in $A \times A$.
   There are eight elements in $A \times A \times A$.
   There are 2^n elements in the Cartesian product of n sets of A.
4. The Cartesian product between any two sets is not commutative.
   Let $A = B = \{1,2\}$, then $A \times B = B \times A$.
   Let $A = \{1,2\}$, $B = \{a,b\}$, then $A \times B \neq B \times A$.
5. Let $A = \{1,2,3,4,5,6,7\}$, $B = \{a\}$. Then $n(A \times B) = 7$.
6. Let $A = \{a,b,c\}$, $B = \{x,y\}$, then $n(A \times B) = 6$. Or let $A = \{a,b,c,d,e,f\}$, $B = \{1\}$, then $n(A \times B) = 6$.
7. (a) $C \times D = \left\{ \begin{array}{l} (H,1),(H,2),(H,3),(H,4),(H,5),(H,6) \\ (T,1),(T,2),(T,3),(T,4),(T,5),(T,6) \end{array} \right\}.$
   (b) $D \times C = \left\{ \begin{array}{l} (1,H),(2,H),(3,H),(4,H),(5,H),(6,H) \\ (1,T),(2,T),(3,T),(4,T),(5,T),(6,T) \end{array} \right\}.$
8. $2^5 \times 3^5 = 7776$.

Sec. 3.2

A. 1. M = milkshakes
H = hamburgers
75 bought only hamburgers.

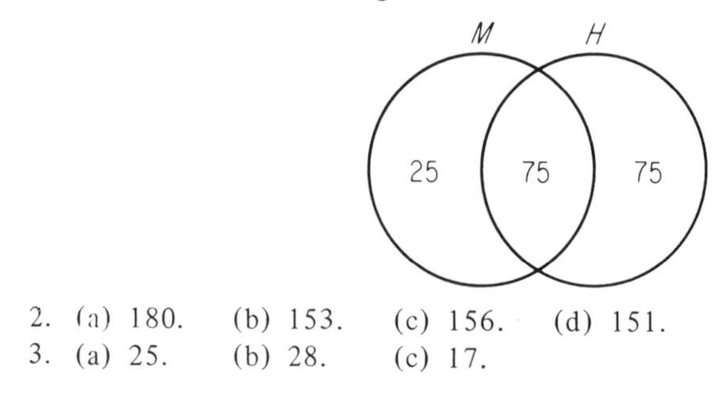

2. (a) 180. (b) 153. (c) 156. (d) 151.
3. (a) 25. (b) 28. (c) 17.

4. (a) 46. (b) 38. 5. 20.
6. (a) 0. (b) 55. (c) 180. (d) 180. (e) 165.
7. (a) 60. (b) 130. (c) 45. (d) 440. (e) 105.
30 instructors are needed.

B. 1. $\mathfrak{U} = A \cup \widetilde{A}$ and $A \cap \widetilde{A} = \emptyset$.
Hence, $n(A \cup \widetilde{A}) = n(A) + n(\widetilde{A})$, but $n(A \cup \widetilde{A}) = n(\mathfrak{U})$.
Thus $n(\mathfrak{U}) = n(A) + n(\widetilde{A})$, $n(\widetilde{A}) = n(\mathfrak{U}) - n(A)$.

2. (b) $n[(A \cup B) \cup C] = n(A \cup B) + n(C) - n[(A \cup B) \cap C]$.
$= n(A) + n(B) - n(A \cap B) + n(C) - n[(A \cup B) \cap C]$.
But $(A \cup B) \cap C = (A \cap C) \cup (B \cap C)$, so
$n(A \cup B \cup C) = n(A) + n(B) + n(C) - n(A \cap B) - [n(A \cap C) + n(B \cap C) - n(A \cap B \cap C)]$
$= n(A) + n(B) + n(C) - n(A \cap B) - n(B \cap C) - n(A \cap C) + n(A \cap B \cap C)$

3. $\widetilde{(A \cup B)} \cup (A \cup B) = \mathfrak{U}$, $(A \cup B) \cap \widetilde{(A \cup B)} = \emptyset$, and $\widetilde{(A \cup B)} = \widetilde{A} \cap \widetilde{B}$.
$n\widetilde{(A \cup B)} + n(A \cup B) = n(\mathfrak{U})$
$n(\widetilde{A} \cap \widetilde{B}) + n(A \cup B) = n(\mathfrak{U})$
$n(\widetilde{A} \cap \widetilde{B}) = n(\mathfrak{U}) - n(A \cup B)$.

Sec. 3.3

A. 1. 6; $xyz, xzy, yxz, yzx, zxy, zyx$.
3. $\underline{9} \times \underline{10} \times \underline{10} \times \underline{10} = 9 \times 10^3 = 9{,}000$.
4. 720. 5. (a) $2^3 = 8$. (b) $2^5 = 32$. (c) 2^{10}. (d) 2^n.
6. (a) $2^2 = 4$. (b) $2^7 = 128$. (c) 2^n.
7. (a) 24. (b) 6. (c) 2.
8. $N = \dfrac{5!}{2!3!} = \dfrac{5 \cdot 4 \cdot 3!}{2!3!} = 10$. 9. 180.
10. (a) $\dfrac{9!}{3!2!4!} = 1260$. (b) 35. (c) 420.

B. 1. (a) 210. (b) 840. (c) 1. (d) 24.
(e) $\dfrac{52!}{47!} = \dfrac{52 \cdot 51 \cdot 50 \cdot 49 \cdot 48 \cdot 47!}{47!} = 52 \cdot 51 \cdot 50 \cdot 49 \cdot 48$.
2. $(n-1)!$ 3. $2 \times 6! = 1440$. 4. 1680.
5. $26 \cdot 25 \cdot 10 \cdot 9 \cdot 8 = 468{,}000$.

Sec. 3.4

A. 1. (a) 1. (b) 4. (c) 6. (d) 4. (e) 1.
(g) $\binom{4}{0,4} + \binom{4}{1,3} + \binom{4}{2,2} + \binom{4}{3,1} + \binom{4}{4,0}$.

(h) $2^4 = \binom{4}{0,4} + \binom{4}{1,3} + \binom{4}{2,2} + \binom{4}{3,1} + \binom{4}{4,0}$.

2. (a) 45. (b) 1,287. 3. (a) 28. (b) 8. 4. 126. 5. $\dfrac{52!}{7!45!}$.

7. (a) 3. (b) 2. (c) 1. 8. 15;5. 9. 28. 10. 10.

B. 1. ${}_nC_{n-1} = \binom{n}{n-1,1} = \dfrac{n!}{1!(n-1)!} = \dfrac{n(n-1)!}{1!(n-1)!} = n$.

2. (a) 35. (b) 35. (c) $\dfrac{52 \cdot 51 \cdot 50 \cdot 49 \cdot 48}{5 \cdot 4 \cdot 3 \cdot 2 \cdot 1}$. (d) $\dfrac{52 \cdot 51 \cdot 50 \cdot 49 \cdot 48}{5 \cdot 4 \cdot 3 \cdot 2 \cdot 1}$.

3. ${}_nC_r \times r! = \dfrac{n!}{r!(n-r)!} \times r!$

$$= \frac{n!}{(n-r)!}$$

$$= {}_nP_r.$$

5. $\dfrac{n!}{n} = \dfrac{n(n-1)!}{n} = (n-1)!$

7. Show that $n \times {}_{n-1}C_{k-1} = k \times {}_nC_1$.

PROOF: $n \times {}_{n-1}C_{k-1} = n \times \dfrac{(n-1)!}{(k-1)!\,[(n-1)-(k-1)]\,!}$

$$= \frac{n!}{(k-1)!\,(n-1-k+1)!}$$

$$= \frac{n!}{(k-1)!\,(n-k)!}$$

$$= \frac{k \cdot n!}{k(k-1)!\,(n-k)!}$$

$$= \frac{k \cdot n!}{k!(n-k)!}$$

$$= k \times {}_nC_k.$$

Sec. 3.5

A. 1. (a) $p^4 + 4p^3q + 6p^2q^2 + 4pq^3 + q^4$.

(b) $(x-y)^7 = x^7 - 7x^6y + 21x^5y^2 - 35x^4y^3 + 35x^3y^4 - 21x^2y^5 + 7xy^6 - y^7$.

(c) $(2a+b^2)^5 = \binom{5}{5,0}(2a)^5 + \binom{5}{4,1}(2a)^4(b^2) + \binom{5}{3,2}(2a)^3(b^2)^2$

$$+ \binom{5}{2,3}(2a)^2(b^2)^3 + \binom{5}{1,4}(2a)(b^2)^4 + \binom{5}{0,5}(b^2)^5$$

$$= 32a^5 + 80a^4b^2 + 80a^3b^4 + 40a^2b^6 + 10ab^8 + b^{10}.$$

(d) $(3r-4t)^3 = 27r^3 - 108r^2t + 144rt^2 - 64t^3$.

2. $(1+0.05)^{20} = \binom{20}{20,0}1^{20} + \binom{20}{19,1}1^{19}(0.05)^1 + \binom{20}{18,2}1^{18}(0.05)^2$.

First term Second term Third term

First term is $\binom{20}{20,0}1^{20} = 1.$

Second term is $\binom{20}{19,1}1^{19}(0.05) = 1.$

Third term is $\binom{20}{18,2}1^{18}(0.05)^2 = 0.475.$

4. ${}_{11}C_7 = \binom{11}{7,4} = 330.$

5. $(10+3)^4 = 10^4 + \binom{4}{3,1}10^3(3) + \binom{4}{2,2}10^2(3)^2 + \binom{4}{1,3}10(3)^3 + 3^4$

$= 10{,}000 + 12{,}000 + 5{,}400 + 1{,}080 + 81$
$= 28{,}561.$

6. (a) −84. (b) 14. (c) 280. (d) −560. 7. 40.

B. 1.

$${}_0C_0$$
$${}_1C_0 \quad {}_1C_1$$
$${}_2C_0 \quad {}_2C_1 \quad {}_2C_2$$
$${}_3C_0 \quad {}_3C_1 \quad {}_3C_2 \quad {}_3C_3$$
$${}_4C_0 \quad {}_4C_1 \quad {}_4C_2 \quad {}_4C_3 \quad {}_4C_4$$
$${}_5C_0 \quad {}_5C_1 \quad {}_5C_2 \quad {}_5C_3 \quad {}_5C_4 \quad {}_5C_5$$

2. 1, 6, 15, 20, 15, 6, 1
1, 7, 21, 35, 35, 21, 7, 1.
3. (a) $n = 10$. (b) $n = 3$.
4. 1, 9, 36, 84, 126, 126, 84, 36, 9, 1.
5. $(1+1)^n = {}_nC_0\,1^n + {}_nC_1\,1^{n-1}(1) + {}_nC_2\,1^{n-2}(1)^2 + \cdots + {}_nC_{n-1}\,(1)^1(1)^{n-1} + {}_nC_n(1)^n$

$2^n = {}_nC_0 + {}_nC_1 + {}_nC_2 + \cdots + {}_nC_{n-2} + {}_nC_{n-1} + {}_nC_n.$

Sec. 4.2

A. 1. (a) $\{(H,H,H),(H,H,T),(H,T,H),(H,T,T),(T,H,H),(T,H,T),(T,T,H),(T,T,T)\}$.
(b) $\{1,2,3,4,5,6\}$.
(c) $\{(1,H),(2,H),(3,H),(4,H),(5,H),(6,H),(1,T),(2,T),(3,T),(4,T),(5,T),(6,T)\}$.
2. (b) $\{2,3,4,5,6,7,8,9,10,11,12\}$.
3. (a) 5/13. (b) 0. (c) 8/13.
4. (a) 2/7. (b) 1/7. (c) 3/7.
5. (a) 3/4. (b) 7/8. (c) 31/32. (d) $\dfrac{2^n - 1}{2^n}$.
6. (a) 1/8. (b) 1/8. (c) 1/8. (d) 7/8. (e) 4/8 = 1/2. (f) 3/8.
7. $S = \{3,4,5,6,7\}$.
$w(3) = 2/12$, $w(4) = 2/12$, $w(5) = 4/12$, $w(6) = 2/12$, $w(7) = 2/12$.
$S = \{2,3,4,5,6,7,8\}$.
$w(2) = 1/16$, $w(3) = 2/16$, $w(4) = 3/16$, $w(5) = 4/16$.
$w(6) = 3/16$, $w(7) = 2/16$, $w(8) = 1/16$.

8. Let x represent the number of red marbles, then $3x$ represents the number of black marbles.

$$3x + x = 16$$
$$4x = 16$$
$$x = 4$$
$$3x = 12$$

(a) $12/16 = 3/4$. (b) 0. (c) 1.

9. (a) $4/52 = 1/13$. (b) $26/52 = 1/2$. (c) $36/52 = 9/13$. (d) $12/52 = 3/13$.

10. Let $w(x) = 1/6$, $w(y) = 2/6$, $w(z) = 3/6$. The measure of a set A, $m(A)$, is the sum of the weights of the elements in the set.

$m(\emptyset) = 0, m(\{x\}) = 1/6, m(\{y\}) = 2/6, m(\{z\}) = 3/6$
$m(\{x,y\}) = 3/6, m(\{x,z\}) = 4/6, m(\{y,z\}) = 5/6$
$m(\{x,y,z\}) = 1$.

B. 1. No.

2. $S = \{s, b\}$ where s represents the pyramid landing on its side, and b represents the pyramid landing on its base. There is no prior way of assigning weights to s and b in the sample space.

3. $S = \{(T),(H,T),(H,H,T),(H,H,H)\}$.
$w((T)) = 1/2, w((H,T)) = 1/4, w((H,H,T)) = 1/8, w((H,H,H)) = 1/8$.
$w((H,H,H)) = 1/8$.

4. $S = \{B,E,O\}$.
$w(B) = 1/6$, $w(E) = 2/6$, $w(O) = 3/6$.

5. (a) 1/6. (b) 3/5. (c) 2/5. (6) 1/6.

Sec. 4.3

A. 1. 1/3.

2. Let E_1 represent the event that an even number appears on the roll of a die and let E_2 represent the event that a number divisible by 3 appears when a die is rolled.

$$E_1 = \{2,4,6\}, E_2 = \{3,6\}, E_1 \cap E_2 = \{6\}.$$
$$\begin{aligned}\Pr(E_1 \cup E_2) &= \Pr(E_1) + \Pr(E_2) - \Pr(E_1 \cap E_2)\\ &= 3/6 + 2/6 - 1/6\\ &= 4/6\\ &= 2/3.\end{aligned}$$

3. (a) $8/52 = 2/13$. (b) $26/52 = 1/2$. (c) 11/13.

4. (a) 29/30. (b) 1/30.

5. 11/26.

6. (a) 2/9. (b) 1/9. (c) 1/9. (d) 1/18. (e) 17/18.

7. (a) 0.5. (b) 0.3.

8. 1 to 2.

9. (a) 3 to 2. (b) 2 to 3.

10. (a) 3/1000. (b) 17/1000. (c) 32/1000. (d) 968/1000.

11. (a) 1/2. (b) 8/9. (c) 5/12. (d) 7/12. (e) 2/9.

Sec. 4.4

A. 1. (a) 1/10. (b) 65/100. (c) 35/100. (d) 5/11. (e) 2/9.

2. (a) 5/6. (b) 11/21.

3.

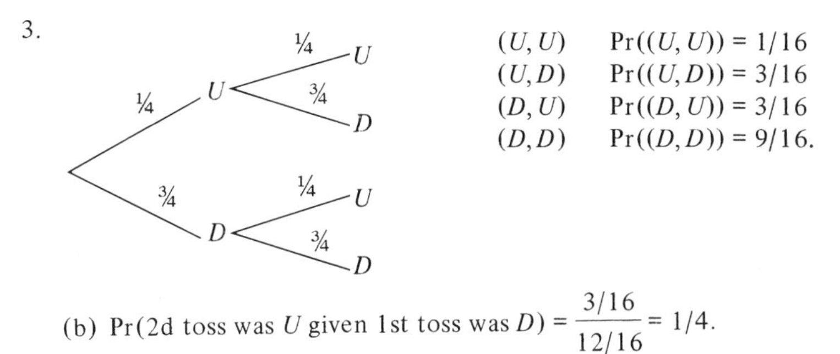

(U,U) $\Pr((U,U)) = 1/16$
(U,D) $\Pr((U,D)) = 3/16$
(D,U) $\Pr((D,U)) = 3/16$
(D,D) $\Pr((D,D)) = 9/16.$

(b) $\Pr(\text{2d toss was } U \text{ given 1st toss was } D) = \dfrac{3/16}{12/16} = 1/4.$

4. (a) Pr(all 3 are blue) = 60/12144
Pr(all 3 are red) = 120/12144
Pr(all 3 are yellow) = 504/12144
Pr(all 3 are white) = 24/12144.

(b) $\dfrac{11436}{12144}$.

5. (a) 1. (b) 0.
6. (a) 3/4. (b) 7/12. (c) 1/3. (d) 2/5. (e) 1/2. (f) 3/7.
7. (a) $\Pr(\text{all 5 spades}) = \left(\frac{13}{52}\right)\left(\frac{12}{51}\right)\left(\frac{11}{50}\right)\left(\frac{10}{49}\right)\left(\frac{9}{48}\right).$
8. 38/132,4. 9. 1/420.

Sec. 4.5

A. 1. No, since $\Pr(A \cap B) \neq \Pr(A) \cdot \Pr(B)$.
2. (a) 1/2. (b) 1/2. (c) 1/4. (d) 3/4.
The events E, F are independent.
3.
$$E = \left\{\begin{matrix}(2,1),(2,2),(2,3),(2,4),(2,5),(2,6),\\(4,1),(4,2),(4,3),(4,4),(4,5),(4,6),\\(6,1),(6,2),(6,3),(6,4),(6,5),(6,6)\end{matrix}\right\}.$$
$$F = \left\{\begin{matrix}(1,1),(1,3),(1,5),\\(2,2),(2,4),(2,6),\\(3,1),(3,3),(3,5),\\(4,2),(4,4),(4,6),\\(5,1),(5,3),(5,5),\\(6,2),(6,4),(6,6)\end{matrix}\right\}.$$
$E \cap F = \{(2,2),(2,4),(2,6),(4,2),(4,4),(4,6),(6,2),(6,4),(6,6)\}$.
$\Pr(E) = 1/2$, $\Pr(F) = 1/2$, $\Pr(E \cap F) = 1/4$.
Since $\Pr(E \cap F) = \Pr(E) \cdot \Pr(F)$, the events E,F are independent.
4. $$\begin{aligned}\Pr(\text{target is hit}) &= (1/3)(2/5) + (1/3)(3/5) + (2/3)(2/5)\\ &= 2/15 + 3/15 + 4/15\\ &= 9/15\\ &= 3/5.\end{aligned}$$
5. (a) Yes. (b) Yes. (c) Yes. (d) No.
6. (a) Yes. (b) Yes. (c) Yes. (d) Yes.

B. 1. **PROOF**: Let A, B be independent events with nonzero probability. (To show A, B are not mutually exclusive.) Since A and B are independent, $\Pr(A \cap B) =$

$\Pr(A) \cdot \Pr(B)$. Also, $\Pr(A) \neq 0$ and $\Pr(B) \neq 0$ by hypothesis. Thus $\Pr(A \cap B) \neq 0$. Since $\Pr(A \cap B) \neq 0$, then $A \cap B \neq \emptyset$. Hence A, B are not mutually exclusive.

2. Let $\Pr(A \mid B) = \Pr(A)$. By definition, $\Pr(A \mid B) = \dfrac{\Pr(A \cap B)}{\Pr(B)}$. Now $\Pr(B) \neq 0$, otherwise $\Pr(A \mid B)$ is not defined. Thus $\Pr(A \mid B) \cdot \Pr(B) = \Pr(A \cap B)$. By hypothesis $\Pr(A) \neq 0$, so $\Pr(B) = \dfrac{\Pr(A \cap B)}{\Pr(A)}$. But by definition $\dfrac{\Pr(A \cap B)}{\Pr(A)} = \Pr(B \mid A)$, therefore, $\Pr(B \mid A) = \Pr(B)$.
3. (a) **PROOF**: Let A and B be independent events. Then

   $\Pr(A \mid B) = \Pr(A)$ and $\Pr(B \mid A) = \Pr(B)$. Also, $\Pr(A \cap B) = \Pr(A) \cdot \Pr(B)$. Now

   $\Pr(A) = \Pr(A \cap B) + \Pr(A \cap \tilde{B})$

   $\Pr(A) = \Pr(A) \cdot \Pr(B) + \Pr(A \cap \tilde{B})$

   $\Pr(A \cap \tilde{B}) = \Pr(A) - \Pr(A) \cdot \Pr(B)$

   $\Pr(A \cap \tilde{B}) = \Pr(A)\,(1 - \Pr(B))$

   $\Pr(A \cap \tilde{B}) = \Pr(A) \cdot \Pr(\tilde{B})$. Thus, A and $\tilde{B}$ are independent.

   (b) $\Pr(\tilde{B}) = \Pr(A \cap \tilde{B}) + \Pr(\tilde{A} \cap \tilde{B})$, by (a) we have

   $\Pr(A \cap \tilde{B}) = \Pr(A) \cdot \Pr(\tilde{B})$

   $\Pr(\tilde{A} \cap \tilde{B}) = \Pr(\tilde{B}) - \Pr(A) \cdot \Pr(\tilde{B})$

   $\Pr(\tilde{A} \cap \tilde{B}) = \Pr(\tilde{B})\,(1 - \Pr(A))$

   $\Pr(\tilde{A} \cap \tilde{B}) = \Pr(\tilde{B}) \Pr(\tilde{A})$. Hence $\tilde{A}$ and $\tilde{B}$ are independent.

Sec. 4.6

A. 1. (a) Not binomial, since the number of trials is not specified.
   (b) Not binomial. (c) Binomial. (d) Not binomial.

2. (a) $b(2;4,1/2) = \binom{4}{2,2}\left(\frac{1}{2}\right)^2\left(\frac{1}{2}\right)^2 = 3/8.$ (b) 1/4. (c) 11/16.

3. $\Pr(\text{exactly 3 baskets in 5 shots}) = b\left(3;5,\frac{8}{10}\right)$

$$= \frac{2048}{10^4} \doteq \frac{2}{10}.$$

4. (a) $b(1;4,1/4) = \binom{4}{1,3}\left(\frac{1}{4}\right)^1\left(\frac{3}{4}\right)^3$

$$= 4\left(\frac{1}{4}\right)\left(\frac{27}{64}\right)$$

$$= \frac{27}{64}.$$

   (b) 27/128. (c) 3/64. (d) 1/256. (e) 175/256.

5. (a) $b(6;6,8/10) = (8/10)^6$. (b) $b(3;6,8/10)$.
   (c) $\Pr(\text{at least 4 arrive on schedule}) = b(4;6,8/10) + b(5;6,8/10) + b(6;6,8/10)$.
   (d) $1 - b(5;6,8/10) - b(6;6,8/10)$. (e) $b(6;6,2/10)$.

B. 1. (a) 3/8. (b) 3/64. (c) 3/64.

2. $b(k;n,p) = \binom{n}{k,n-k} p^k(1-p)^{n-k}$

$= \binom{n}{n-k,k} (1-p)^{n-k}p^k$

$= b(n-k;n,1-p).$

3. (a) 6/7. (b) 1/7.

Sec. 4.7

A. 1. (a) 1/7. (b) 1/4. (c) 1/2. 3. 19/60. 5. 0.425.

Sec. 4.8

A. 1. 2.

2. If $E(x)$ represents the expected number of men, then $E(x) = 2.8$.

3. No. Give player $14 if a 2 comes up.

4. $E(x) = \frac{1}{1000}(2495) - 5\left(\frac{999}{1000}\right)$

$= \frac{2495}{1000} - \frac{4995}{1000}$

$= \frac{-2500}{1000}$

$= -2.5$. The tickets are overpriced, since it is not a fair game.

5. No.

6. (a) $1 (the game is in favor of the player).
(b) −$1. (c) $4. (d) $k + 1$. (e) $4. (f) −$3. (g) $k(1)$.

B. 1. By definition of expected value, $E(x)$, we have $E(x) = a_1p_1 + a_2p_2 + \cdots + a_np_n$. Since each outcome is multiplied by k, each probability will be multiplied by k. *Note*: x represents an arbitrary outcome in the experiment. Thus,

$$\begin{aligned} E(kx) &= a_1(kp_1) + a_2(kp_2) + \cdots + a_n + (kp_n) \\ &= k(a_1p_1) + k(a_2p_2) + \cdots + k(a_np_n) \\ &= k(a_1p_1 + a_2p_2 + \cdots + a_np_n) \\ &= kE(x). \end{aligned}$$

2. Let the outcomes of an experiment be $a_1, a_2, \ldots, a_n$ where $p_1, p_2, \ldots, p_n$ are the respective probabilities for these outcomes.

$$\begin{aligned} E(k + x) &= (a_1 + k)p_1 + (a_2 + k)p_2 + \cdots + (a_n + k)p_n \\ &= a_1p_1 + kp_1 + a_2p_2 + kp_2 + \cdots + a_np_n + kp_n \\ &= (a_1p_1 + a_2p_2 + \cdots + a_np_n) + (kp_1 + kp_2 + \cdots + kp_n) \\ &= E(x) + k(p_1 + p_2 + \cdots + p_n). \end{aligned}$$

But

$$p_1 + p_2 + \cdots + p_n = 1$$

hence

$$E(k + x) = E(x) + k.$$

Sec. 5.2

A. 1. (a) Discrete. (b) Continuous. (c) Continuous. (d) Discrete.
(e) Discrete.

B. 1. (a)

Interval	Frequency
95- 99	1
100–104	3
105–109	2
110–114	3
115–119	4
120–124	2
125–129	2
130–134	2
135–139	4
140–144	0
145–149	2
150–154	3
155–159	1
160–164	1

(b)
(c)

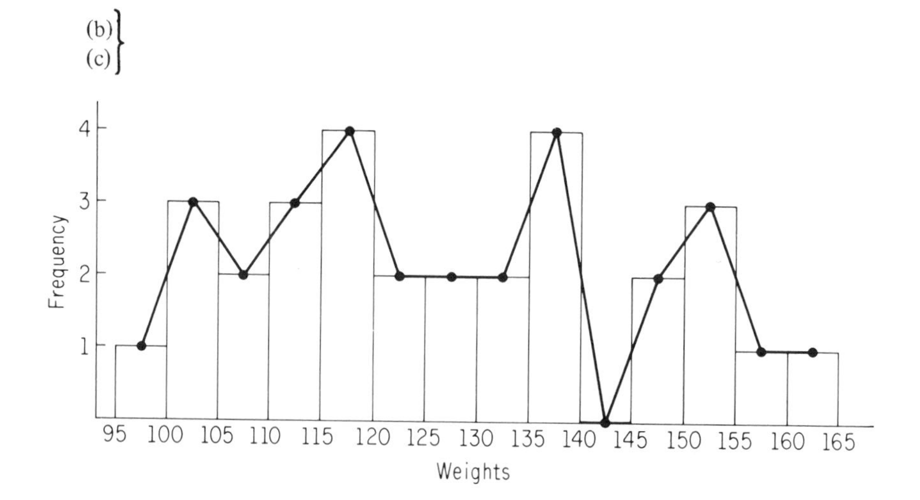

2. (a)

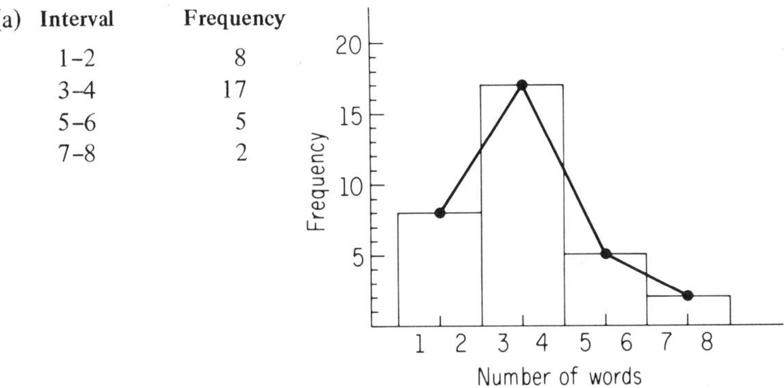

Interval	Frequency
1–2	8
3–4	17
5–6	5
7–8	2

(b)

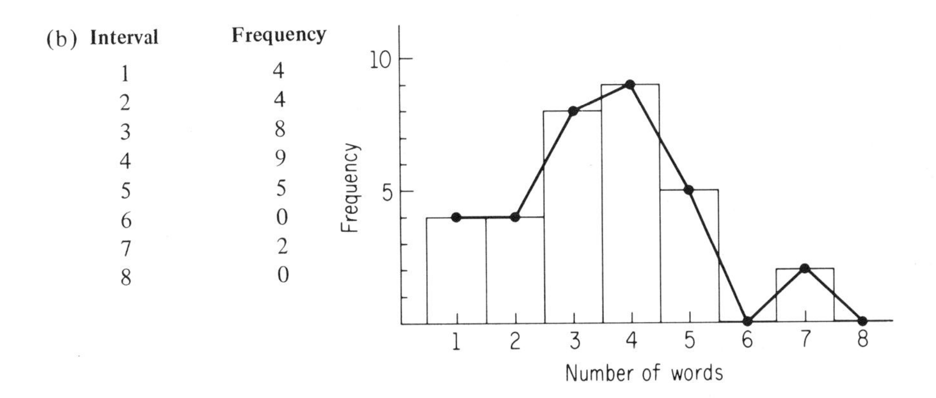

Interval	Frequency
1	4
2	4
3	8
4	9
5	5
6	0
7	2
8	0

3.

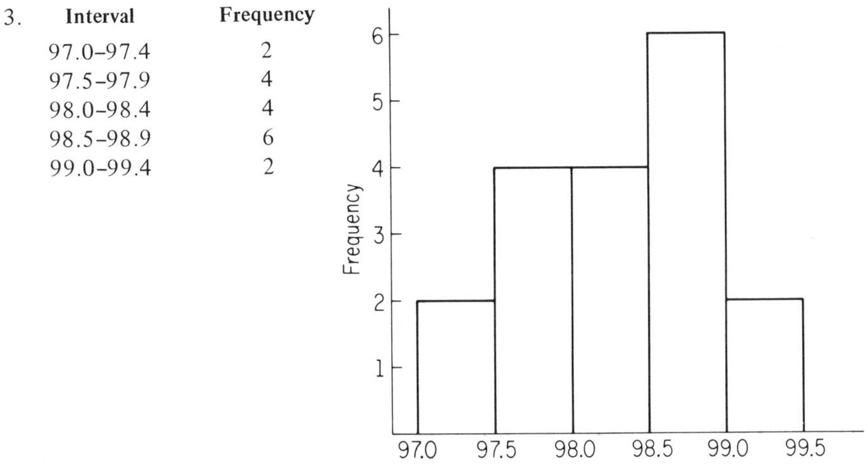

Interval	Frequency
97.0–97.4	2
97.5–97.9	4
98.0–98.4	4
98.5–98.9	6
99.0–99.4	2

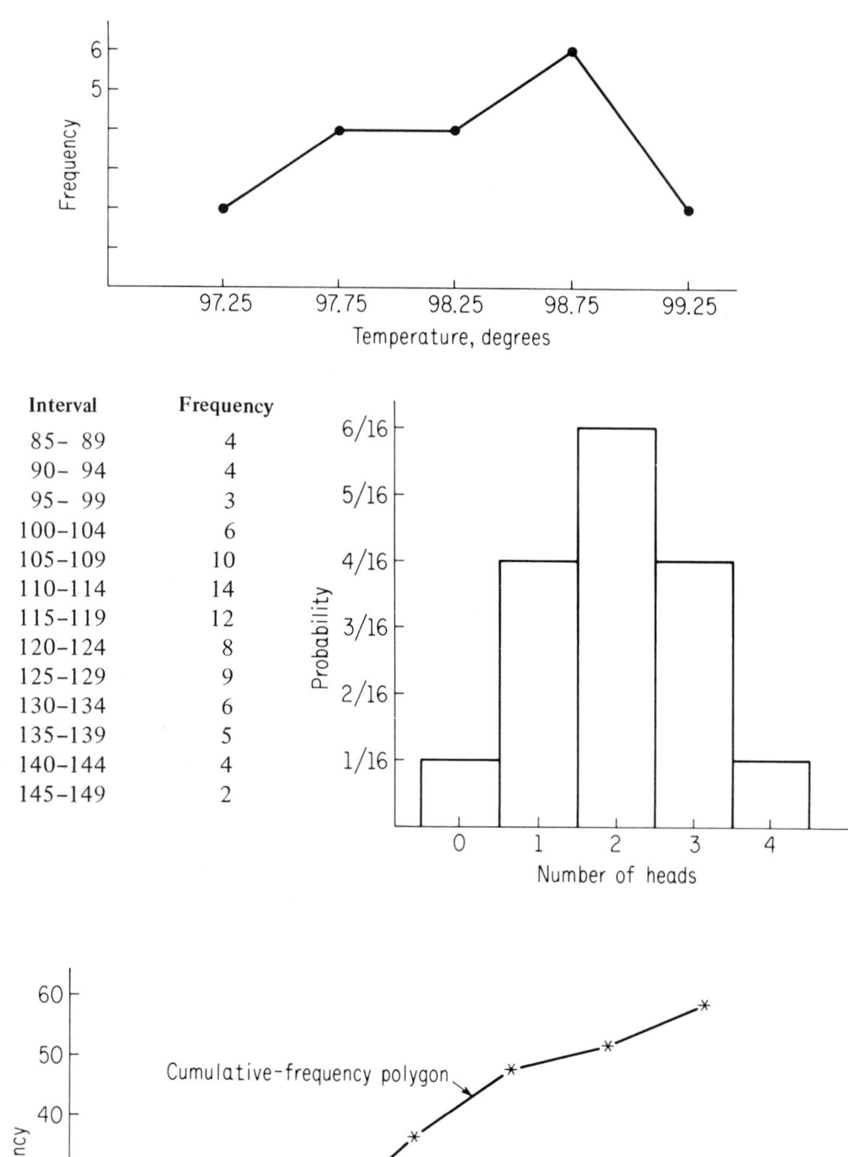

6.

Interval	Frequency
85– 89	4
90– 94	4
95– 99	3
100–104	6
105–109	10
110–114	14
115–119	12
120–124	8
125–129	9
130–134	6
135–139	5
140–144	4
145–149	2

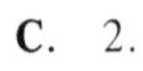
C. 2.

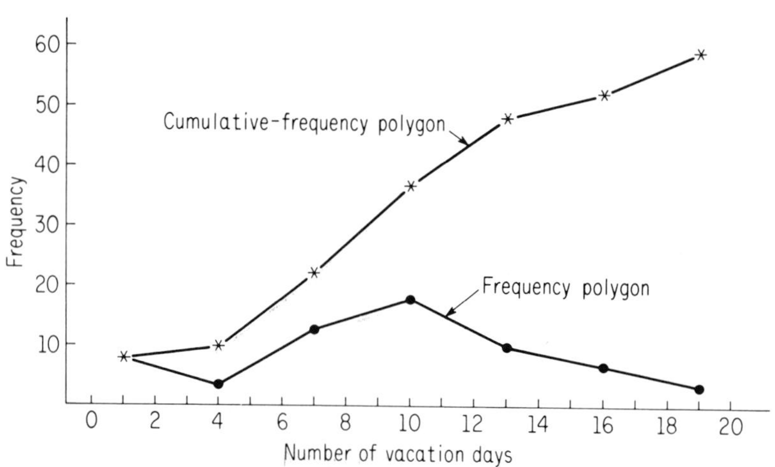

Sec. 5.3

A. 1. 68 in.

2. $\bar{x}_w = \dfrac{x_1 w_1 + x_2 w_2 + \cdots + x_n w_n}{w_1 + w_2 + \cdots + w_n}$, $\quad \bar{x} = \dfrac{x_1 + x_2 + \cdots + x_n}{n}$

$\bar{x}_w = \dfrac{4(2) + 3(3) + 2(4) + 5(1)}{2 + 3 + 4 + 1}$, $\quad \bar{x} = \dfrac{2 + 3 + 4 + 1}{4}$

$\bar{x}_w = \dfrac{30}{10}$ $\quad \bar{x} = \dfrac{10}{4}$

$\underline{\bar{x}_w = 3}$ $\quad \underline{\bar{x} = 2.5}$

3. Letters a, e are the modes for the given data.
4. (a) Mean is 7. (b) Numbers 4, 10 are the modes. (c) Median is 7.
5. (a) 10,029 is the mean. (b) 10,000 is the mode. (c) 10,000 is the median.
6. (a) Mean, 5.7. Median, 5.5. (b) Mean, 5.7. Median, 5.5.
7. Mean is approximately 5.8. Median is 5.5.
8. (a) Mean is 5 2/3. Median is 5.5. (b) Mean is 8 2/3. Median is 8.5. (c) Mean is 22 2/3. Median is 22. (d) Mean of b is 3 more units than mean a. Mean of c is 4 times mean of a.
9. Samples: (2,4), (2,6), (2,8), (4,2), (4,6), (4,8), (6,2), (6,4), (6,8), (8,2), (8,4), (8,6).

   The average of all the sample means is 5. The average of the numbers 2, 4, 6, 8 is the same as the average of all the sample means of size two.
10. If the mean, mode, and median are approximately the same then the data is uniformly dispersed about the mean.

B. 1. The mean for the data in Table 5.1, Sec. 5.2 is 78. The mean found by using the data in Table 5.2, Sec. 5.2 is 78.3. The means found by the two methods are approximately the same.

2. Let $x_1, x_2, \ldots, x_n$ be the given data with mean $\bar{x} = \dfrac{x_1 + x_2 + \cdots + x_n}{n}$. Let $\bar{x}_a$ be the mean of the given data when the constant k is added to each entry of the given data. (We must show $\bar{x}_a = \bar{x} + k$.) By definition,

$$\bar{x}_a = \frac{(x_1 + k) + (x_2 + k) + \cdots + (x_n + k)}{n}$$

$$= \frac{x_1 + x_2 + \cdots + x_n + nk}{n}$$

$$= \frac{x_1 + x_2 + \cdots + x_n}{n} + \frac{nk}{n}$$

$$= \bar{x} + k.$$

3. Let $x_1, x_2, \ldots, x_n$ be the given data with mean of $\bar{x} = \dfrac{x_1 + x_2 + \cdots + x_n}{n}$. Let

$\overline{x}_c$ be the mean of the given data when each entry is multiplied by the constant k. (We must show $\overline{x}_c = k\overline{x}$.) By definition,

$$\overline{x}_c = \frac{kx_1 + kx_2 + \cdots + kx_n}{n}$$

$$= k\,\frac{x_1 + x_2 + \cdots + x_n}{n}$$

$$= k\overline{x}.$$

4. Mean is 79.

Sec. 5.4

A. 1. (a) 219.3. (b) 3,244. (c) 57.
2. (a) Mean 4, standard deviation 2.366. (b) 2.366. (c) 11.83.
(d) Adding a constant to each entry does not affect the standard deviation. Multiplying each entry by a constant multiplies the standard deviation by that constant.
3. $\frac{99}{100}, \frac{24}{25}$. 5. (a) 64. (b) 1,067. (c) 33. 6. 3. 8. 20.
9. (a) (i) 13 (ii) 13. (b) (i) 3.74 (ii) 3.74. 10. (a) 13. (b) 14.96.

Sec. 5.5

A. 1. 3.59 percent. 2. (a) 38.3 percent. (b) 68.26 percent. 4. 12.
5. 0.3643. 6. 0.6458. 7. 83. 9. (a) 1.88. (b) 1.96. (c) 2.57.

B. 1. Mathematics, 21.5 percent, English, 71 percent, Social Studies, 56 percent.
2. The tolerance space is from 4.98 to 5.02 inches corresponding to a z interval of from $z = -2$ to $z = 2$.
$\Pr(z < -2) + \Pr(z > 2) = .0456$ representing 4.56 percent rejects. Thus 674 of the 15,000 parts will be rejected.
3. Approximately 5 1/2 years.

Sec. 5.6

A. 1. $\overline{x} = 10/3$, $s = 1.49$.
2. (a) $b(3;10,1/3) + b(4;10,1/3) \doteq 0.37$ (b) 0.42. 3. 0.5752.
4. $\overline{x} = np$ $\quad s = \sqrt{npq}$ $\quad z = \frac{x - \overline{x}}{s}$, $\Pr(k \leqslant 5) \doteq \Pr(z < 1.72)$

$\overline{x} = 10(0.3)$ $\quad s = \sqrt{10(.3)(.7)}$ $\quad z = \frac{5.5 - 3}{1.45}$, $\Pr(z < 1.72) = 0.9573$

$\overline{x} = 3.$ $\quad s = 1.45.$ $\quad z = 1.72.$

5. 0.0694. 6. (a) 0.8714. (b) 0.0057.

7. Assume his claim true, then

$$\bar{x} = np$$
$$\bar{x} = (400)(0.9)$$
$$\bar{x} = 360.$$

Now $s = \sqrt{npq}$ $\qquad z = \dfrac{341.5 - 360}{6} = -\dfrac{18.5}{6} \doteq -3.1.$

$s = \sqrt{400(0.9)(0.1)}$
$s = \sqrt{36}$
$s = 6.$

So $\Pr(z < -3.1) = 0.0010$. This probability indicates that only 0.1 percent of the time would he obtain this result if his accuracy is 90 percent. His claim is exaggerated on the basis of the data.

8. 0.2578. 9. (a) 0.5557. (b) 0.0087.

Sec. 5.7

A. 1. (a) 5. (b) $\sqrt{2.5}$. (c) 5. (d) $\dfrac{\sqrt{2.5}}{2}$.

2. 2.5. The average of the sample variances is the population variance i.e, $E(s^2) = \sigma^2$.

3. (a)

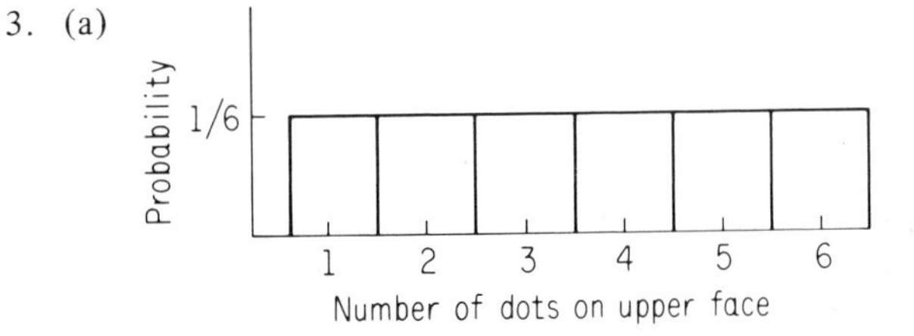

(b)

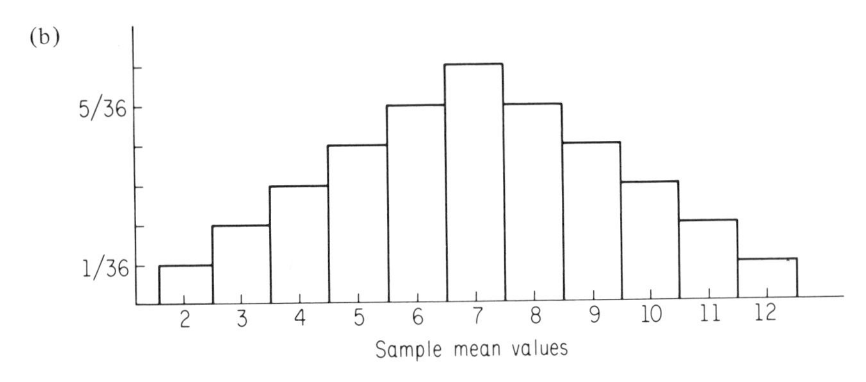

4. 0.0280, 0.0105. 5. 0.0062. 6. 0.0013, no.

7. Maximum 0.762 in., minimum 0.738 in.

B. 1. (a) 0.6826. (b) 61. 2. 256.

3. 0.0062. The average nicotine content is greater than 17.7.

Sec. 5.8

A. 1. (1) H_0: $\mu = 8000$. (2) H_1: $\mu \neq 8000$. (3) Level of significance is 0.05.
(4) $z > 1.96$ or $z < -1.96$.
(5) $z_{OBS} = \dfrac{7500 - 8000}{1600/\sqrt{100}} = -\dfrac{500}{160} = -3.13.$
(6) Since $z = -3.13$ is less than $z = -1.96$, reject H_0.

2. H_0: $\mu = 2.50$, H_1: $\mu > 2.50$
$z_{OBS} = 4.0$. Reject H_0.

3. H_0: $\mu = 74$, H_1: $\mu \neq 74$
$z_{OBS} = -1.5$. Accept H_0.

4. H_0: $\mu = 15$, H_1: $\mu < 15$.
$z_{OBS} = -1.25$. From the data, the manufacturer is meeting the standard. Accept H_0.

5. H_0: $\mu = 450$, H_1: $\mu > 450$.
$z_{OBS} = 2.5$. Reject H_0, i.e., the property taxes are significantly higher than the national average.

6. H_0: $\mu = 20$, H_1: $\mu > 20$.
$z_{OBS} = 3.53$. Reject H_0. The claim that the fertilizer gives higher crop production is justified by the information given.

7. H_0: $\mu = 48$, H_1: $\mu \neq 48$
$z_{OBS} = -1.25$ for science majors.
$z_{OBS} = 1.67$ for humanities majors. Accept H_0.

8. H_0: $\mu = 4.5$, H_1: $\mu \neq 4.5$
$z_{OBS} = -2.14$. Reject H_0.

9. H_0: $\mu = 250$, H_1: $\mu > 250$
$z_{OBS} = 2.5$. Reject H_0. The driving distance has been improved on the basis of the given data.

10. H_0: $\mu = 45$, H_1: $\mu < 45$
$z_{OBS} = -1.5$. Accept H_0. The data indicates that the cost of production has not been significantly reduced.

Sec. 6.1

A. 1. On the real line as shown below, mark off the unit distance and also $\sqrt{2}$. Let B be the point whose coordinate is one. At B construct a line perpendicular to the real line and mark off a distance of $\sqrt{2}$. Let the end point be C. Draw AC. By the Pythagorean theorm the distance AC is equal to $\sqrt{3}$ units. Now mark off AC on the real line. A similar procedure can be used for finding $-\sqrt{5}$, since $\sqrt{5} = \sqrt{1^2 + 2^2}$.

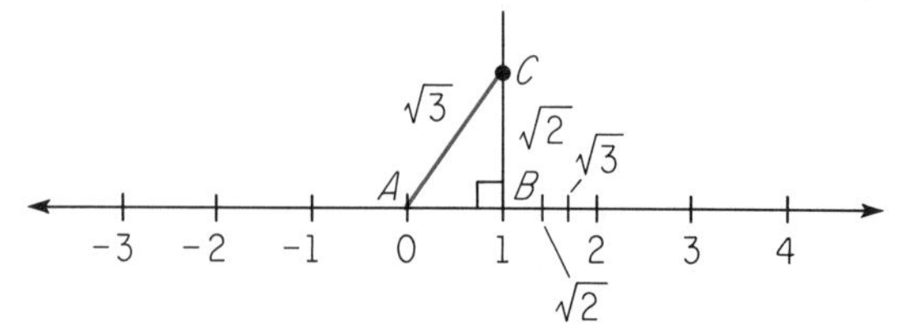

2.

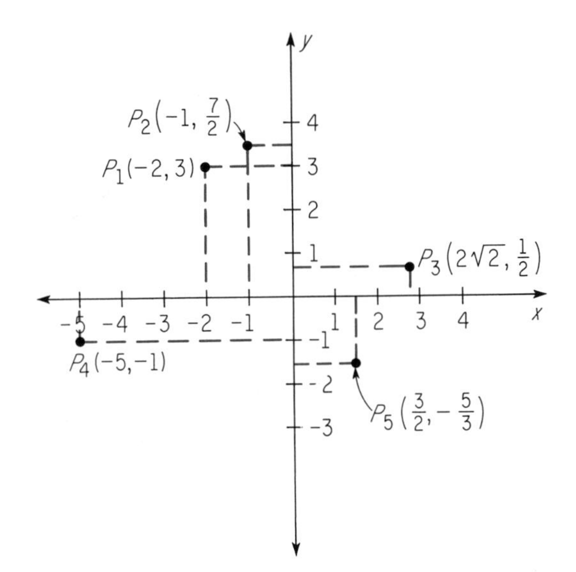
y
x
$P_2(-1, \frac{7}{2})$
$P_1(-2, 3)$
$P_3(2\sqrt{2}, \frac{1}{2})$
$P_4(-5, -1)$
$P_5(\frac{3}{2}, -\frac{5}{3})$
4
3
2
1
-1
-2
-3
-5 -4 -3 -2 -1
1 2 3 4

3. (b)

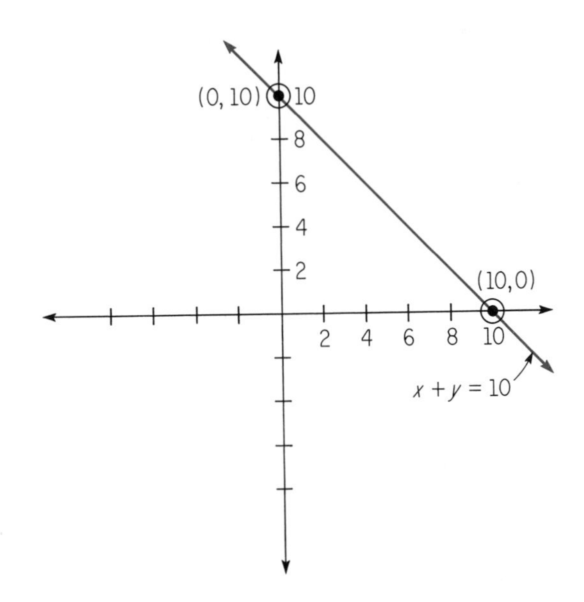
(0, 10)
10
8
6
4
2
(10,0)
2 4 6 8 10
$x + y = 10$

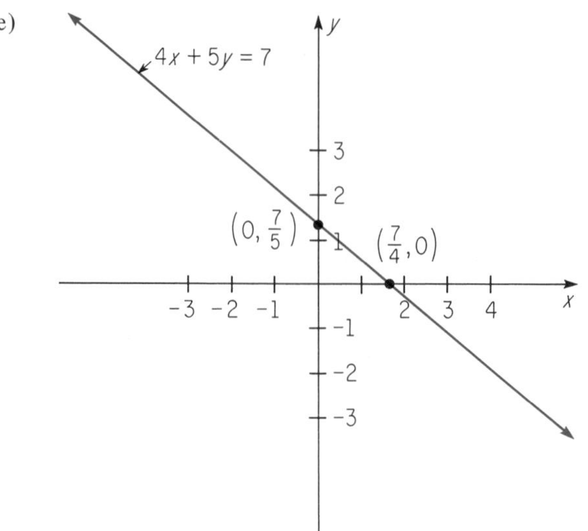

4. The y-coordinate for the x-intercept is 0. Substituting zero for y in the equation, we obtain $-2x = 6$ so that $x = -3$. Thus the coordinates of the x-intercept are $(-3,0)$. Similarly, the coordinates of the y-intercept are $(0,2)$. Using the coordinates of the x- and y-intercepts, we find that the slope is given by $m = \dfrac{2-0}{0-(-3)}$, thus $m = \dfrac{2}{3}$.

5. (a) The slope of the line determined by $(3,0)$ and $(0,1)$ is $m = \dfrac{1-0}{0-3}$. Thus $m = -\dfrac{1}{3}$. Using the slope-intercept form we obtain $y - 1 = -\dfrac{1}{3}(x - 0)$. Hence $3y - 3 = -x$, and equivalently $x + 3y = 3$ which is the desired form.

   (c) $m = 1, x - y = 0$. (e) $m = -\dfrac{2}{3}, 2x + 3y = -7$.

6. The coordinates of the y-intercept are $(0,5/3)$. There is no x-intercept. The coordinates of another point on the line are $(1,5/3)$. Using these two points we find that the slope of the line determined by the equation $3y = 5$ is zero.

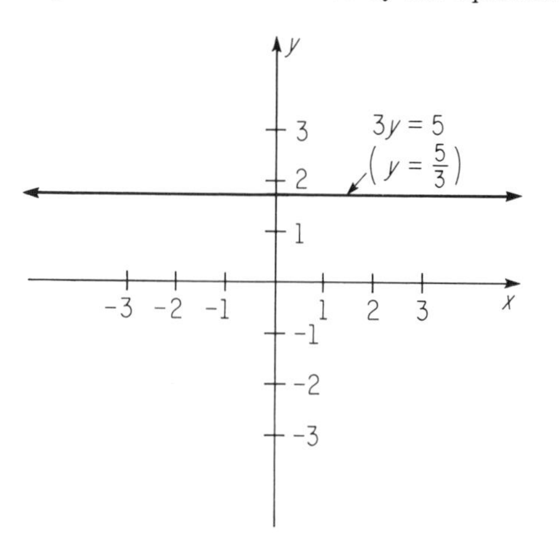

7.

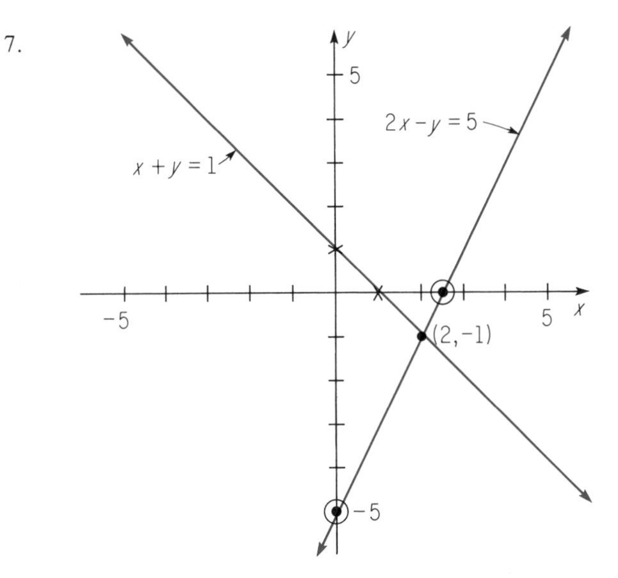

The coordinates of the point where the lines intersect is (2,−1). The coordinates of this point do satisfy each of the given equations. The point of intersection is the only point that will satisfy both of the given equations.

9. **Reasons**
   (a) Additive identity axiom.
   (b) Multiplication property of equality.
   (c) Distributive law of multiplication over addition.
   (d) Addition property of equality.
   (e) Associative law of addition.
   (f) Additive inverse axiom.
   (g) Additive identity axiom.

11.

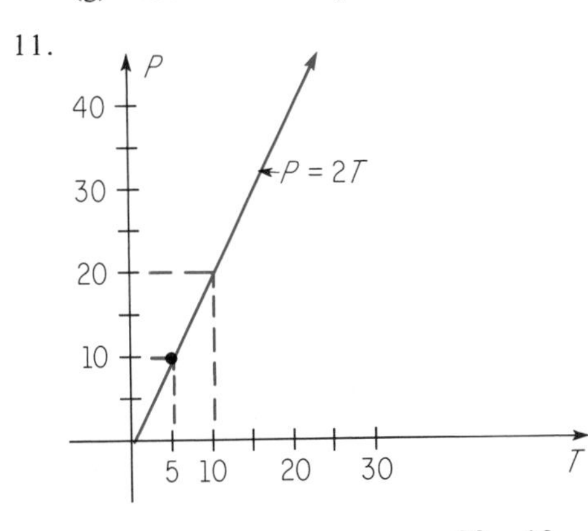

The slope of the line is $m = \dfrac{20 - 10}{10 - 5}$. Thus $m = 2$. Using the point slope form the equation of the line is

$$P - 10 = 2(T - 5)$$
$$P - 10 = 2T - 10$$
$$P = 2T.$$

Sec. 6.2

A. 1. (a) The solution set to the given system of equations is $\{(2,1)\}$. The equations are independent and the system of equations is consistent. In the above ordered pair solution, the first component is the value of x and the second component is the y-value.

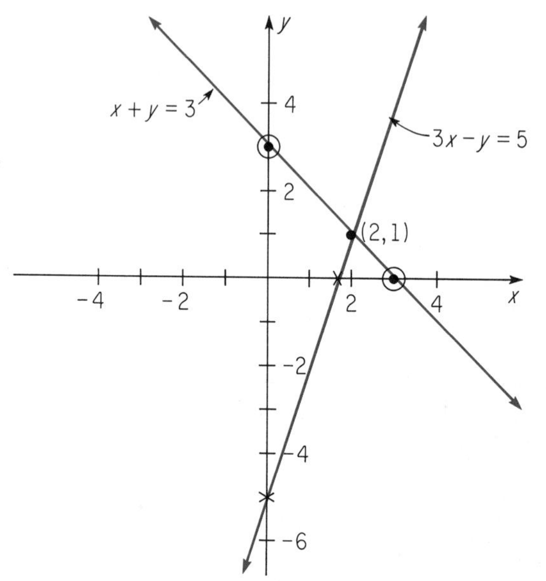

(c) From the graph, the solution set appears to be $\{(4,-1)\}$. However, this does not satisfy both equations. The exact solution, obtained by other methods is $(17/4,-3/4)$. (This solution is difficult to obtain by graphing.) The equations are independent and the system of equations is consistent.

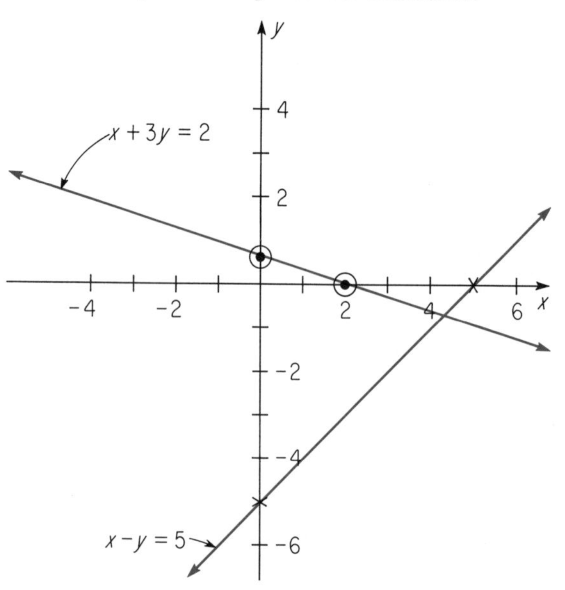

(d) The solution set to the given system of equations is empty. (The lines do not intersect.) The equations are dependent and the system of equations is inconsistent.

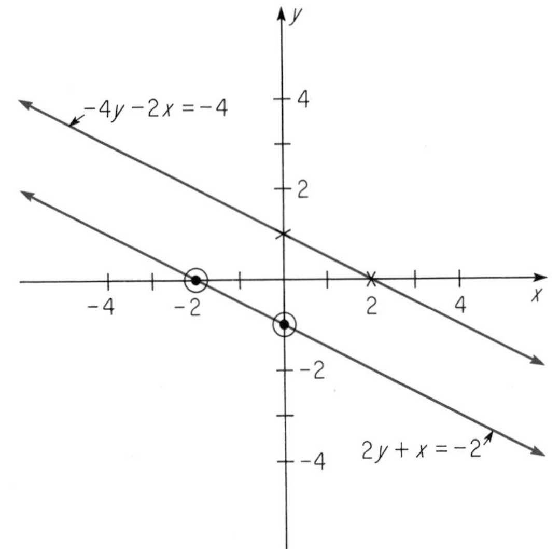

2. (a) $\begin{cases} 2x - y = 9 \\ x - 5y = 0. \end{cases}$

First, solve $x - 5y = 0$ for x which gives $x = 5y$. Then substitute for x in the equation $2x - y = 9$ to obtain

$$\begin{aligned} 2(5y) - y &= 9 \\ 10y - y &= 9 \\ 9y &= 9 \\ y &= 1. \end{aligned}$$

Now, since $y = 1$ and $x = 5y$ we obtain $x = 5$. Thus the solution set to the given system is $\{(5,1)\}$.

(b) $\begin{cases} 2x + 3y = -6 \\ 3x - 2y = 12. \end{cases}$

We can solve the equation $2x + 3y = -6$ for x to obtain

$$\begin{aligned} 2x &= -3y - 6 \\ x &= -\frac{3}{2}y - 3. \end{aligned}$$

We then substitute for x into the equation $3x - 2y = 12$ to obtain

$$\begin{aligned} 3\left(-\frac{3}{2}y - 3\right) - 2y &= 12 \\ -\frac{9}{2}y - 9 - 2y &= 12 \\ -\frac{13}{2}y &= 21 \\ -13y &= 42 \\ y &= -\frac{42}{13} \end{aligned}$$

For $y = -\frac{42}{13}$ we have

$$x = -\frac{3}{2}\ -\frac{42}{13} - 3$$

$$= +\frac{63}{13} - 3$$

$$= +\frac{24}{13}.$$

Thus the solution set to the original system of equations is $\left(\frac{24}{13}, -\frac{42}{13}\right)$.

(f) $\begin{cases} 4x - 3y + 3 = 3x - y + 6 \\ 5x - y = 2x - 4y + 2. \end{cases}$

To solve the given system of equations, we can solve $4x - 3y + 3 = 3x - y + 6$ for x to obtain

$$x - 2y = 3$$
$$x = 2y + 3.$$

We then simplify the equation $5x - y = 2x - 4y + 2$ to obtain $3x + 3y = 2$ and then substitute for x in equation $3x + 3y = 2$ to obtain

$$3(2y + 3) + 3y = 2$$
$$6y + 9 + 3y = 2$$
$$9y = -7$$
$$y = -\frac{7}{9}.$$

For $y = -\frac{7}{9}$, we have

$$x = 2\left(-\frac{7}{9}\right) + 3$$

so that $x = \frac{13}{9}$. Thus the solution set to the system of equations is

$$\left\{\left(\frac{13}{9}, -\frac{7}{9}\right)\right\}.$$

3. (a) $\begin{cases} 2y - 5x = -7 \\ -3y + 7x = 5. \end{cases}$

The following systems of equations are all equivalent

$$\begin{cases} 2y - 5x = -7 \\ -3y + 7x = 5 \end{cases} \qquad \begin{cases} -5x + 2y = -7 \\ 7x - 3y = 5 \end{cases}$$

$$\begin{cases} x - \frac{2}{5}y = \frac{7}{5} \\ 7x - 3y = 5 \end{cases} \qquad \begin{cases} x - \frac{2}{5}y = \frac{7}{5} \\ -\frac{1}{5}y = -\frac{24}{5} \end{cases} \qquad \begin{cases} x - \frac{2}{5}y = \frac{7}{5} \\ y = 24 \end{cases} \qquad \begin{cases} x = 11 \\ y = 24. \end{cases}$$

Hence $\{(x, y) | 2y - 5x = -7 \text{ and } -3y + 7x = 5\} = \{(11, 24)\}$.

(c) $\begin{cases} 2x - 3y = 1 \\ x + 5y = 9. \end{cases}$

The following systems of equations are all equivalent

$$\begin{cases} 2x - 3y = 1 \\ x + 5y = 9 \end{cases} \qquad \begin{cases} x + 5y = 9 \\ 2x - 3y = 1 \end{cases}$$

$$\begin{cases} x + 5y = 9 \\ -13y = -17 \end{cases} \qquad \begin{cases} x + 5y = 9 \\ y = \dfrac{17}{13} \end{cases} \qquad \begin{cases} x = \dfrac{32}{13} \\ y = \dfrac{17}{13} \end{cases}$$

Hence the solution set $\{(x,y)|2x - 3y = 1, x + 5y = 9\} = \left\{\left(\dfrac{32}{13}, \dfrac{17}{13}\right)\right\}$.

(f) The $\{(x,y)|x = -y, x - y = 0\} = \{(0,0)\}$.

Sec. 6.3

A. 1. (a) $\alpha + \beta = (-3,5) + (3,2)$ (b) $(-3,-2)$. (c) $(-3,-2)$.

$\alpha + \beta = (-3 + 3, 5 + 2)$

$\alpha + \beta = (0,7)$.

(d) $\alpha + (-2)\beta = (-3,5) + (-2)(3,2)$ (e) $\sqrt{34}$. (f) $\sqrt{13}$. (g) 7.

$\alpha + (-2)\beta = (-3,5) + (-6,-4)$

$\alpha + (-2)\beta = (-3 + (-6), 5 + (-4))$

$\alpha + (-2)\beta = (-9,1)$.

2. (a) 0. (b) $\sqrt{107}$. (c) $\sqrt{30}$. 4. $a \cdot b = 0, b \cdot a = 0, a \cdot b = b \cdot a$, yes.

5. Let $\alpha = (a_1 a_2, \ldots, a_n)$, then $-\alpha = (-a_1, -a_2, \ldots, -a_n)$.

$$|\alpha| = \sqrt{a_1^2 + a_2^2 + \cdots + a_n^2}$$
$$|-\alpha| = \sqrt{(-a_1)^2 + (-a_2)^2 + \cdots + (-a_n)^2}$$
$$|-\alpha| = \sqrt{a_1^2 + a_2^2 + \cdots + a_n^2}.$$

Therefore $|\alpha| = |-\alpha|$.

8. Let the river current velocity be represented by the vector $\left(0, -\dfrac{5}{2}\right)$ and the wind velocity be represented by $(5,0)$. The resultant vector of the sailboat will be the vector sum of the aforementioned vectors which is $\left(5, -\dfrac{5}{2}\right)$. The magnitude of the resultant vector is

$$\sqrt{\frac{125}{4}} \quad \text{or} \quad \frac{5\sqrt{5}}{2}.$$

9. $a \cdot b = (4,-7) \cdot (b_1, b_2)$

$= 4b_1 - 7b_2$.

But $a \cdot b = 0$, so $4b_1 - 7b_2 = 0$, thus $b_1 = \dfrac{7}{4} b_2$.

B. 2. (a) $\gamma = (5,-8)$. (b) $\gamma = (1,2)$.

3. The vector sum is the vector $(0,0)$ with magnitude zero.

Sec. 6.4

A. 1. The dimension of matrix A is 3×4.

$$a_{32} = 2, a_{13} = -1, a_{24} = -1, a_{33} = 3, a_{21} = 0.$$

2.
$$B = \begin{bmatrix} -1 & 1 & 5 & 2 & 1 \\ -2 & 1 & 4 & 2 & -3 \\ -1 & 2 & 0 & 3 & 0 \\ 3 & 5 & 5 & -1 & 5 \end{bmatrix}_{4 \times 5}.$$

5. (a)
$$AB = \begin{bmatrix} 3 & -\frac{3}{7} & 0.6 \\ 7 & -1 & 1.4 \\ -5 & \frac{5}{7} & -1.0 \end{bmatrix}_{3 \times 3}.$$
(b) $BA = [1]_{1 \times 1}$. (c) No.

6. (a)
$$AB = \begin{bmatrix} -21 & -29 \\ 32 & 24 \\ 19 & 62 \end{bmatrix}.$$
(b)
$$A((2)B) = \begin{bmatrix} -42 & -58 \\ 64 & 48 \\ 38 & 124 \end{bmatrix}.$$
(c)
$$(A(2))B = \begin{bmatrix} -42 & -58 \\ 64 & 48 \\ 38 & 124 \end{bmatrix}.$$

7.
$$\left[\begin{array}{ccc|ccc} 1 & -3 & 2 & 1 & 0 & 0 \\ 0 & 5 & -4 & 0 & 1 & 0 \\ -1 & 4 & 7 & 0 & 0 & 1 \end{array}\right] \underset{\sim}{R_1 + R_3} \left[\begin{array}{ccc|ccc} 1 & -3 & 2 & 1 & 0 & 0 \\ 0 & 5 & -4 & 0 & 1 & 0 \\ 0 & 1 & 9 & 1 & 0 & 1 \end{array}\right]$$

$$\underset{\sim}{\frac{1}{5}R_2} \left[\begin{array}{ccc|ccc} 1 & -3 & 2 & 1 & 0 & 0 \\ 0 & 1 & -\frac{4}{5} & 0 & \frac{1}{5} & 0 \\ 0 & 1 & 9 & 1 & 0 & 1 \end{array}\right] \underset{\sim}{-R_2 + R_3} \left[\begin{array}{ccc|ccc} 1 & -3 & 2 & 1 & 0 & 0 \\ 0 & 1 & -\frac{4}{5} & 0 & \frac{1}{5} & 0 \\ 0 & 0 & \frac{49}{5} & 1 & -\frac{1}{5} & 1 \end{array}\right]$$

$$\underset{\sim}{\frac{5}{49}R_3} \left[\begin{array}{ccc|ccc} 1 & -3 & 2 & 1 & 0 & 0 \\ 0 & 1 & -\frac{4}{5} & 0 & \frac{1}{5} & 0 \\ 0 & 0 & 1 & \frac{5}{49} & -\frac{1}{49} & \frac{5}{49} \end{array}\right] \underset{\sim}{\frac{4}{5}R_3 + R_2} \left[\begin{array}{ccc|ccc} 1 & -3 & 2 & 1 & 0 & 0 \\ 0 & 1 & 0 & \frac{4}{49} & \frac{9}{49} & \frac{4}{49} \\ 0 & 0 & 1 & \frac{5}{49} & -\frac{1}{49} & \frac{5}{49} \end{array}\right]$$

$$\underset{\sim}{-2R_3+R_1}\left[\begin{array}{ccc:ccc} 1 & -3 & 0 & \frac{39}{49} & \frac{2}{49} & -\frac{10}{49} \\ 0 & 1 & 0 & \frac{4}{49} & \frac{9}{49} & \frac{4}{49} \\ 0 & 0 & 1 & \frac{5}{49} & -\frac{1}{49} & \frac{5}{49} \end{array}\right] \underset{\sim}{3R_2+R_1} \left[\begin{array}{ccc:ccc} 1 & 0 & 0 & \frac{51}{49} & \frac{29}{49} & \frac{2}{49} \\ 0 & 1 & 0 & \frac{4}{49} & \frac{9}{49} & \frac{4}{49} \\ 0 & 0 & 1 & \frac{5}{49} & -\frac{1}{49} & \frac{5}{49} \end{array}\right]$$

$$AB = \begin{bmatrix} 1 & 0 & 0 \\ 0 & 1 & 0 \\ 0 & 0 & 1 \end{bmatrix} \quad BA = \begin{bmatrix} 1 & 0 & 0 \\ 0 & 1 & 0 \\ 0 & 0 & 1 \end{bmatrix}.$$

If matrix A is multiplied by matrix B (or matrix B by matrix A) the result is a 3×3 identity matrix, i.e.,

$$AB = BA = \begin{bmatrix} 1 & 0 & 0 \\ 0 & 1 & 0 \\ 0 & 0 & 1 \end{bmatrix}.$$

Each matrix is the multiplicative inverse of the other.

10. (a) $\begin{bmatrix} 1 & 5 & 3 & -2 \\ 0 & 1 & 1 & -\frac{9}{7} \\ 0 & 0 & 1 & -\frac{34}{7} \end{bmatrix}$. (b) $\begin{bmatrix} 1 & \frac{1}{4} & -\frac{5}{4} \\ 0 & 1 & -\frac{11}{3} \\ 0 & 0 & 1 \\ 0 & 0 & 0 \end{bmatrix}$. (c) $\begin{bmatrix} 1 & -\frac{5}{2} & -\frac{1}{2} & 0 \\ 0 & 1 & \frac{1}{9} & -\frac{2}{3} \end{bmatrix}$.

Sec. 6.5

A. 1. (b) $\begin{cases} x_1 + 2x_2 + 3x_3 = 4 \\ -x_1 + 3x_2 - 3x_3 = 1 \\ 2x_1 + 4x_2 - 6x_3 = 2. \end{cases}$ The augmented matrix for the given system of equations is

$$\left[\begin{array}{ccc:c} 1 & 2 & 3 & 4 \\ -1 & 3 & -3 & 1 \\ 2 & 4 & -6 & 2 \end{array}\right].$$

Beginning with the augmented matrix and performing elementary row operations, we obtain the following equivalent matrices:

$$\begin{bmatrix} 1 & 2 & 3 & \vdots & 4 \\ -1 & 3 & -3 & \vdots & 1 \\ 2 & 4 & -6 & \vdots & 2 \end{bmatrix} \underset{-2R_1+R_3}{\overset{R_1+R_2}{\sim}} \begin{bmatrix} 1 & 2 & 3 & \vdots & 4 \\ 0 & 5 & 0 & \vdots & 5 \\ 0 & 0 & -12 & \vdots & -6 \end{bmatrix}$$

$$\underset{-\frac{1}{12}R_3}{\overset{\frac{1}{5}R_2}{\sim}} \begin{bmatrix} 1 & 2 & 3 & \vdots & 4 \\ 0 & 1 & 0 & \vdots & 1 \\ 0 & 0 & 1 & \vdots & \frac{1}{2} \end{bmatrix} \overset{-3R_3+R_1}{\sim} \begin{bmatrix} 1 & 2 & 0 & \vdots & \frac{5}{2} \\ 0 & 1 & 0 & \vdots & 1 \\ 0 & 0 & 1 & \vdots & \frac{1}{2} \end{bmatrix}$$

$$\overset{-2R_2+R_1}{\sim} \begin{bmatrix} 1 & 0 & 0 & \vdots & \frac{1}{2} \\ 0 & 1 & 0 & \vdots & 1 \\ 0 & 0 & 1 & \vdots & \frac{1}{2} \end{bmatrix}.$$

Thus, the solution set to $\{(x_1,x_2,x_3) \mid x_1 + 2x_2 + 3x_3 = 4, -x_1 + 3x_2 - 3x_3 = 1, 2x_1 + 4x_2 - 6x_3 = 2\} = \{(1/2, 1, 1/2)\}$.

(c) The system of equations is inconsistent.

(d) $x_1 = 0, x_2 = -1, x_3 = 2, x_4 = -3$.

2. (a) The augmented matrix is $\begin{bmatrix} 1 & 1 & \vdots & 3 \\ \frac{1}{2} & \frac{1}{2} & \vdots & -1 \end{bmatrix}$.

$$\begin{bmatrix} 1 & 1 & \vdots & 3 \\ \frac{1}{2} & \frac{1}{2} & \vdots & -1 \end{bmatrix} \overset{-\frac{1}{2}R_1+R_2}{\sim} \begin{bmatrix} 1 & 1 & \vdots & 3 \\ 0 & 0 & \vdots & -\frac{5}{2} \end{bmatrix}.$$

The matrix

$$\begin{bmatrix} 1 & 1 & \vdots & 3 \\ 0 & 0 & \vdots & -\frac{5}{2} \end{bmatrix}$$

corresponds to the equations

$$\begin{cases} x_1 + x_2 = 3 \\ \quad 0 = -\frac{5}{2}. \end{cases}$$

Since $0 = -\frac{5}{2}$ is a contradiction, the solution set to the given system is empty and the system of equations is inconsistent.

(b) The equations are dependent. There are an infinite number of solutions. The solutions are of the form $(x_1,x_2) = (c, 3-c)$ for any real number c.

(c) $(x_1,x_2,x_3) = (1,-1,0)$. (e) The system of equations is inconsistent.

3. (b)

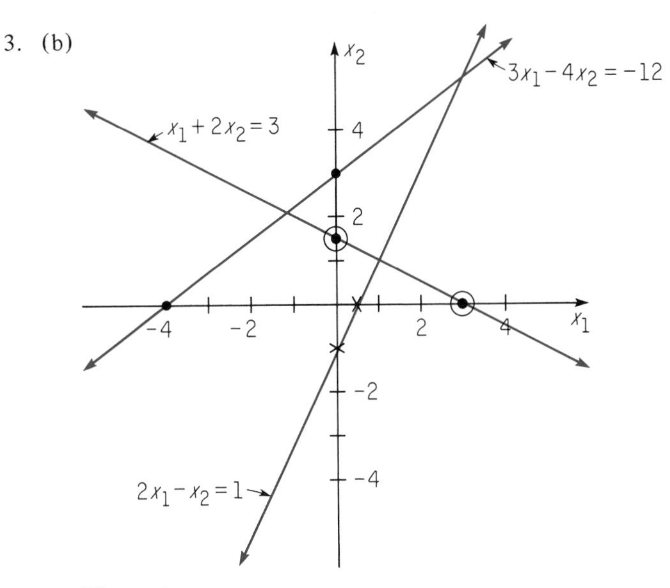

The system

$$\begin{cases} x_1 + 2x_2 = 3 \\ 2x_1 - x_2 = 1 \\ 3x_1 - 4x_2 = -12 \end{cases}$$

has no unique solution, for if it did, the three lines would all intersect in a common point.

(c) The system of equations is consistent and there are an infinite number of solutions.

4. (a) $(x_1,x_2) = (-2,-3)$. (b) $(x_1,x_2,x_3) = (2,-1,1)$.
(d) The system of equations is inconsistent. (f) $(x_1,x_2,x_3,x_4) = (1,2,1,0)$.

Sec. 7.1

A. 1. (a) $\frac{1}{2}x + 3 > -\frac{3}{2}x - 9$.

The given inequality is equivalent to

$$2x > -12$$
$$x > -6.$$

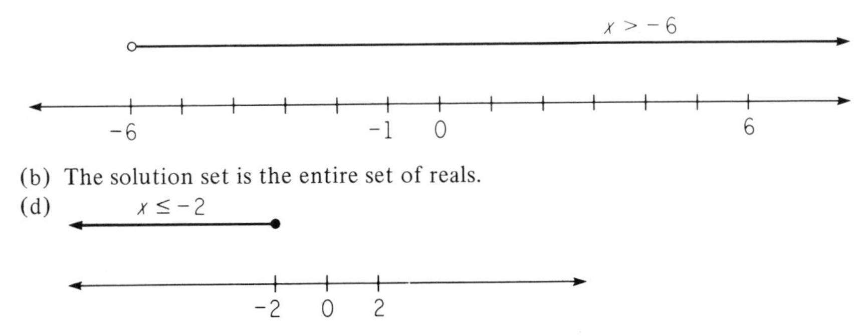

(b) The solution set is the entire set of reals.

(d)

3. (a) $3 < 2x - 1 < 5$
$4 < 2x < 6$
$2 < x < 3.$

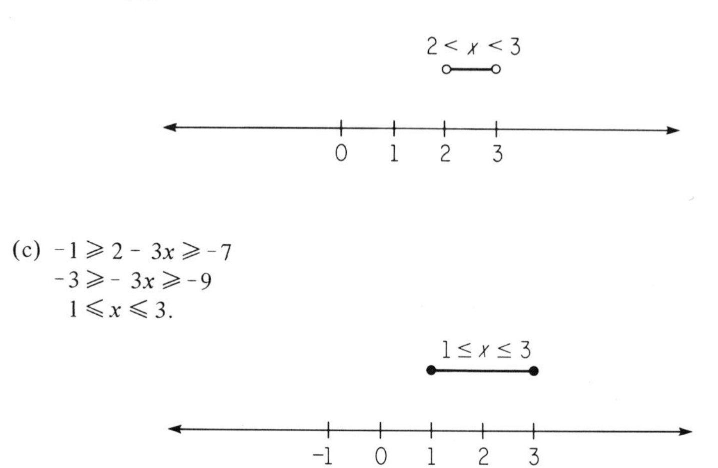

(c) $-1 \geqslant 2 - 3x \geqslant -7$
$-3 \geqslant -3x \geqslant -9$
$1 \leqslant x \leqslant 3.$

Sec. 7.2

A. 1.

$-3x - 2y = -6$

3. (a) $2x + y < 2x + y - 3$ is equivalent to $0 < -3$. But $0 < -3$ is false so the solution set for the given inequality is empty.

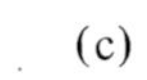

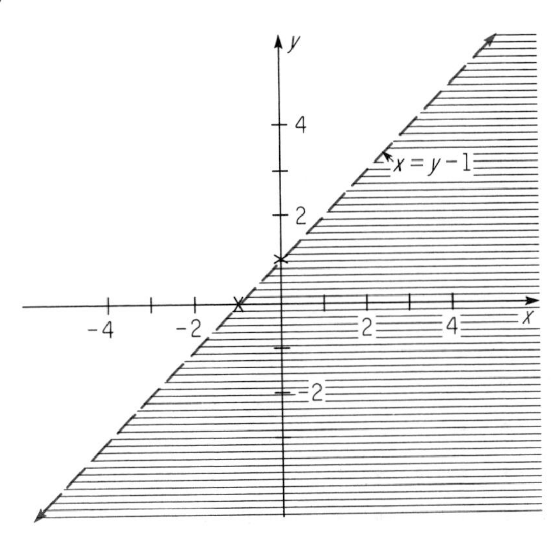

(e)

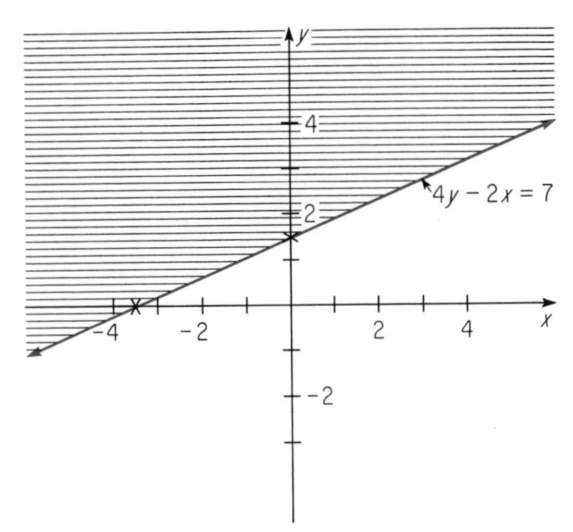

5.

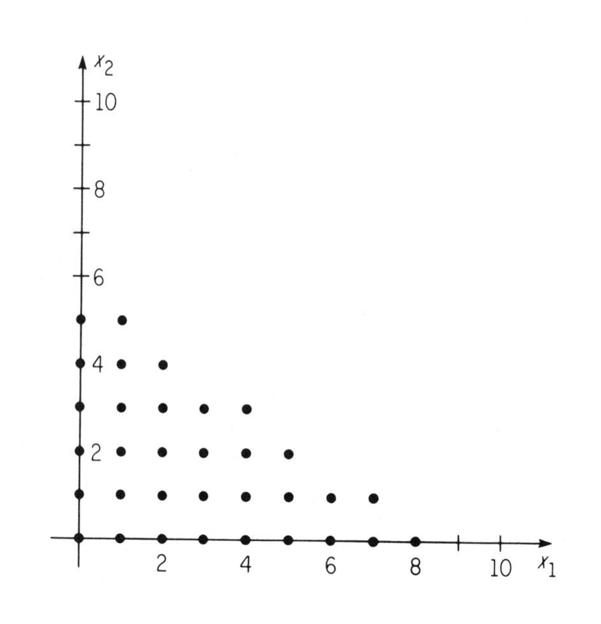

6. (a)

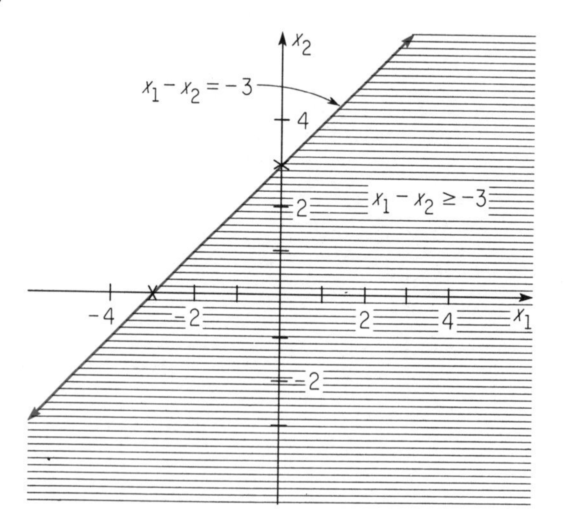

(c)

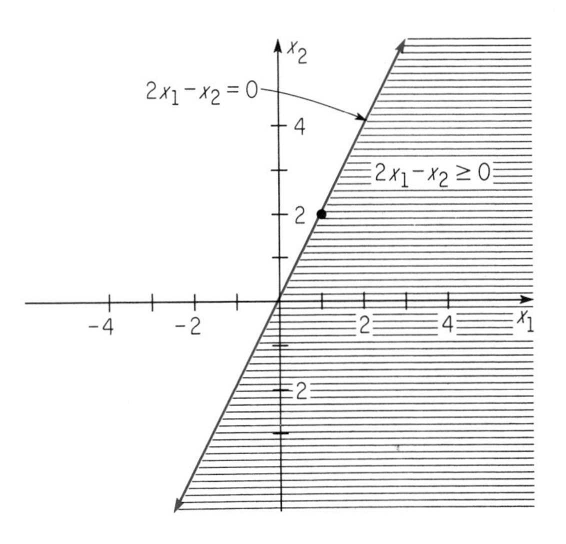

(e)

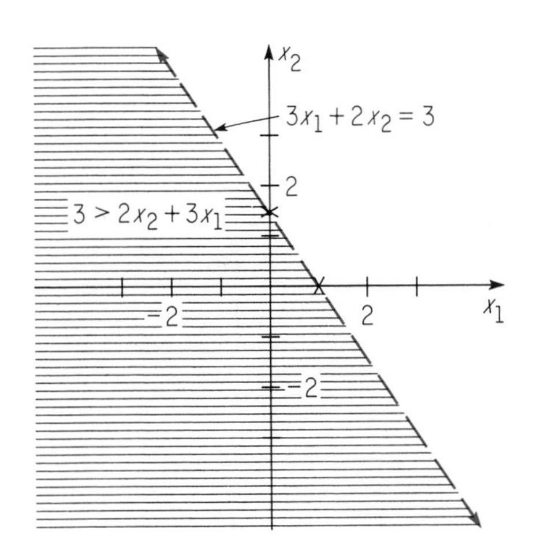

7. (a)

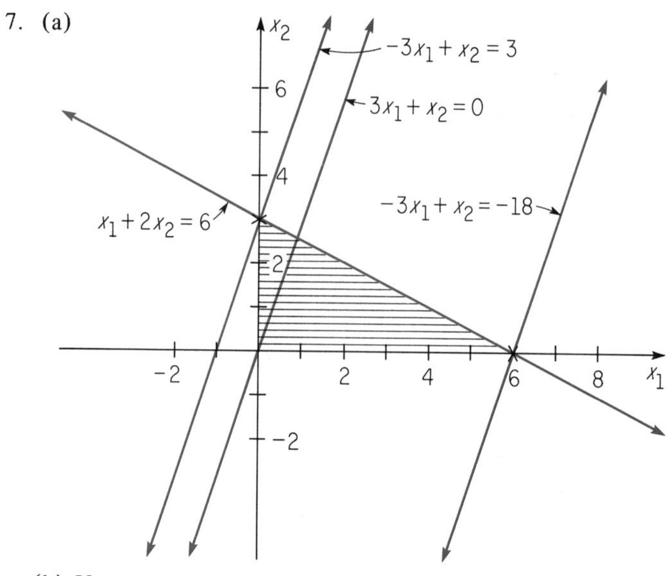

(b) Yes.

(d) (6,0) will make $-3x_1 + x_2$ a minimum, and (0,3) will make $-3x_1 + x_2$ a maximum.

8.

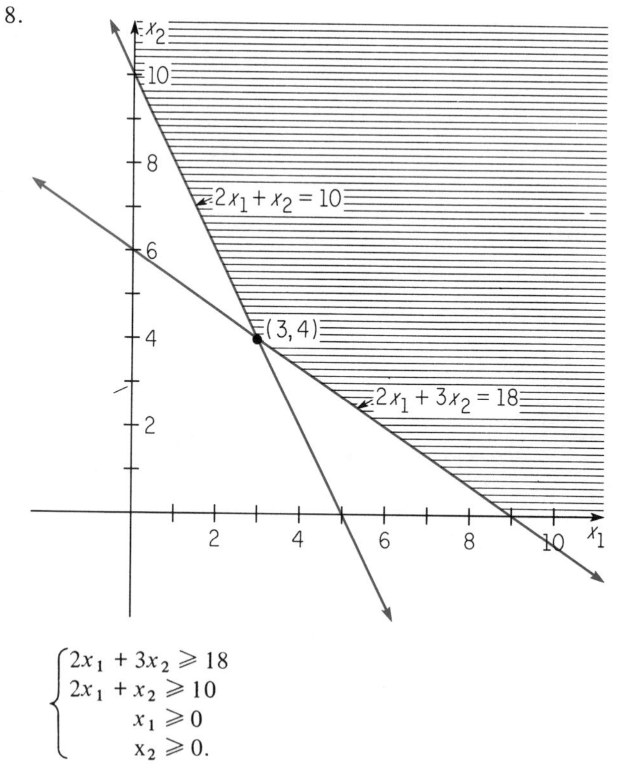

$$\begin{cases} 2x_1 + 3x_2 \geqslant 18 \\ 2x_1 + x_2 \geqslant 10 \\ x_1 \geqslant 0 \\ x_2 \geqslant 0. \end{cases}$$

Total production time can be symbolized as $x_1 + x_2$, which is to be minimized. The production time is minimized when 3 hours is used to produce sweaters and 4 hours is used to produce blouses.

Sec. 7.3

A. 1. (a)

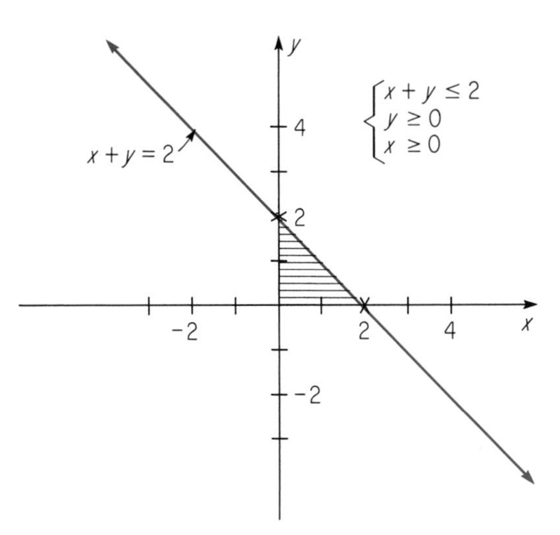

(c)

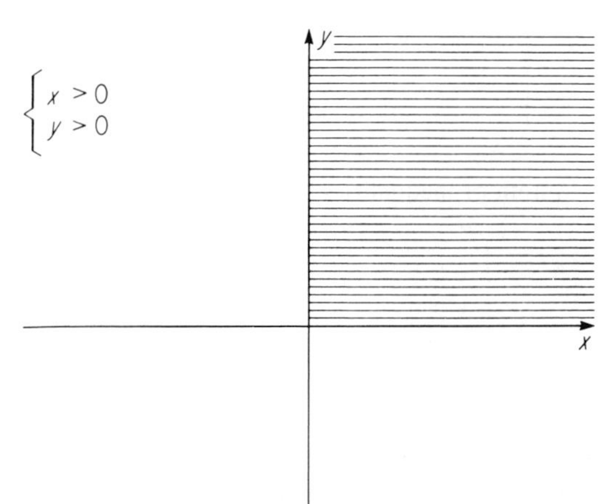

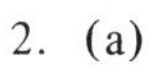

2. (a)

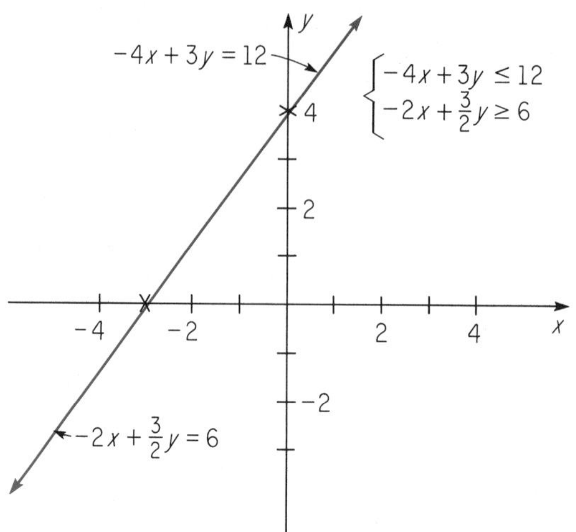

The solution set to the system of inequalities is $\{(x, y) \mid -4x + 3y = 12,\ x \text{ and } y$ real numbers$\}$.

3. (b)

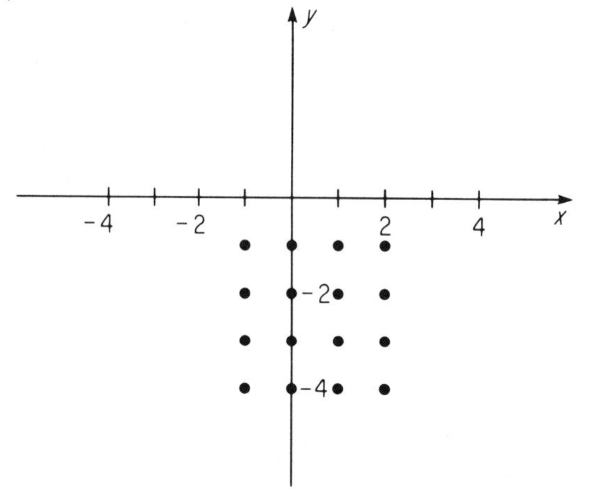

The solution set to the system of inequalities

$$\begin{cases} y < 0 \\ x < 3 \\ x > -2 \\ y > -5 \end{cases}$$

is

$$\begin{Bmatrix} (-1,-1),(-1,-2),(-1,-3),(-1,-4) \\ (0,-1),(0,-2),(0,-3),(0,-4) \\ (1,-1),(1,-2),(1,-3),(1,-4) \\ (2,-1),(2,-2),(2,-3),(2,-4) \end{Bmatrix}.$$

4. (a)

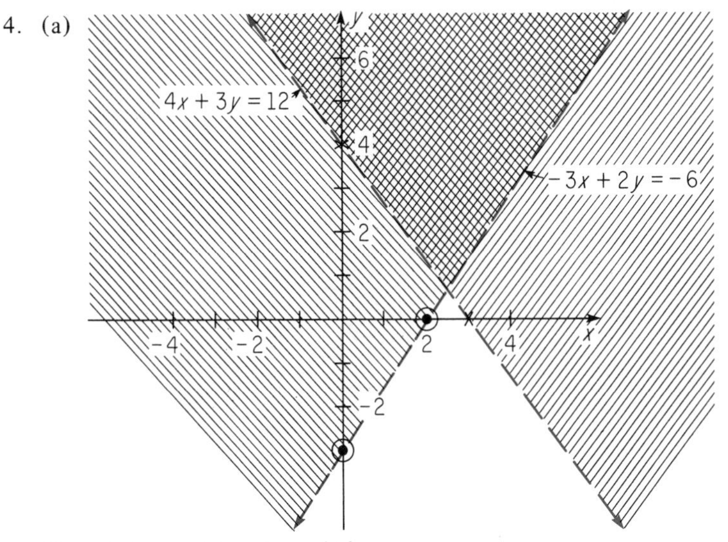

The solution set is unbounded.

5. (c) A system of linear inequalities that represents the given shaded region is

$$\begin{cases} 3x + 2y \leqslant 12 \\ x + y < 6 \\ 2y - x \geqslant -6 \\ 3x + 4y \leqslant 12. \end{cases}$$

6. Let x_1 represent the number of super sound radios built in one day and let x_2 represent the number of modern sound radios built in one day. The inequalities are:

$$\begin{cases} x_1 \geqslant 0 \\ x_2 \geqslant 0 \\ 5x_1 + 10x_2 \leqslant 500. \end{cases}$$

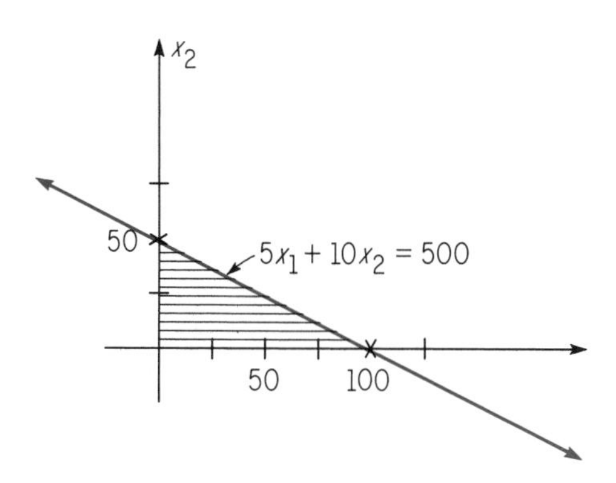

Sec. 7.4

A. 1. (a) Convex. (b) Not convex. (c) Convex. (f) Not convex.

2. (a)

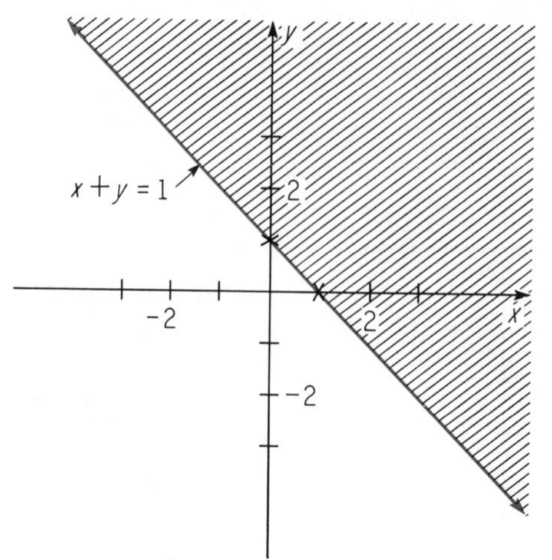

The solution set is convex and it is not bounded.

(c)

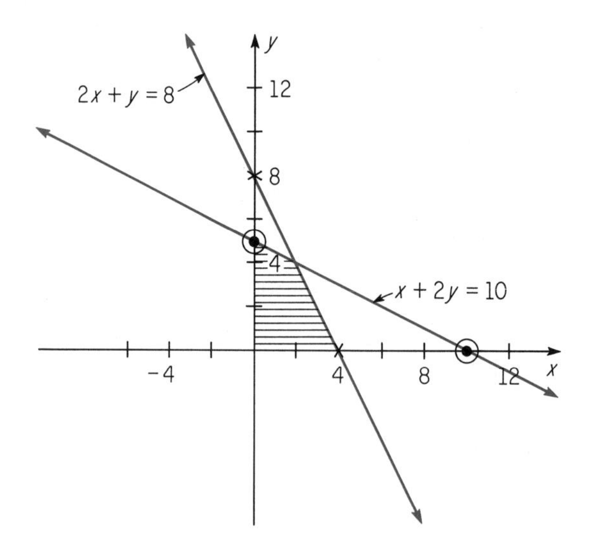

The solution set is convex and it is bounded.

3.

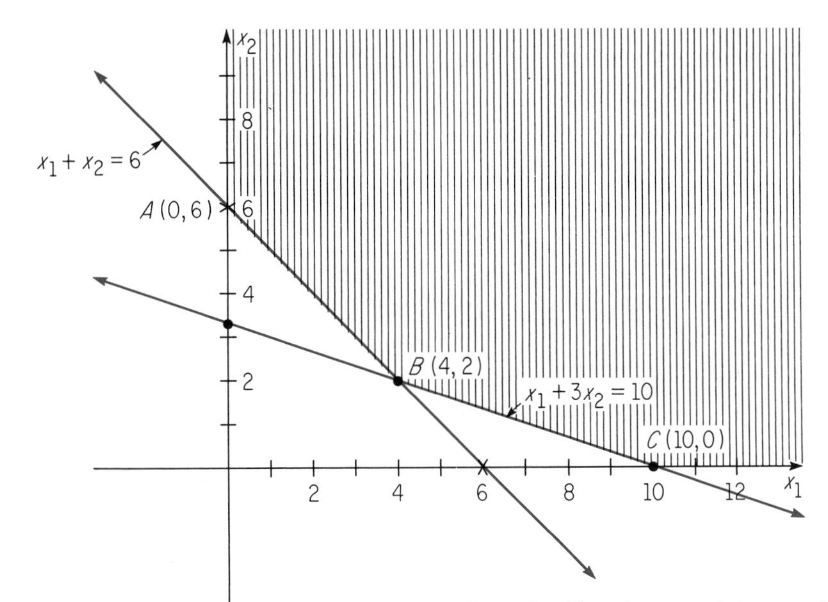

The coordinates of one of the vertices A, B or C will make g a minimum. By substituting the coordinates of each of the vertices into the objective function g, we find the point A whose coordinates are (0,6) makes g a minimum.

6. (a)

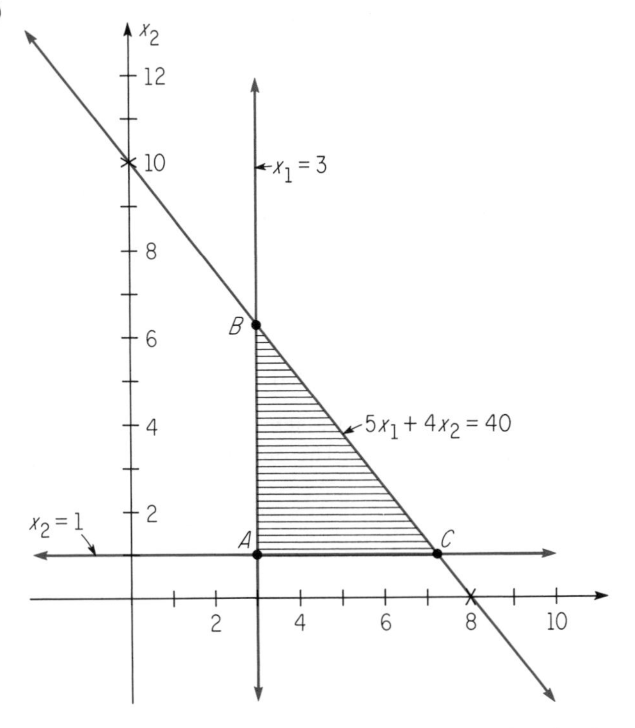

The coordinates of the vertices A, B, C are $(3, 1)$, $\left(3, \frac{25}{4}\right)$, $\left(\frac{36}{5}, 1\right)$ respectively.

7. To keep the cost minimum and maintain normal temperature 10 units of gas should be consumed and 30 units of oil.

Sec. 7.5

A. 1. The constraint set is

$$\begin{cases} 6x + 12x \leqslant 72 \\ 10x_1 + 3x_2 \leqslant 72 \\ x_1 \geqslant 0 \\ x_2 \geqslant 0 \end{cases}$$

where x_1 represents the number of industrial plants and x_2 represents the number of small estates.

(a) The objective function $f = 60{,}000\,x_1 + 60{,}000\,x_2$ will be a maximum when $x_1 = 4$ and $x_2 = 4$.

(b) The objective function $f = 25{,}000\,x_1 + 50{,}000\,x_2$ will be a maximum when $x_1 = 7.2$ and $x_2 = 0$. In practical terms the real estate firm should build seven industrial plants and no small estates.

(c) The objective function $f = 40{,}000\,x_1 + 60{,}000\,x_2$ will be a maximum when $x_1 = 4$ and $x_2 = 4$.

3. The constraint set is

$$\begin{cases} 3x_1 + x_2 \leqslant 400 \\ x_1 + 2x_2 \leqslant 200 \\ x_1 \geqslant 0 \\ x_2 \geqslant 0 \end{cases}$$

where x_1 represents the number of pounds of the first mixture and x_2 represents the number of pounds of the second mixture. The objective function $f = 65x_1 + 75x_2$ will be a maximum when $x_1 = 100$ and $x_2 = 50$.

4. The constraint set is

$$\begin{cases} 200x_1 + 100x_2 \geqslant 1700 \\ 100x_1 + 200x_2 \geqslant 1600 \\ x_1 \geqslant 0 \\ x_2 \geqslant 0 \end{cases}$$

where x_1 represents the number of days factory A operates and x_2 represents the number of days factory B operates. The objective function $f = x_1 + x_2$ will be a minimum when $x_1 = 6$ and $x_2 = 5$.

Index